Volume **51**

Advances in
CHROMATOGRAPHY

Volume **51**

Advances in
CHROMATOGRAPHY

EDITORS

Eli Grushka · Nelu Grinberg

CRC Press
Taylor & Francis Group
Boca Raton London New York

CRC Press is an imprint of the
Taylor & Francis Group, an **informa** business

CRC Press
Taylor & Francis Group
6000 Broken Sound Parkway NW, Suite 300
Boca Raton, FL 33487-2742

First issued in paperback 2019

© 2014 by Taylor & Francis Group, LLC
CRC Press is an imprint of Taylor & Francis Group, an Informa business

No claim to original U.S. Government works

ISBN-13: 978-1-4665-6965-2 (hbk)
ISBN-13: 978-0-367-37974-2 (pbk)

Visit the Taylor & Francis Web site at
http://www.taylorandfrancis.com

and the CRC Press Web site at
http://www.crcpress.com

Contents

Contributors

Dror Avisar
Hydrochemistry Laboratory
    Department of Geography and
    Human Environment
Tel Aviv University
Tel Aviv, Israel

V. G. Berezkin
A. V. Topchiev Institute of
    Petrochemical Synthesis
Russian Academy of Sciences
Leninsky Prospekt, Moscow

Barbara Bojko
Department of Chemistry
University of Waterloo
Waterloo, Ontario, Canada

Marios Chrysanthakopoulos
Department of Pharmaceutical
    Chemistry
School of Pharmacy
University of Athens
Athens, Greece

Igal Gozlan
Hydrochemistry Laboratory
    Department of Geography and
    Human Environment
Tel Aviv University
Tel Aviv, Israel

David. F. Grant
Department of Pharmaceutical Sciences
University of Connecticut
Storrs-Mansfield, Connecticut

Lowell H. Hall
Department of Chemistry
Eastern Nazarene College
Quincy, Massachusetts

L. Mark Hall
Product Development
Hall Associates Consulting
Quincy, Massachusetts

Dennis W. Hill
Department of Pharmaceutical Sciences
University of Connecticut
Storrs-Mansfield, Connecticut

Tzipporah M. Kormos
Department of Pharmaceutical Sciences
University of Connecticut
Storrs-Mansfield, Connecticut

Heng Liang
Separation Science Institute
Key Laboratory of Biomedical
    Information Engineering of
    Education Ministry
Xi'an Jiaotong University
Xi'an, China

Qian Liu
Separation Science Institute
Key Laboratory of Biomedical
    Information Engineering of
    Education Ministry
Xi'an Jiaotong University
Xi'an, China

Antonio Martín-Esteban
Department of Environment
INIA
Madrid, Spain

Liang-Fei OuYang
Separation Science Institute
Key Laboratory of Biomedical
    Information Engineering of
    Education Ministry
Xi'an Jiaotong University
Xi'an, China

Janusz Pawliszyn
Department of Chemistry
University of Waterloo
Waterloo, Ontario, Canada

S. N. Shtykov
Saratov State University
Saratov, Russia

E. G. Sumina
Saratov State University
Saratov, Russia

Anna Tsantili-Kakoulidou
Department of Pharmaceutical
    Chemistry
School of Pharmacy
University of Athens
Athens, Greece

Fotios Tsopelas
Laboratory of Inorganic and Analytical
    Chemistry
School of Chemical Engineering
National Technical University of Athens
Athens, Greece

1 Nonequilibrium Thermodynamics in Nonlinear Chromatography and Electrophoresis
Theory and Applications

Heng Liang, Qian Liu, and Liang-Fei OuYang

CONTENTS

1.1 INTRODUCTION

Since the 1940s, separation scientists [1–3] have become accustomed to understanding the separation processes of chromatography or electrophoresis from the point of view of equilibrium thermodynamics or dynamics, which are two separate worlds. Giddings was convinced of thermodynamic limitations in separation science [1]. He pointed out that one must apply and manipulate external work and heat and allow dilution in a thermodynamically consistent way. Although Giddings [1] recognized early that nonequilibrium thermodynamics [4–6] is an elegant discipline describing the separation processes, he considered that it provides too little substance and too much conceptual difficulty. It is unquestionable that nonequilibrium thermodynamics has played an essential role in both the physical and biological sciences [7]. As the elegant framework of separation theories, is nonequilibrium thermodynamics capable of unifying the phenomena of equilibrium (classical) thermodynamics and dynamics involved in the separation processes? Can nonequilibrium thermodynamic separation theory (NTST) [8–17] merge the related principles of fluid mechanics and modern control science to understand or control the separation processes of nonlinear chromatography or electrophoresis more easily, intuitively, and effectively? The answers to these two questions should be positive as a result of the efforts of Prof. Heng Liang, his co-workers, and students over the past 15 years. Of course, the researches concerning the applications of nonequilibrium thermodynamics in nonlinear chromatography or electrophoresis are still open issues. The developments [8–17] of NTST do not reach the deserved level that nonequilibrium thermodynamics should gift separation science.

The current chromatographic theories [1–3], which are based on partial differential equations of convection–diffusion with Eulerian presentation, have been successfully inosculated with the modern control science, especially in multicolumn chromatographic systems. However, it is necessary to notice the key time points of the origin of chromatography theories (1940) [18], cybernetics (1948) [19], and linear nonequilibrium thermodynamics (1960s) [4]. It hints that the current theories of chromatography could be affected a little by the quintessential concepts of nonequilibrium thermodynamics [4] and cybernetics [19] on understanding or controlling the separation processes systematically. During the period, nonequilibrium thermodynamics and intersecting sciences were also being initially developed, and it was impossible for these subjects to filter into the framework of separation theories at that time. It provides a reliable justification to develop NTST through merging the related concepts and principles of nonequilibrium thermodynamics, fluid mechanics, and modern control science [8].

Giddings [1], Guiochon [2], Morbidelli [3,20], and others are the outstanding representatives in developing contemporary separation theories. With the convection–diffusion equations, Giddings, Guiochon, Morbidelli, and their colleagues mainly contributed theories and applications in linear chromatography, nonlinear chromatography, and the optimization and control of multicolumn chromatographic systems, respectively. The contemporary separation theories and their applications are too tremendous and splendid to exceed the range of this review.

Linear and nonliear chromatography such as frontal analysis (FA) [21,22] and multicolumn continuous chromatography (MCC) [3,23], capillary electrophoresis (CE) [24], and the isoelectric focusing (IEF) [25,26] such as the moving neutralization boundary (MNB) [27] on immobilized pH gradients (IPG) strips were considered as objective separation systems that we used in developing NTST.

During the 1940s to the 1960s, Svensson [28–31] originally took cognizance of the relationship between electrophoresis and moving boundary. He adopted the partial differential equations of convection–diffusion (Svensson–Tiselius' differential equation) to describe the steady states of IEF at the isoelectric points of carrier ampholytes (CAs). The partial differential equations of convection–diffusion in electrophoresis and moving boundary used the same concepts as that in nonlinear chromatography, such as local equilibrium assumption and Lagrangian coordinates [12–15]. Therefore, those concepts can be used in the time-varying IPG–MNB system. As pointed out by Righetti in his reviews [25,26] of IEF and IPG, Svensson in his PhD thesis made an original contribution in relating "electrophoresis" to a "moving boundary" as early as in 1946 [28]. Svennson in 1961 and 1962 advanced the classic IEF theory.

The original idea of the recursion equation [13–15] benefited from our studies of the matters of common concern between electrophoresis and chromatography, for example, partial differential equations of convection–diffusion, nonequilibrium thermodynamics [4,32] in separation processes, as well as Lagrangian or Lagrangian–Eulerian coordinates [12–14,33,34]. The applications of concepts of nonequilibrium thermodynamics [4,32] would create a wide space to develop the theories of analytical or preparation electrophoresis. It can be traced to the original research concerning the Gaussian distribution of the ampholytes at the stationary state by Kauman (1957) who was directed by Prigogine [4,35,36]. In addition, the Lagrangian coordinates [12–14,33,34] would be preponderant in dealing with the complex time-varying physicochemical processes when the images in time sequences are just obtained in both IPG–MNB system and single-molecule imaging system [15].

In this chapter, we review the theories and applications concerning nonequilibrium thermodynamics in nonlinear chromatography and electrophoresis. We do not intend to give a comprehensive review of the current separation theoretical research, but only focus on some of our researches on the theories and applications of nonequilibrium thermodynamics in separation science. In intersecting sciences, we review the related concepts and principles used in NTST, such as nonequilibrium thermodynamics, local thermodynamic equilibrium assumption, Onsager reciprocal relations, entropy balance equation, irreversibility in separation processes, Eulerian and Lagrangian description, time-variant system, Markovian representations, operator splitting, finite-difference, time-domain, and state space representation. We emphasize that separation processes involve the linear nonequilibrium thermodynamic processes, rather than classical dynamics or equilibrium thermodynamics, which are separated. In this regard, we presented some new concepts, such as entropy nature of separation, integral optimizing functional of separation efficiency, the coexistence of Clausius's heat death and Darwin's evolutionism in separation processes, and the framework of NTST published as a review in *Journal of Chromatography A* by Liang 13 years ago, titled "Frameworks of separation theories from two separate worlds: dynamics and thermodynamics" [8]. We introduced the novel concepts (Lagrangian description, time-variant system, and local thermodynamic equilibrium assumption) into IPG–MNB electrophoresis systems [27] and single-molecule imaging system [37,38]. We presented thermodynamic state recursion equations to simulate nonideal, nonlinear chromatographic processes, nonlinear transport chromatographic processes, and the Varicol processes of MCC based on the local thermodynamic path (LTP) as the cover image of *Journal of Separation Science*, 2012, Vol.35, No.12, see [112] and Lagrangian description. Simulation softwares for MCC based on the 0-1 model were also introduced. We emphasize that NTST only extends some concepts and principles of the current chromatographic theories through felicitously using some ideas of nonequilibrium thermodynamics, control science, and fluid mechanics. Of course, separation process control is more important than understanding or describing them. We ought to adopt the concepts of state space through thermodynamic state recursion approach to control separation processes.

1.2 RELATED CONCEPTS AND PRINCIPLES IN INTERSECTING SCIENCES

In 1945, Schrödinger noted the significance of "negative entropy" to life [39]. In 1948, Wiener discovered the relationship between entropy and information in founding cybernetics [19]. In 1949, Shannon founded information theory and used the formula to measure information, which is the same as the measurement entropy used by Boltzmann [40]. In 1950, Brillouin systematically demonstrated and developed the thought that information is equated to negative entropy flow. In other words, gain in entropy always means loss of information and nothing more [41]. After noting the equivalence relationship between information and negative entropy, we will answer the question, how do we control separation processes to gain more information (or negative entropy) through separation systems?

With the point of view of nonequilibrium thermodynamics, one should understand the separation processes better and design and control separation systems more easily for a complex separation system. However, to achieve this objective, we have to face some concepts and principles that are difficult to comprehend in intersecting sciences. In this section, the hard concepts and principles are introduced one by one.

1.2.1 NONEQUILIBRIUM THERMODYNAMICS

Nonequilibrium thermodynamics is also called the thermodynamics of irreversible processes [4,5,32]. It deals with systems that are not in thermodynamic equilibrium. They evolve with time. They are continuously subject to the flux of matter and energy from other systems. In this chapter, nonequilibrium thermodynamics that we mention only limits near-equilibrium thermodynamics [42], which is the study of processes of systems said to be close to equilibrium. The near-equilibrium processes are characterized by Onsager reciprocal relations [43,44]. Contrarily, nonlinear (far-from-equilibrium) thermodynamics [45] of irreversible processes (e.g., an oscillating reaction) developed by Ilya Prigogine are not included in this chapter.

Most separation scientists know well the equilibrium thermodynamics and dynamics of chromatographic processes [1,46]. However, there are two fundamental differences between equilibrium thermodynamics and nonequilibrium thermodynamics. One difference lies in the behavior of inhomogeneous systems (e.g., the inhomogeneous spacial distributions of solute bands along the column axial direction). It requires for their study knowledge of flux rates that are not considered in equilibrium thermodynamics of homogeneous systems. This difference directly results in the essential and internal relations among the local equilibrium hypothesis, mesoscopic description, and time-varying control, when the concepts of nonequilibrium thermodynamics were adopted in separation science. It is discussed in more detail in Sections 1.2.2 and 1.2.7. Another fundamental difference is the difficulty in defining entropy in macroscopic terms for systems not in thermodynamic equilibrium [4,47]. For example, the integral optimizing functional of the separation efficiency for column separations involves the calculation of entropy and the inhomogeneous distributions of two components [8,10]. Some concepts of particular importance for nonequilibrium thermodynamics include time rate of entropy production [43,44] and entropy flow [4].

The entropy production and some flows of nonequilibrium thermodynamic steady states are nonzero, but there is no time variation. The stationary state includes the occurrence of unpredictable and experimentally unreproducible fluctuations in the state of the system. The fluctuations are due to the system's internal subprocesses and the exchange of matter or energy with the system's surroundings that create the constraints that define the process. The fluctuation cannot be reproduced with a significant level of probability. Fluctuations about stable stationary states are extremely small except near critical points [45]. The stable stationary state has a local maximum of entropy and is locally the most reproducible state of the system. The concepts of steady states and fluctuations should be used on the online control of multicolumn chromatographic systems.

1.2.2 Local Thermodynamic Equilibrium Assumption

Local thermodynamic equilibrium of matter [6,48,49] means that the system can be spatially and temporally divided into "cells" or "micro-phases" of small (infinitesimal) size, in which classical thermodynamic equilibrium conditions for matter are fulfilled to good approximation. When these "cells" are defined, one admits that matter and energy may pass freely between contiguous "cells," slowly enough to leave the "cells" in their respective individual local thermodynamic equilibria with respect to intensive variables [6,48–50]. With the assumption of the local equilibrium, nonequilibrium thermodynamics [4,47] extended classical thermodynamics to open systems and treated the state parameters of thermodynamics as field variables, that is, as continuous space- and time-dependent fields such as mass density $\rho(r,t)$, entropy $s(r,t)$, or temperature $T(r,t)$, and so on. The local thermodynamic equilibrium assumption is materialized in NTST [8,12–14]. For example, the local thermodynamic equilibrium assumption ensures that the equilibrium isotherms can be localized in the two corresponding solute cells, one (SCm) in the mobile phase and the other (SCs) in the stationary phase with the Lagrangian–Eulerian description (LED) [14].

1.2.3 ONSAGER RECIPROCAL RELATIONS

Following Onsager [43,44], let us extend our considerations to thermodynamically nonequilibrium systems. Such locally defined gradients of intensive macroscopic variables are called "thermodynamic forces." They "drive" flux densities, perhaps misleadingly often called "fluxes," which are dual to the forces. Onsager reciprocal relations consider the stable near-steady thermodynamically nonequilibrium regime, which has dynamics linear in the forces and flux densities. In stationary conditions, such forces and associated flux densities are by definition time invariant, as are the system's locally defined entropy and rate of entropy production. When an open system is in conditions that allow it to reach a stable, stationary, thermodynamically nonequilibrium state, it organizes itself so as to minimize total entropy production defined locally. Onsager [43,44] showed that in the regime where both the flows are small and the thermodynamic forces vary slowly, there will be a linear relation between them, parametrized by a matrix of coefficients conventionally. Onsager's reciprocal relations affirmed that thermodynamics could provide available information in the neighborhood of equilibrium. The direct value of the combination of the entropy production and Onsager's reciprocal relations resulted in the description of linear nonequilibrium thermodynamics.

1.2.4 ENTROPY BALANCE EQUATION

Only in equilibrium states or during reversible processes is entropy a well-defined function of state. However, with the local equilibrium hypothesis, entropy remains a valuable state function even in nonequilibrium situations. To describe the evolution of many-body systems, the balance equation of entropy plays a central role in near equilibrium. Let us consider a system consisting of n components in a viscous medium without chemical reactions. For a volume cell, by inserting the conservation equations of mass, energy, and motion to the Gibbs relation, a general format of entropy balance equation can be obtained [5]:

$$\rho \frac{ds}{dt} = -\text{div } J_s + \sigma \tag{1.1}$$

where

$$J_s = \frac{1}{T} \cdot \left(J_q - \sum_{i=1}^{n} \mu_i J_i \right) \tag{1.2}$$

$$\sigma = -\frac{1}{T^2} \cdot J_q \,\text{grad}\, T - \frac{1}{T} \cdot \sum_{i=1}^{n} \left(T \,\text{grad}\, \frac{\mu_i}{T} - F_i \right) \\ - \frac{1}{T} \cdot \Pi \,\text{grad}\, v \tag{1.3}$$

where ρ is the local total density; s is the local entropy per unit mass; J_s, J_q, and J_i are the entropy flow, the heat flow, and the diffusion flow of component i per unit area and unit time, respectively; σ is the entropy production per unit volume and unit time; T is the local absolute temperature; μ_i is the chemical potential of component i; F_i is the force per unit mass exerted on component i; Π is viscous pressure tensor; and v is the center of mass velocity. The set of conservation laws (mass, energy, and momentum), entropy balance equation, the separation state equations [8,10], and phenomenological equations may be said to be complete, which may be solved under the proper initial and boundary conditions for a material [5,35].

The entropy balance equation (Equation 1.1) is only available for small volume cells, which are so small that the thermodynamic properties of the system vary little over each cell but large enough for the cell to be treated as macroscopic thermodynamic subsystems in contact with their surroundings. Therefore, we cannot only define thermodynamic variables and functions in the cells but also wish to vary them from cell to cell in such a way that the thermodynamic variables can be described in continuous space and time-dependent fields [32]. Thus, solute systems in a cell form an open system of thermodynamics, because the exchange of heat and matter between the cell and its surrounding can be found in Equations 1.1 and 1.2. In practice, many fields (the gradients of chemical potential, external fields, and temperature) [1] promote separations unexceptionally, which results in two effects—entropy flux and entropy production, which are closely related to each other in the entropy balance equation [5]. The entropy flux corresponds to the relative migration of the mass center of different solutes partly due to the differences of solutes themselves, for example, electric charge, mass, size, shape, hydrophobicity, and affinity, partly due to the properties of the mobile phase and stationary phase or applied fields. Here, we stress the matching between the characters of solutes and applied fields (including mobile phase and stationary phase), and we can describe them suitably in the diffusion flow (J_i) of component i, and heat flow (J_q) in Equations 1.1 through 1.3. Separation scientists have known well the gradient of temperature due to Joule heating in CE [51] and also noted the influence of adsorption heat on elution band profiles in nonlinear liquid chromatography [52]. Diffusion flow and heat flow always coexist [53] in electroseparation methods, which are related to electrophoreses, electroosmosis or electroendosmosis, partition, and so on. Increasing relative migration among solutes and releasing more heat from solute systems imply that separation systems put more negentropy flow into solute systems. In this aspect, all separation conditions as the separation surroundings are equivalent to Maxwell's Demon [40] in that they supply the solute system with negative entropy flow and make the solute system more ordered (better separation efficiency). The entropy production inside solute systems is related to the irreversible processes, which contain three contributions due to heat conduction, diffusion, and viscous flow in general separation processes, and correspond to related factors contributing to band spreading. With the entropy balance equation, the equations of state and phenomenological equations and a group of conservation laws (mass, energy, and momentum) and the separation state function were obtained between the integral separation efficiency and operation parameters to describe the big separation system. The more negative the integral optimizing functional of separation efficiency (ΔS_s), $\Delta S_s < 0$ is, the more information solute systems gain and the better the separation efficiency is [8,9]. It is obvious that effective separation systems can supply solute systems with more negentropy, which was called the amount of information by Wiener [40]. The amount of information in separations corresponds to the quantity of separated solutes, which is incarnated in the number of separated peaks and their quantity (namely the number of separated solute moles) [10].

1.2.5 IRREVERSIBILITY IN SEPARATION PROCESSES

The first and second terms on the right-hand side of Equation 1.1 are, respectively, the divergence of an entropy flux and an entropy production as a source term. Equation 1.2 shows that for open systems, the entropy flow consists of two parts: one is the "reduced" heat flow and the other is connected with the diffusion flows of matter. Equation 1.3 demonstrates that the entropy production contains three different contributions. The first term on the right-hand side of Equation 1.3 arises from heat conduction, the second from diffusion, and the third is connected to gradients of velocity field, giving rise to viscous flow. The structure of the expression for σ is that of a bilinear form. It consists of the sum of products of two factors

$$\sigma = \sum_{g=1} Y_g \cdot X_g \geq 0 \tag{1.4}$$

where Y_g are the thermodynamic fluxes (J_q, J_i, Π, etc.), which are "driven" by X_g, the conjugate thermodynamic forces ($-\dfrac{1}{T^2}\operatorname{grad}T$, $-\operatorname{grad}\dfrac{\mu_i}{T}$, $\operatorname{grad} v$, etc.).

In the entropy balance equation (Equations 1.1), on one hand, entropy production is always a nonnegative quantity relating the various irreversible processes occurring in a system. On the other hand, entropy flows, which supply the system with the entropy through its surroundings, can be positive, zero, and negative. The thermodynamic forces are the source of causing the evolution of complex systems (toward order or disorder). We noted that one thermodynamic force always causes two opposite kinds of effects simultaneously. One is in entropy flow, and this could put negative entropy flows into object systems to make the system more ordered through matter and energy or informational exchange, and this effect incarnates the idea of Darwin's evolutionism. The other one is in entropy production, which always increases the entropy of object systems to make the system more disordered, and this effect incarnates the idea of Clausius's heat death of the universe.

The irreversibility of separation processes can be explained from a few aspects: (1) The actualizing of any separation process necessarily depends on the applied fields (e.g., chemical potential, electrical, sedimentation, and temperature graduations) provided by separation systems. In fact, the fields are irreversible surroundings for solute systems [1]. (2) In the applied fields, the different kinds of solute band could be separated on the separation pathway. Thus, the solute systems are provided with negative entropy flow (negentropy) by separation systems (as surroundings), which make the solute system evolve from disorder to order and obtain information [9]. (3) Just for the sake of separating, the irreversible processes caused by the applied fields are adopted in any separation. It follows that irreversible processes accompany entropy production with nonnegative values, which correspond to the factors making solute zone spreading and information losses. Obviously, both separation among bands and spreading of bands originate from the irreversibility simultaneously [8,9].

The irreversibility of separation processes has to be described by the entropy balance equation instead of the second law from equilibrium thermodynamics. If the methods of dynamics are only used to deal with separation processes, irreversibility in separation processes as a restriction condition (the entropy balance equation) is lost, which leads to difficulty in establishing the relationship between operation parameters and separation efficiency quantitatively and systematically. Qualitatively thinking the entropy balance equation, we could relate entropy to separation efficiency or relate flow entropy to migration different between two solute zones or relate the entropy production to solute zone's spreading. Up to now, the complexity of nonlinear chromatography makes entropy balance equation just as the frameworks of separation theories, but not as a universal and quantitative equation of separation theories.

1.2.6 EULERIAN AND LAGRANGIAN DESCRIPTIONS

In fluid dynamics, there are the two views, Eulerian description and Lagrangian description [54,55], of the flow field. The Eulerian description describes fluid motion at fixed locations and considers velocities and other properties of fluid particles to be functions of time and of fixed space coordinates. It can be imagined that one sits on the bank of a river to watch the water pass at the fixed location. It is a way of looking at fluid motion that focuses on specific locations in the space through which the fluid flows as time passes. In the Eulerian description, the velocities of the flow are given at fixed points in space as time varies.

By contrast, the Lagrangian description considers the positions of fluid particles and their other properties to be functions of time and of their permanent identifications, such as their initial positions or any set of material functions of fluid particles. In the Lagrangian description, the fluid dynamics is formulated in terms of trajectories of fluid elements. The Lagrangian description is for the observer to follow an individual fluid parcel as it moves through space and time. It is a way of looking at fluid motion where the observer follows an individual fluid parcel as it moves through space and time. It is can be visualized as sitting in a boat and drifting down a river. Plotting the position of an individual parcel through time gives the pathline of the parcel. The Lagrangian description is chosen to account for the possible changes of the shape over time.

The Eulerian grid and the Lagrangian grid are the two disparate kinds of grid of domain discretization in the coordinates of fluid field. Eulerian grid is fixed on the space. The shape and volume of the mesh cell remain unchanged in the entire process of the computation. Eulerian grid dominates the area of computational fluid dynamics. It also dominates the current nonlinear chromatographic theories based on the convection-diffusion equation with the Eulerian description [1–3]. As its disadvantages, it is difficult to analyze the time history of field variables at a fixed point on the material. And it is also difficult to determine the positions of free surfaces, deformable boundaries, and moving material interfaces accurately. The Lagrangian grid is attached on the material, the grid moves with the material. The deforming material causes the deforming mesh accordingly. The controlled time step can become small to be efficient for the time marching. Remeshing is possible, but it is expensive. Many numerical methods for fluid flow simulation use representations of the flow in terms of Lagrangian elements as opposed to Eulerian fields [56]. Solutes (e.g., proteins and compounds) as separation objects and solvents (e.g., the buffers of mobile phase) must be distinguished when the Eulerian description or the Lagrangian description is adopted.

The equivalency of Eulerian and Lagrangian descriptions in a one-dimensional case was rigorously proved [57]. However, from a numerical computation point of view, they are not equivalent. Computationally, in the numerical solutions of the Eulerian coordinates, there is excessive numerical diffusion since fluid particles move across space–grid interfaces fixed in space. Although great improvement has been made, numerical diffusion still exists, resulting in inaccuracy. Another disadvantage of the Eulerian coordinates is that a grid generation, which can be time consuming, needs priority to flow computation to satisfy with boundary conditions on solid boundaries [58]. From fluid dynamics point of view, the finite-difference methods, finite-element methods, as well as Lagrangian methods comprise a special group of fluid computational algorithms combined by the discretization method of the initial value mathematical problem, but they possess a number of different characters for the numerical description of evolution of a complex physical process [59–61]. The Lagrangian description, as an alternative to the more common Eulerian description, offers a new series of perspectives and addresses for a broad audience of physicists with interest in fields such as plasma and fluid dynamics, semiconductor and astrophysics, and so on [62]. The Lagrangian method of solving fluid equations turns out to be a powerful tool for compressible media in general [63,64]. Even if Lagrangian description is getting increasing attention in fluid dynamics and related areas as Lagrangian codes [61], it is not explored in chromatographic theories although Eulerian methods were synchronously introduced when chromatography theory was presented in the 1940s. According to our researches [11,12], the evolution of solute zones in column separation processes can be considered as the "migration" and "deformation" motions of dynamic interface [61] of solute zones in nonlinear chromatography processes based on Lagrangian description. It is more important that Lagrangian description gives more information thorough it tells the history and material functions of each fluid particle. This method plays a key role in constructing of theory of the optimal control of nonlinear chromatography processes through fining an optimal thermodynamic path in ergodic state space of the variables with state space representation (see Section 1.2.11). Recently, Lagrangian description was used in chromatographic and electrophoretic theories by the efforts of our group [11–17].

1.2.7 TIME-VARIANT SYSTEM

A time-variant system is a system that is neither time-invariant nor stationary. The characteristics of its output depend explicitly on time. There are many well-developed techniques for dealing with the response of linear time-invariant systems, such as Laplace and Fourier transforms. However, these techniques are not strictly valid for time-varying systems [65].

The state space equations can be solved for time-variant systems, but the solution is significantly more complicated than the time-invariant case. Unlike the time-invariant case, we cannot define this as a simple exponential. In time-variant systems, the most important thing is to determine the state-transition matrix. For nonlinear chromatographic processes, authors determined the state-transition matrix, $\mathbf{S}_i^k \rightarrow \mathbf{S}_i^{k+1}$, through an LTP [12–14], where $\mathbf{S}_i^k$ and $\mathbf{S}_i^{k+1}$ are the state vectors of space distribution of solute i at kth stage and $k + 1$th stage, respectively.

Specially, in nonlinear chromatography, the space- and time-dependent fields $(u_{j,k})$ generally include the chemical potential between mobile and stationary phases due to the space- and time-dependent solvent strength under gradient elution [8,66,67], or due to the space- and time-dependent solute concentration in nonlinear chromatography [68], or due to time-varying electric field in CE [69]. Accordingly, we can define local external force fields that act on the given volume cell j of the solute zone i at the time t_k. These local external force fields include local strong eluent concentration $(\varphi_{i,j,k})$ of mobile phase, local chemical potential $(\mu_{i,j,k})$, local electrical field strength $(E_{i,j,k})$, and so on. One can use any thermodynamic laws with local thermodynamic variables in the identifiable volume cells [4,6]. With the local equilibrium assumption, we can define a set of local thermodynamic variables in the given volume cell j of solute zone i at the time t_k in separation processes, such as local isotherm parameters $(a_{i,j,k}$ and $b_{i,j,k})$, local diffusion coefficient $(D_{i,j,k})$, local lumped mass transfer coefficient $(k_{fi,j}^k)$, local molar concentration $(c_{i,j,k})$, local migration velocity $(v_{i,j,k})$, local thermodynamic distribution constant $(K_{i,j,k})$, local capacity factor $(k'_{i,j,k})$, and so on. We should point out that the time-variant systems of nonlinear chromatography include the systems whose key time-variant parameters, such as $a_{i,j,k}$, $b_{i,j,k}$, $D_{i,j,k}$ $k_{fi,j}^k$, and $k'_{i,j,k}$ (without $c_{i,j,k}$), are time-variant.

1.2.8 Markovian Representation

A Markov process is a time-varying random phenomenon with the Markov property, or memorylessness. A stochastic process with the Markov property is conditional on the present state of the system. Its future and past are independent [70]. The Markov property holds for a certain random process. A Markov chain, which would be defined for a discrete set of times (i.e., a discrete-time Markov chain), means a Markov process that has a discrete (finite or countable) state space.

Markovian representations in modern sciences and technologies are widely applied [71]. Markovian representations are of potential value in developing nonlinear chromatographic theories. The optimal control problems concerning the space-time varying multiparameters, multistage, and multiple objective decisions in nonlinear chromatography processes can be considered as finding the optimal state trajectory in ergodic state space of control variables [13]. To describe the evolution of solute zones in nonlinear chromatography with the Mavkovian characters, the form of explicit scheme concerning thermodynamic states on a time sequence is necessary for some optimal control approaches based on discrete-time states. In this case, we are able to solve for the variables at the next time level only with the local variables at most recent time level under the local actions of space-time-varying external force fields that are just indicated by recent values of local physicochemical parameters [12–14], such as local concentration, local equilibrium isotherms, local velocity along characteristic direction, local axial dispersion coefficient, local operation parameters, and so on.

1.2.9 OPERATOR SPLITTING

Operator splitting formulation has been developed for the multiphysicochemical phenomena to coexist in one complex space-time evolution process [34,64,72,73]. It was used in many numerical methods [72] for the systems of convection–diffusion equations, in which convective and diffusive forces are accounted for in separate sub-steps as an important design principle. One splits a time evolution into partial steps to separate the effects of convection and diffusion. To reduce the influence of temporal splitting errors in operator splitting methods and to allow for the use of large splitting steps, the corrected operator splitting method was introduced [72,73] for general systems of convection–diffusion equations with the ability of correctly resolving the nonlinear balance between the convective and diffusive forces to compensate for the potentially damaging splitting error. The corrected operator splitting exhibits the property of resolving accurately internal layers with steep gradients, gives very little numerical diffusion, and, at the same time, permits the use of large time steps. The motivation for adopting the operator splitting methods lies in that it is easy to combine the Lagrangian description and the discretizations of the parabolic diffusive step and the hyperbolic convection step, and the felicitous combination just coincides with the principle of thermodynamic states in LTP. The operator splitting methods will give a powerful and efficient numerical method. The differences from an original real medium indicate that the computational particle methods require "tuning" with the continuity axiom [33] in the process of numerical calculation as a "know-how" operation. And the accuracy of the auxiliary numerical procedures is used in the programs of particle methods. It also saves time when we deal with problems solved by the state recursion methods [11–17]. The accuracy of particle methods increases nonlinearly with the growth in the number of Lagrangian mesh nodes. The good algorithms of operator splitting formulation as particle-in-cell methods [34] are usually economical. In software implementation, particle methods are easily adapted to computers with parallel architecture [34].

1.2.10 FINITE-DIFFERENCE TIME-DOMAIN

Finite-difference time-domain is a popular computational time-domain method. It belongs to a general class of grid-based differential time-domain numerical modeling methods. The descriptor finite-difference time-domain was originated by Allen Taflove on electromagnetic compatibility [74,75]. In general, certain classes of simulations can require many thousands of time steps for completion.

The 0-1 model [14] on nonlinear chromatography has the characteristics of the finite-difference time-domain. Every numerical modeling technique has strengths and weaknesses, and the 0-1 model is no different. It is a time-domain technique (Lagrangian description). It is intuitive, so users can easily understand how to use it and know what to expect from this model. It is also a systematic approach to deal with the nonlinear behavior naturally. With the 0-1 model, the complete space distribution of solutes along the column axial direction can be obtained at each given time under the continual and time-varying sample concentrations with Markov characteristics. This type of display is useful in understanding what is going on in the model and to help ensure that the model is working correctly. For example, in the 0-1 model, the four subprocesses (e.g., the virtual net migration process, the virtual net diffusion process, the virtual net interphase mass transfer process, or the virtual net distribution process) in a LTP are designed to obtain the local mathematical relationships for the recursions of two adjacent states of solute space distributions. Being an accurate and robust time-domain technique, the 0-1 model can directly calculate the complete space distribution of solute peaks along the axial direction of the columns.

1.2.11 STATE SPACE REPRESENTATION

To implement the state space representation [76–79] of nonlinear chromatography, we should present a mathematical description in state space form. We should choose the smallest set of state variables that fully describes the evolution of the time-variant system. We next compute the solution of the state space equation. The solution to this system is the state vector that depends on time and which contains enough information to completely determine the trajectory of state evolution of the nonequilibrium thermodynamic system. The components of state vectors are called the state variables. We can use matrix notation to rewrite the equations of convective diffusion in nonlinear chromatography in Eulerian description. Of course, we may create a new state space description based on the state recursion with the LED [11–16].

The preparative chromatography systems [2,46,80] can be considered the system of control engineering. A system of MCC [23], such as a simulated moving bed (SMB) [81], is a highly complex engineering system with nonlinearity, nondeterminacy, large time delay, close coupling, distribution parameters, and miscellaneity. Therefore, the state space representation of a nonlinear chromatography should be a mathematical model as a set of input, output, and state variables related by a time sequence of state of the solute spatial distributions. If Lagrangian description is adopted in the fundamental theories of nonlinear chromatography, as a time-domain approach, the state space representation can provide a convenient and compact way to model and analyze systems with multiple inputs and outputs. With inputs and outputs, the equations of state recursions are written down to encode all the information about the MCC. Unlike the classical methods of partial differential equations with Eulerian description, the use of the state space representation is not limited to systems with time-invariant cases, simple initial conditions, or single-column systems. For MCC, the state space contains all information of the time-varying operation parameters, productive rate, purity grade, and costs. At the mesoscopic level, the collection of ordered array of the local concentrations in the mobile phase solute cells and the stationary phase solute cells at a given time should be chosen as the smallest possible subset of system variables to represent the space distribution of solutes in columns at any given time. The most general state space representation of preparative chromatography has been written with the state vector, output vector, input (or control) vector, state matrix, input matrix, and output matrix. In this general formulation, all matrices are allowed to be time-variant. The time variable t can be discrete. The time variable is usually indicated as k. Therefore, our state space model of preparative chromatography belongs to the discrete-time-variant systems (see Section 1.7).

The theory models of preparative chromatography, which are based on nonequilibrium thermodynamics, should have controllability and observability. Controllability provides that the input brings any initial state to any desired final state. Observability is a measure for how well internal states of a system can be inferred by knowledge of its external outputs. Observability provides that knowing an output trajectory provides enough information to predict the initial state of the system. Preparative chromatography should belong to the multiple-input and multiple-output systems in the state space representation. This state space realization is called controllable canonical form because the resulting model is guaranteed to be controllable. Because the control enters a chain of integrators, it has the ability to move every state. This state space realization is called observable canonical form. Every state as a step in separation path has an effect on the output of preparative chromatography system. The strictly proper transfer function can then be transformed into a canonical state space realization. We have chosen thermodynamic state recursion equations to elucidate how the output also depends directly on the input and controllable operations.

1.3 NATURE OF NONEQUILIBRIUM THERMODYNAMICS OF SEPARATION SCIENCE

In 1998, one of authors reviewed the frameworks of separation theories based on nonequilibrium thermodynamics, cybernetics, information theory, and systematology [8]. We understood that the irreversibility of separation processes is the source of both the separation and mixing of solute systems. Clausius's heat death (band spreading of one component) and Darwin's evolutionism (separation among different components) simultaneously coexist in separation processes. By using the entropy balance equation (Equation 1.1), we revealed the evolution that separation systems undergo to make solute systems evolve from Boltzmann's disorder to order. The integral optimization functional [10] was found on the part of the mixed entropy change of the solute system between the final state and the initial one relating directly to the net separation. The more negative the integral optimization function is, the more information solute systems gain and the better the integral separation efficiency of big separation systems. According to the entropy balance equation and the conditions of practical separation processes, a separation state function can be discovered to show how separation systems described by macroscopic physicochemical parameters affect the integral optimization functional. With the methods of modern cybernetics, we will be able to optimize and time-varyingly control separation processes through controlling the operation parameters in separation processes with the separation state function.

The conflict between dynamics and thermodynamics focuses on the second law of thermodynamics, because the second law cannot be deduced from the framework of classical dynamics. Prigogine and Stengers [35] pointed out that the Markov chain corresponding to the reality in natural world would be given only when the second law of thermodynamics is taken as a selection principle.

Separation processes are certainly irreversible, and the solute systems are open systems (in each mesoscopic and small volume cell) in thermodynamics [9,82]. Thus, nonequilibrium thermodynamics that fully reflects the idea of systematicism can be adopted to describe separation processes and to establish the function relations between various operation parameters and separation efficiency [9,82]. With NTST [8–17], the principles of modern cybernetics can be used to properly optimize and time-varyingly control separation processes integrally.

1.3.1 Coexistence of Clausius's Heat Death and Darwin's Evolutionism in Separation Processes

Thermodynamics challenged dynamics in focusing on the irreversibility of processes. Entropy law, which relates to irreversible processes, has already exceeded the category of conservation laws. The irreversibility prescribes the arrow of time to indicate the evolution in the isolated system, which is always in the direction of making a system more disordered [35]. After the irreversibility of processes was admitted, Darwin's evolutionism and Clausius's heat death theory seem to be opposite, but they are unified on the intrinsic randomness of complex systems. Only in the open system is it possible for the environment to put the negative flow of entropy into the system to make it more ordered.

Separation processes involve two different diffusions having two opposite contributions to separation efficiency in the applied fields: one is the band spreading of one component and another is the separation among the bands of different components. Therefore, separations among components were only achieved in nonequilibrium states and irreversible processes although they are in near thermodynamic equilibrium region. In actual separation processes, we found not only Clausius's heat death for the inevitable band spreading (more disorder), but also Darwin's evolutionism for the separating among bands (more order). How do the two opposite sides actually unify in the separation processes?

Nonequilibrium thermodynamics can describe the evolutionary behaviors of complex systems with state parameters. From the viewpoint of nonequilibrium thermodynamics, we can find that separations are processes in which different solutes evolve from disorder to order in applied fields (separation among different components) and simultaneously each solute itself evolves from order to disorder (band spreading of one component). The two opposite processes originate simultaneously from the irreversibility of separation processes, which are described quantitatively by the items of entropy flows and entropy production. In separation processes, a portion of negentropy flow is used up by entropy productions. Thus, the results of separation processes are determined by both the negentropy flows and the entropy productions. Obviously, the entropy balance equation adequately incarnates the dual nature of the irreversibility through the opposite contributions of the constructive action of negentropy flows and the destructive action of entropy productions. In the entropy balance equation, thermodynamic forces (X_g) are thermodynamic variables, and the thermodynamic fluxes (Y_g) are kinetic variables (as shown in Equation 1.4) [48].

Thus, all terms of Equation 1.1 have dual characters of dynamics and thermodynamics, and the two terms on the right-hand side of Equation 1.1 indicate the entropy changes of the solute system determined by Darwin's evolutionism (entropy flow) and Clausius' heat death (entropy produce).

1.3.2 ENTROPY NATURE OF SEPARATION WITH A SIMPLE MODEL

We can use a simple model (as shown in Figure 1.1) to clarify the entropy nature of separation [83]. The entropy increase (ΔS) and the entropy increase per unit time ($\Delta S/\Delta t$) of the solute zone in the processes of chromatography and electrophoresis can be chosen as new criteria of separation efficiency. ΔS and $\Delta S/\Delta t$ were expressed as

$$\Delta S = 2n_i R \ln\left[1 - \left(\frac{\Phi_i}{2}\right)\right] \qquad (1.5)$$

$$\frac{\Delta S}{\Delta t} = 2n_i R \cdot \left[\frac{\Phi_i}{(\Phi_i - 2)\Delta t}\right] = 2n_i R \cdot \left[\frac{\Phi_i'}{\Delta t}\right] = 2n_i R \cdot \varphi_i \qquad (1.6)$$

where n_i is the total moles of component i, R is the gas constant, Φ_i is the separation ratio of component i, $\Phi_i = n_{s,i}/n_i$ is the ratio of the moles of separated component i ($n_{s,i}$) to its total moles (n_i), Φ' is corrected separation ratio, φ_i is corrected separation rate to indicate the separation power of a separation system for component i, Δt is the separation time. In Figure 1.1 C_{01} and C_{02} are initial concentrations of the solutes 1 and 2. The assumption s are made, $n_1 = n_2$, and the concentration ($C_{s,i}$) in the space that separated solute i occupies is two times of the initial concentration. Equation 1.5 indicates that the entropy increase of the solute system is directly proportional to the logarithm of the difference between one and one half of the substantial separation ratio. Equation 1.6 indicates that the entropy increase per unit time is directly proportional to the corrected separation rate of separation system. The entropy increase of separations and the entropy increase per unit time of the solute zone are the important bridge between separation efficiency of chromatography or electrophoresis and operating parameters, especially when nonequilibrium thermodynamics is adopted [8,9,12,16].

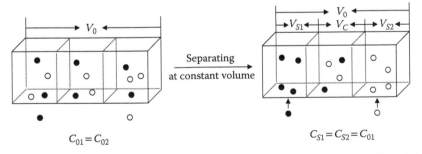

FIGURE 1.1 The nature of two components' separation at constant volume. (From Liang et al., *Journal of Beijing Institute of Technology*, 6, 137–143, 1996. With permission.)

1.3.3 INTEGRAL OPTIMIZING FUNCTIONAL OF SEPARATION EFFICIENCY

According to a virtual thermodynamic path as shown in Figure 1.2, the integral optimizing functional of separation efficiency (ΔS_s) [10], which is the mixed entropy change between initial and final states directly associated with separation efficiency, was presented to indicate integral separation efficiency for any zone separation method by further considering the entropy change from the uniform to the arbitrary distribution of solute zones, as shown in Figure 1.3 [10]. Physically, as a system property, ΔS_s is equal to the amount of information that the solute system obtains from its separation surrounding or separation system. It can be quantitatively related to

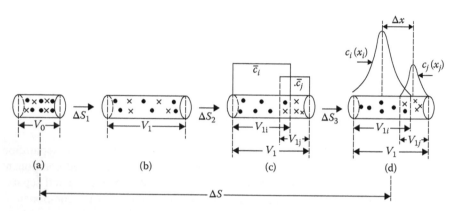

FIGURE 1.2 The spreading and separating of the two-component system in an imaginary separation path. They are respectively the process of net volume change from (a) to (b), the net separation from (b) to (c), and the net distribution change from (c) to (d). The mixed entropy changes of each process and the whole process are ΔS_1, ΔS_2, ΔS_3, and ΔS, respectively. (From Liang and Lin, *Journal of Chromatography A*, 841, 133–146, 1999. With permission.)

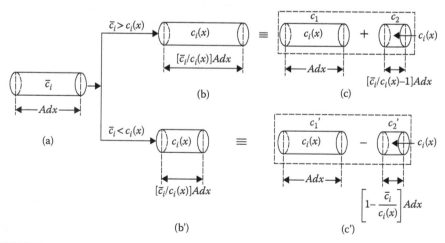

FIGURE 1.3 Schematic for the calculation of the mixed entropy from the average concentration $\bar{c}$ to the actual concentration distribution $c_i(x)$ of component i in a volume cell (Adx). (From Liang and Lin, *Journal of Chromatography A*, 841, 133–146, 1999. With permission.)

the irreversibility of separation processes in the entropy balance equation [4,32] and corresponds to the extent of Boltzmann order. In separation science, ΔS_s corresponds to the quantity of separated solutes. For any arbitrary distribution of solute zones, ΔS_s can be calculated directly from the distributions and relative positions of solute zones. Thus, ΔS_s can be calculated directly from the separation results of chromatography and electrophoresis to indicate separation efficiency integrally and quantitatively. For example, for Gaussian distributions of zones, ΔS_s can be calculated directly from the standard deviations (σ) of the peaks, the distance between the centers of gravity of adjoining zones (Δx) and the number of moles of the solutes (n) or the peak height (h), as shown in Figure 1.4. Equation 1.7 can be used to calculate ΔS_s

$$\Delta S_s = \sqrt{2\pi}\,\sigma_i r_h A R \ln\left[\frac{6\sigma_i}{3(\sigma_i+\sigma_j)+\Delta x}\right] + \sqrt{2\pi}\,\sigma_j A R \ln\left[\frac{6\sigma_j}{3(\sigma_i+\sigma_j)+\Delta x}\right]$$

$$+ AR \int_{-3\sigma_i}^{\Delta x-3\sigma_j} r_h \cdot \exp\left(\frac{-x^2}{2\sigma_i^2}\right)\left[\left(\frac{-x^2}{2\sigma_i^2}\right)+\ln\left(\frac{\sqrt{2\pi}}{6}\right)\right]\cdot dx \tag{1.7}$$

$$+ AR \int_{3\sigma_i-\Delta x}^{3\sigma_j} \exp\left(\frac{-x^2}{2\sigma_j^2}\right)\left[\left(\frac{-x^2}{2\sigma_j^2}\right)+\ln\left(\frac{\sqrt{2\pi}}{6}\right)\right]\cdot dx$$

where σ_i and σ_j are the standard deviations of the solutes i and j, respectively; r_h is the ratio (h_i/h_j) of peak heights with $h_j = 1$; and A is the cross-sectional area perpendicular to the separation path.

A quasi-inverse relation between ΔS_s and separation pureness of solutes (φ) was found numerically. In any effective separation process, ΔS_s is always a negative value, and the more negative ΔS_s is, the better the separation efficiency is. The computer simulation supported the above characters of ΔS_s. The discovery of ΔS_s is a part of the NTST, which could be used to integrally optimize and time-varyingly control the complete separation systems.

1.3.4 FRAMEWORK OF NONEQUILIBRIUM THERMODYNAMIC SEPARATION THEORY

In the framework of NTST [8], the solute systems consist of two or more components. If only a one-component system is chosen (plate theory) [84–86], the separation is not involved principally. A separation system means the whole separation

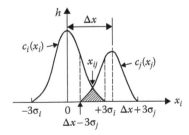

FIGURE 1.4 The distribution of a two-component zone. x_{ij} is the intersection point of the elution curves of solutes i and j. (From Liang and Lin, *Journal of Chromatography A*, 841, 133–146, 1999. With permission.)

surroundings contribute to the separation of the solute system except the solute system itself. Therefore, we also name the separation system as the environment of the solute system. It is in correspondence with the concepts of the system and environment in thermodynamics [4,5]. The assemblages of the separation systems and solute systems compose the big separation systems, whose various properties and interactions can be described by using certain macroscopic physicochemical parameters. The essence of the integral optimization is just the optimization of the big separation system.

Dynamics and thermodynamics are two complementary explanations of nature, which can be unified by a noncanonical transformation. We must abandon the usual formulation of dynamics to reach the thermodynamic description [35,36,87]. In the framework of NTST, we only insert a new element, the irreversibility of separation processes, into the framework of contemporary separation theories. Thus, there is no conflict in any levels between contemporary separation theories and NTST, where the former specializes in dealing with the details of separation processes and the latter acts as an integral framework for whole separation processes. In nonequilibrium thermodynamics, the entropy balance equation contains three kinds of (mass, energy, and momentum) conservation equations with no conflict [5], in which the mass conservation equation only deals with each component in the solute systems. Obviously, we can use all achievements of contemporary separation theories on conservation equations to enrich NTST with no conflict. Thus, NTST cannot replace contemporary separation theories. Figure 1.5 shows the frameworks of contemporary separation theories and NTST. In one word, NTST emphasizes the two opposite actions of the irreversibility in separation processes. Thermodynamic forces make solute bands not only spread but also separate. Integral optimization means that the opposite actions of thermodynamic forces must be considered in one theoretical framework. NTST can be used to integrally optimize and time-varyingly control separation processes with the methods of modern cybernetics. When NTST puts contemporary separation theories back together again, separation science will achieve the unification between art and science on the basic levels of natural science. One of the functions of NTST would provide the optimized and time-varying constraints of various parameters in separation processes to make the black box, we assumed, transfer to a white box (a gray box at some time) in separation science.

1.3.5 Nonequilibrium Thermodynamic Separation Theory in Capillary Electrophoresis

In 1997, one of the authors and coworkers firstly presented a separation model in CE based on the entropy equation of nonequilibrium thermodynamics [9]. We determined the entropy flow of the solute system, which is composed of both energetic and material exchange terms relating to capillary cooling and relative migrations among solute zones. It is just the CE separation system as an exterior surrounding, which contributes to the enhanced separation efficiency. The more the CE system (except the solute system) provides the solute system with negative entropy flow, the better the separation efficiency of the CE system. We also determined six thermodynamic forces and their thermodynamic fluxes corresponding to six irreversible

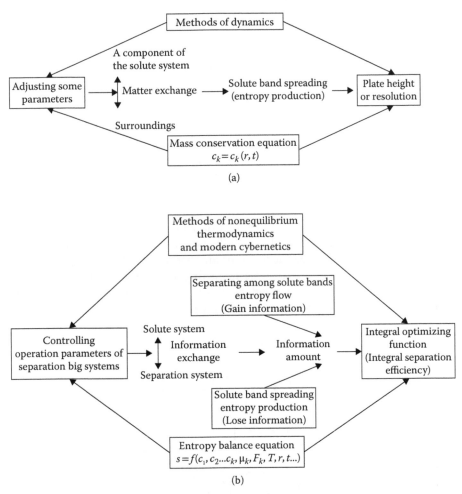

FIGURE 1.5 Frameworks of (a) contemporary separation theories and (b) nonequilibrium thermodynamic separation theory. (From Liang and Lin, *Journal of Chromatography A*, 828, 3–17, 1998. With permission.) (c_k the local density of component k, T the absolute temperature, r space vectors, t time, s the local entropy of solute system per unit volume, μ_k chemical potential of component k, and F_k external force per unit mass acting on component k.)

processes: heat conduction, four kinds of diffusion (electrical field, axial concentration gradient, electrophoretic dispersion, and wall adsorption), and viscous flow. Entropy production is thus composed of the six terms corresponding to time-dependent CE efficiency loss factors. The bigger the entropy production is, the greater the loss of separation efficiency is. The objective functions were built based on the entropy equation of solute systems developed between CE separation efficiency (ΔS_s) and the optimizing parameters (electrical strength, coolant temperature; the composition and concentration of buffer; the radius, length, and wall adsorption of the capillary; the concentration, charge, molecular weight, and conformation of solutes; injection conditions, etc.). The more negative ΔS_s is, the better the separation efficiency is. The significance of this research lies in the fact that it enriched current separation theories and extended the application fields of nonequilibrium thermodynamics.

Plate theory focuses on factors causing zone broadening using the dynamic methods. However, it considers other factors that impel the separation between two adjoining zones at the same time to a lesser extent. In fact, the factors of the two sides are present simultaneously in the separation processes. Thus, the factors involved on both sides should be taken into account in one theoretical frame. We should assemble the factors of the two sides in one theoretical frame.

The entropy equation occupied an important position in studying the linear phenomena of nonequilibrium thermodynamics. The use of entropy as a general criterion of separation has been proposed by De Clerk and Cloete [88] and Stewart [89]. We have confirmed that the entropy, entropy change, and entropy change rate of solute systems make up a group of general criteria for determining separation efficiency [10,82,83]. In the electrophoretic separation of solute systems, three main thermodynamic phenomena exist, electrokinetic effects, diffusion, and heat transfer, which belong to the category of linear nonequilibrium thermodynamics. Thus, the entropy equation could be used as a basis for establishing an original separation theory of CE based on nonequilibrium thermodynamics [4,49]. The objective function of the mono-component system (ΔS_H) for CE can be written as

$$
\begin{aligned}
\Delta S_H &= \Delta S_{e,1} + \sum_{j=1}^{6} \Delta S_{i,j} - n_i R \ln\left(\frac{6l^{\frac{1}{2}}}{\Delta l_0}\right) \\
&= \pi R_1^2 \left[4\sqrt{2D_i}\left(\frac{FE\Theta_i}{\alpha}\right)^{-\frac{3}{2}} l^{\frac{3}{2}} \right] \\
&\quad \cdot \left\{ -\left(\frac{\Lambda C_b E^2}{T_c}\right) + \left[\frac{(\Lambda C_b R_1)^2 E^4}{8k_1 T_c}\right] \right. \\
&\quad + \left(\frac{\Lambda_i z_i^2 F^2 E^2}{T_c^2}\right) + \left[\frac{R_i^2 T_c}{(A_0 M_i^{R_{cl}})}\right] B^2 C_i \\
&\quad + \left(\frac{\kappa R T_c}{C_b}\right) \Delta l_0 \left(\frac{FE\Theta_i}{\alpha}\right) B^2 C_i^2 \\
&\quad + \left(\frac{D_i^{\text{surf}} R}{n_i'^2}\right) B_1^2 C_i \\
&\quad \left. + \frac{a'(F\Theta_i \Lambda C_b \varepsilon R_1)^2 E^6}{\left[4k_1^2 \alpha^2 T_c^4 \ln(\varepsilon/RT_c)\right]} \right\} \\
&\quad - n_i R \ln\left(\frac{6l^{\frac{1}{2}}}{\Delta l_0}\right)
\end{aligned}
\tag{1.8}
$$

where $\Delta S_{e,1}$ is the contribution to the entropy flow (ΔS_e) from heat transportation (energetic exchange) between the solute system and its surroundings. $\Delta S_{i,j}$ are all contributions to entropy production for heat conduction ($\Delta S_{i,j=1}$), electrical field ($\Delta S_{i,j=2}$), axial concentration gradient ($\Delta S_{i,j=3}$), electrophoretic dispersion ($\Delta S_{i,j=4}$), wall adsorption ($\Delta S_{i,j=5}$), and viscous flow ($\Delta S_{i,j=6}$). n_i is the mole number of solute i. R is the gas constant, l and R_1 are the effective length and inner radius of the capillary, respectively. Δl_0 is the initial width of the sample plug. D_i is total diffusion coefficient of solute i. k_1 is the thermal conductivity of buffer. F is Faraday constant, E is electrical field strength. $\Theta_i = Z_i / M_i^{R_{c,i}}$, Θ_i electrophoretic characterization factor of solute i, which is only determined by the solute electrophoretic parameters, the electronic charge (Z_i), molecular weight (M_i), and conformation parameter ($R_{c,i}$) of the solute i [90]. $\alpha = k' \exp(\varepsilon / RT_c)$, α a constant relating to viscosity of buffer, k' other constant relating to the viscosity of buffer, ε the activation energy of buffer, T_c the temperature of cooling, Λ and Λ_i are the equivalent conductances of buffer and solute i, respectively. z_i the electronic charge per unit mass of the solute i, κ a constant relating to electrophoretic dispersion of solutes. B and B_1 are constants relating to the axial diffusion of the solutes and it on the internal face of capillary, respectively. C_i is the concentration of the solute i, D_i^{surf} is the diffusion coefficient of the solute i on the internal face of capillary wall. n' is a constant in Freundlich adsorption equation, a' is a constant in the viscosity equation. C_b is the concentration of buffer electrolyte solution.

For the multicomponent solute system, the objective function (ΔS_S) was given as

$$
\begin{aligned}
\Delta S_S = {} & \sum_{i=1}^{2} \Delta S_{e,i} + \sum_{j=1}^{6} \Delta S_{i,j} - \Delta S_V \\
= {} & -\left[\frac{(n_1 + n_2)}{(72D)} \right] s' \left(\frac{FEl}{\alpha} \right) \Delta\Theta_{1,2} \\
& + \pi R_1^2 \left[4\sqrt{2D} \left(\frac{FE\Theta_i}{\alpha} \right)^{-\frac{3}{2}} l^{\frac{3}{2}} \right] \\
& + (1/2) \left(\frac{FE}{\alpha} \right)^{-1} l^2 \Delta\Theta_{1,2} \Theta_i^{-2} \cdot \left\{ -\left(\frac{\Lambda C_b E^2}{T_c} \right) + \left[\frac{(\Lambda C_b R_1)^2 E^4}{8 k_1 T_c} \right] \right. \\
& + \sum_{i=1}^{2} \left(\frac{\Lambda_i z_i^2 F^2 E^2}{T_c^2} \right) + \sum_{i=1}^{2} \left[\frac{R^2 T_c}{(A_0 M_i^{R_{c,i}})} \right] B^2 C_i \\
& + \sum_{i=1}^{2} \left(\frac{\kappa R T_c}{C_b} \right) \Delta l_0 \left(\frac{FE\Theta_i}{\alpha} \right) B^2 C_i^2 \\
& + \sum_{i=1}^{2} \left(\frac{D_i^{\text{surf}} R}{n_i'^2} \right) B_1^2 C_i \\
& \left. + \frac{a' \left(F\Theta_i \Lambda C_b \varepsilon R_1 \right)^2 E^6}{\left[4 k_1^2 \alpha^2 T_c^4 \ln \left(\frac{\varepsilon}{RT_c} \right) \right]} \right\} \\
& - \sum_{i=1}^{2} n_i R \ln \left\{ \left(\frac{l}{\Delta l_0} \right) \left[6\sqrt{2\alpha D / (Fl\Theta_i)} + \frac{\Delta\Theta_{1,2}}{\Theta_i} \right] \right\}
\end{aligned}
$$

(1.9)

where ΔS_V is the entropy change due to the volume change only. n_1 and n_2 are the mole numbers of solute $i = 1$ and solute $i = 2$, respectively. s' is the transfer entropy density per unit mass of the solute i. $\Delta\Theta_{1,2} = \Theta_2 - \Theta_1$, $\Delta\Theta_{1,2}$ is the difference between the solutes 2 and 1 in the electrophoretic characterization factors.

The mixed entropy change of the solute system is a bridge between the separation efficiency of CE and optimizing parameters in the entropy equations of CE shown in Equations 1.8 and 1.9. It is the outstanding characteristics of the developed nonequilibrium thermodynamic separation model. In Equation 1.8, ΔS_H indicates the broadening of solute's zone under given CE operation conditions. $\Delta S_{e,1}$ indicates the mixed entropy reduction of the mono-component solute system due to the heat transferred from the solute system with cooling. It denotes equivalently the level of reduction of the solute's diffusion due to the thermal diffusion of solute zones. $\sum_{j=1}^{6} \Delta S_{i,j}$ in Equation 1.8 is the contribution of ΔS_H due to the entropy production, which corresponds to six irreversible processes in CE separation, heat conduction, four kinds of diffusion (electrical field, axial concentration gradient, electrophoretic dispersion, and wall adsorption), and viscous flow. Under typical CE operating conditions, $\Delta S_{i,j} > 0$ and $\sum_{j=1}^{6} \Delta S_{i,j} > 0$. Thus, the terms of the entropy production correspond to time-dependent efficiency loss factors in CE. The bigger the entropy production is, the greater the loss of separation efficiency is. The last term on the right-hand side of Equation 1.8 indicates the contributions of ΔS_H due to the solute's migration distance (l) and injecting conditions (Δl_0). Therefore, Equation 1.8 reveals a quantitative relationship between the separation efficiency of the CE system for given solutes and optimizing parameters.

The fact that $\Delta S_{e,1}$ in Equation 1.8 is negative compelled us to introduce the concept of "negative plate height" to indicate that the more negative the term in the contribution of ΔS_H is, the better the separation efficiency is. It is important in optimizing operational conditions to understand negative plate height. In Equation 1.9, ΔS_S is the mixed entropy change in the net separation process only (see Figure 1.6). Thus, $\Delta S_S < 0$ in any effective separations. $\sum_{i=1}^{2} \Delta S_{e,i}$ in Equation 1.9 is the contribution of ΔS_S due to the entropy flow, which relates to the capillary cooling and relative migrations of the solute. Thus, $\Delta S_{e,i} < 0$ and $\sum_{i=1}^{2} \Delta S_{e,i} < 0$ in any effective separations, it just shows that the CE system, as exterior surroundings, contributes to enhancing the separation efficiency of the solute system. The more the CE system provides the solute system with a negative entropy flow, the better the separation efficiency of the CE system. $\sum_{j=1}^{6} \Delta S_{i,j}$ in Equation 1.9 is the contribution of ΔS_S due to the entropy production. They have the similar meanings in Equation 1.8. ΔS_V in Equation 1.9 shows the contributions of ΔS_S due to the injecting conditions (Δl_0) and the change of total volume all solutes occupy (V_t). Therefore, Equation 1.9 reveals a quantitative relationship between CE separation efficiency (ΔS_S) and optimizing parameters (electrical strength, coolant temperature, injecting conditions, the composition and concentration of the buffer, the radius, length and wall adsorption of the capillary, the concentration, charges, molecular weight and conformation of solutes, and so on). The more negative ΔS_S is, the better the separation efficiency is. It is thus obvious that Equation 1.9 reflects quantitatively what Giddings said [1], separation

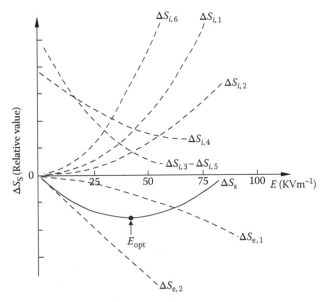

FIGURE 1.6 Schematic of influence of field strength on the separation efficiency (ΔS_S) for the two-component system, under typical experimental conditions of CE. These curves were generated using Equation 1.9. The optimal field strength, E_{opt} is also shown. (From Liang et. al, *Journal of Chromatography A*, 763, 237–251, 1997. With permission.)

is the art and science of maximizing separative transport (making $\sum_{i=1}^{2} \Delta S_{e,i}$ more negative) relative to dispersive transport (decreasing $\sum_{j=1}^{6} \Delta S_{i,j}$) in CE. For the multi-component system, each term of their objective function has a similar meaning in the two-component system.

In principle, we can optimize any operation parameter in the objective functions with the nonequilibrium thermodynamic separation model. In this chapter, only injection sample length (Δl_0) and field strength (E) are discussed with experimental data. Figure 1.6 is based on Equation 1.9 for the two-component system; it shows how eight kinds of factors influence separation efficiency (ΔS_S) as E increases. In Figure 1.6, as E increases, the terms of the viscous flow ($\Delta S_{i,j=6}$), heat conduction ($\Delta S_{i,j=1}$), and electrical field ($\Delta S_{i,j=2}$) increase with difference power of E. Thus, these three kinds of factors lose separation efficiency (making ΔS_H increase) more greatly as E increases. On the contrary, the terms of the axial concentration gradient ($\Delta S_{i,j=3}$), electrophoretic dispersion ($\Delta S_{i,j=4}$), and wall adsorption ($\Delta S_{i,j=5}$) decrease as E increases, which is mainly due to the decrease of solute migration time. Heat transportation ($\Delta S_{e,1}$) is negative across the whole scope of field strength. Thus, it is a factor that enhances separation efficiency (making ΔS_H decrease) due to the fact that more Joule heat is taken off from the solute system as E increases. The objective function (ΔS_H) is the sum of those seven effects ($\Delta S_{e,1}$, $\sum_{j=1}^{6} \Delta S_{i,j}$), the optimal field strength (E_{opt}), which can be derived by using Equation 1.8. $\Delta S_{e,2}$ corresponds to the relative migration between two adjoining solute zones. The terms of energy transportation ($\Delta S_{e,1}$) and

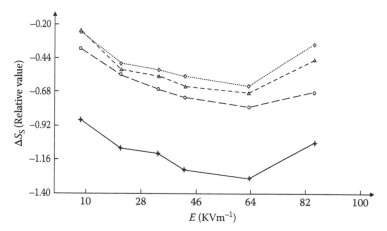

FIGURE 1.7 Dependence of separation efficiency (ΔS_S) on field strength for the experimental data between SDS proteins in non-gel sieving of capillary electrophoresis. $\Diamond$, phosphorylase B and β-galactosidase; $\triangle$, ovalbumin and serum albumin; $\circ$, serum albumin and phosphorylase B; $+$, trypsin inhibitor and carbonic anhydrase. (From Liang et. al, *Journal of Chromatography A*, 763, 237–251, 1997. With permission.)

mass transportation ($\Delta S_{e,2}$) are negative across the whole scope of field strength. $\Delta S_{e,1}$ and $\Delta S_{e,2}$ are more negative, which correspond to the enhanced separation system as E increases. This is why high field strength is adopted as the driving force for CE separation with fine cooling (making $\Delta S_{e,1}$ more negative). The objective function (ΔS_S) is the sum of those eight effects ($\sum_{i=1}^{2} \Delta S_{e,i}$ and $\sum_{j=1}^{6} \Delta S_{i,j}$). The optimal field strength (E_{opt}) is shown in Figure 1.6, and corresponds to the minimum of ΔS_S for the best separation efficiency, and it can be derived by using Equation 1.9. Figure 1.7 shows the relationship between ΔS_S and E from our experimental results for the separation between two adjoining SDS proteins in non-gel sieving of CE [82]. The similar shape between Figures 1.6 and 1.7 reflects that the experimental data support our theoretical prediction of Equation 1.9. Experimental data that support Equation 1.9 can be found in the literature [91]. Analogously, the cooling temperature, the concentration of buffer, the radius and length of capillaries, and so on, could be optimized with the objective functions in the nonequilibrium thermodynamic separation model.

1.3.6   TIME-VARYING MIGRATION IN MNB ON IMMOBILIZED pH GRADIENT STRIPS

MNB is an important foundation to understand and improve IEF [27]. However, there are obstacles to carry out the theoretical studies of MNB on IPG strips due to the unknown local concentrations of CAs on commercial IPG strips and the time-varying phase boundary velocities. Aiming at the problems in the IPG–MNB systems, we introduce a recursion approach to extend the current MNB theories into the space-time-varying MNB system [16,17].

The recursion approach [16,17] emphasizes the localizations of physicochemical parameters in the discrete time intervals and local positions in Lagrangian coordinates, such as local and relative equivalent concentrations of CAs, the local β-phase OH concentrations, the local phase boundary velocity, local relative judgment expression, and so on. The boundary position recursion equation in a complete time sequence was presented to quantitatively predict the MNB position-time curves by distinguishing three kinds of titration cases according to the NaOH concentrations in rehydration buffers on IPG–MNB strips. The theoretical position-time curves were satisfactorily validated by the corresponding images of boundary migrations achieved from the IPG–MNB experiments with some typical NaOH concentrations-bromophenol blue-rehydration buffers on pH 4–7 IPG strips [16].

For ammonia-rehydration buffers [17], the time-varying migration processes of MNB on IPG strip are determined by both time-varying dissociation equilibria of ammonia and position-varying pH environments formed by immobilized CAs on the IPG strip. Thus, the local dissociation equilibria of ammonia and the position-varying pH are introduced into the recursion equation of position of MNB migrations. The theoretical position-time curves and the velocity-time curves of MNB migrations obtained by the recursion approach were satisfactorily validated by a series of images of boundary migrations from the IPG–MNB experiments by using rehydration buffers with different ammonia concentrations on pH 3–6 IPG strips [17]. The results achieved herein have evident significance to a quantificational understanding of the mechanism of MNB and IEF with the point of view of nonequilibrium thermodynamics.

1.3.6.1 Recursion Equations for NaOH–IPG–MNB System

The velocities of the boundary displacement are not constant, but are related to the boundary positions at the length direction of IPG strip due to the space-time-varying-pH gradients [16]. The positive direction of the x coordinates (as shown in Figure 1.8) is from the anode to the cathode along the IPG strip. The time that the boundary

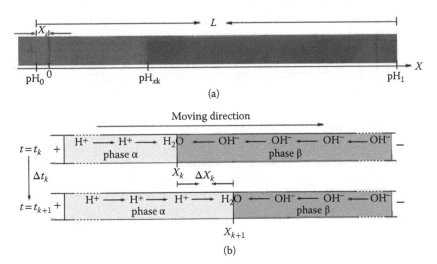

FIGURE 1.8 (See color insert.) Diagrams of Lagrangian coordinates and the boundary migrations at two adjacent local times in NaOH–IPG–MNB system. (From Liang et al., *Electrophoresis*, 30, 3134–3143, 2009. With permission.)

moves is divided into a series of time intervals, $\Delta t_k = t_{k+1} - t_k$. Thus, the local migration distance, Δx_k, of the MNB can be expressed as discrete recursion format [11–16]

$$\Delta x_k = x_{k+1} - x_k \tag{1.10}$$

According to the local equilibrium assumption of nonequilibrium thermodynamics [6,48,49], Figure 1.8 shows the adjacent local positions, x_k and x_{k+1}, of the boundary at two adjacent times, t_k and t_{k+1}. The local equilibrium assumption ensures that all physicochemical parameters or relationships, which are tenable in classical thermodynamics and kinetics, also hold water in the mesoscopic matter cells in the local space-time regions (Δx_k, Δt_k) with the Lagrangian description [54,55].

Since the gel strips used in the experiments have IPG, the different titration relations will be obtained when different NaOH concentrations are added into rehydration buffers. They can be described as the three kinds of different situations. Firstly, if $C_{A'(x_k),\mathrm{act}} - C_{B'(x_k),\mathrm{act}} \geq C_{\mathrm{alk}}$, which means that the added NaOH concentrations, C_{alk}, in rehydration buffers are less than or equal to the difference between the concentrations of the acidic Immobiline A and the basic Immobiline B, the situation can be considered as titrating acidic Immobiline A with NaOH and the basic Immobiline B. Therefore, the local concentrations of H⁺ in the β phase, $C^{\beta}_{\mathrm{H}^+(x_k),\mathrm{act}}$, can be expressed as

$$C^{\beta}_{\mathrm{H}^+(x_k),\mathrm{act}} = K_A \frac{\left[C_{A'(x_k),\mathrm{act}} - \left(C_{\mathrm{alk}} + C_{B'(x_k),\mathrm{act}} \right) \right]}{\left(C_{\mathrm{alk}} + C_{B'(x_k),\mathrm{act}} \right)} \tag{1.11}$$

As a pair of conjugate bases and acids, the local concentrations of OH⁻ in the β phase, $C^{\beta}_{\mathrm{OH}^-(x_k),\mathrm{act}}$, is given by

$$C^{\beta}_{\mathrm{OH}^-(x_k),\mathrm{act}} = \left(\frac{K_W}{K_A} \right) \cdot \frac{\left(C_{\mathrm{alk}} + C_{B'(x_k),\mathrm{act}} \right)}{\left[C_{A'(x_k),\mathrm{act}} - \left(C_{\mathrm{alk}} + C_{B'(x_k),\mathrm{act}} \right) \right]} \tag{1.12}$$

Secondly, if $C_{A'(x_k),\mathrm{act}} > C_{\mathrm{alk}} > C_{A'(x_k),\mathrm{act}} - C_{B'(x_k),\mathrm{act}}$, the acidic Immobiline A will be partially neutralized by NaOH, then basic Immobiline B will be titrated with the residual Immobiline A. In this case, the local concentrations of OH⁻ in the β phase, $C^{\beta}_{\mathrm{OH}^-(x_k),\mathrm{act}}$, is given by

$$C^{\beta}_{\mathrm{OH}^-(x_k),\mathrm{act}} = \left(\frac{K_W}{K_B} \right) \frac{\left[C_{B'(x_k),\mathrm{act}} - \left(C_{A'(x_k),\mathrm{act}} - C_{\mathrm{alk}} \right) \right]}{\left(C_{A'(x_k),\mathrm{act}} - C_{\mathrm{alk}} \right)} \tag{1.13}$$

Thirdly, if $C_{\mathrm{alk}} \geq C_{A'(x_k),\mathrm{act}}$, the acidic Immobiline A will be completely neutralized, and the partial NaOH and the basic Immobiline B will be left in the β phase of the IPG gel. In this case, $C^{\beta}_{\mathrm{OH}^-(x_k),\mathrm{act}}$ can be obtained by

$$C^{\beta}_{\mathrm{OH}^-(x_k),\mathrm{act}} = C_{\mathrm{alk}} - C_{A'(x_k),\mathrm{act}} + \sqrt{C_{B'(x_k),\mathrm{act}} \cdot K_W / K_B} \tag{1.14}$$

Therefore, the actual local concentrations of OH⁻ in the β phase, $C^{\beta}_{\mathrm{OH}^-(x_k),\mathrm{act}}$, can be obtained by Equations 1.11 through 1.14. Lastly, the local concentrations of H⁺ in the α phase, $C^{\alpha}_{\mathrm{H}^+(x_k),\mathrm{act}}$, can be directly calculated by

$$C^{\alpha}_{H^+(x_k),\,act} = 10^{-pH_{xk}} \qquad (1.15)$$

The discrete form of recursion equation of boundary displacement of the space-time-varying MNB can be obtained by

$$x_{k=n} = \sum_{k=1}^{n} \left[\frac{\left(m^{\alpha}_{H^+,\,act} \cdot \gamma_{H^+} \cdot C^{\alpha}_{H^+(x_k),\,act} - m^{\beta}_{OH^-,\,act} \cdot \gamma_{OH^-} \cdot C^{\beta}_{OH^-(x_k),\,act} \right)}{\left(\gamma_{H^+} \cdot C^{\alpha}_{H^+(x_k),\,act} - \gamma_{OH^-} \cdot C^{\beta}_{OH^-(x_k),\,act} \right)} \cdot (i/\kappa) \right] \cdot \Delta t_k \quad (1.16)$$

With Equation 1.16, the discrete position-time curves (x_k, t_k) of the boundary movement can be obtained by the theoretical prediction. The local concentrations of H^+ and OH^- ions are calculated in the local space-time regions $(\Delta x_k, \Delta t_k)$ of the IPG–MNB system with Equations 1.11 through 1.13.

The three time sequences of images of boundary movement of neutralization reactions on 4–7 pH IPG strips are shown in Figure 1.9. The experimental position-time curves (x_k, t_k) can be accurately obtained to explore the time-varying velocity of the moving boundary of IPG–MNB systems.

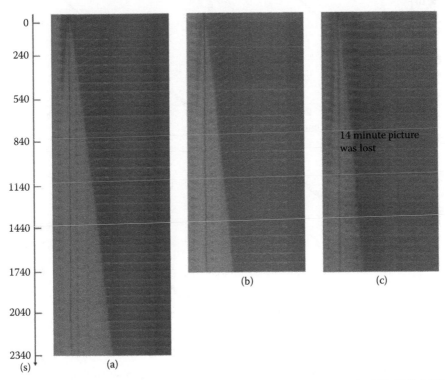

FIGURE 1.9 (**See color insert.**) Images of the moving boundaries in the NaOH–IPG–MNB systems with 4–7 pH IPG strips. (a) 5 mmol·L^{-1}, (b) 10 mmol·L^{-1}, (c) 12 mmol·L^{-1} with strong alkali rehydration buffers (C_{alk} NaOH + 0.1 mol L^{-1} NaCl + 0.1% (w/v ethanol) bromophenol blue). The IEF experiments were carried on a **PROTEAN** IEF Cell with limitative current ($I_1 = 5 \times 10^{-5}$ A) and temperature 25°C.

The experimental and theoretically predicted position-time curves of the moving boundary are shown in Figure 1.10. The recursion equation (Equation 1.16) of boundary displacement is nonlinear and space-time varying, although they are two quasi-lines for the two lower NaOH concentrations (2 and 5mmol·L^{-1}). The local velocity of phase boundary of the lower NaOH concentration (2 mmol·L^{-1}) is bigger than the one of higher NaOH concentration (mmol·L^{-1}) by comparing the local boundary displacements (e.g., 7.682 mm in Figure 1.10a and 7.664 mm in Figure 1.10b) at the same time $t_k = 1500$ seconds. As increasing C_{alk} (e.g., 10 and 12 mmol·L^{-1}) the position-time curves become the curved ones (as shown in Figure 1.10c and d) from the above quasi-lines. And the curve in Figure 1.10d has more down-bend trend than the one in Figure 1.10c by comparing the special points, (x_k, t_k), such as (3.092 mm, 540 seconds) and (6.803 mm, 1500 seconds) with $C_{alk} = 10$ mmol·L^{-1}, as well as (2.510 mm, 540 seconds) and (5.779 mm, 1500 seconds) with $C_{alk} = 12$ mmol·L^{-1}. Similarly, theoretical

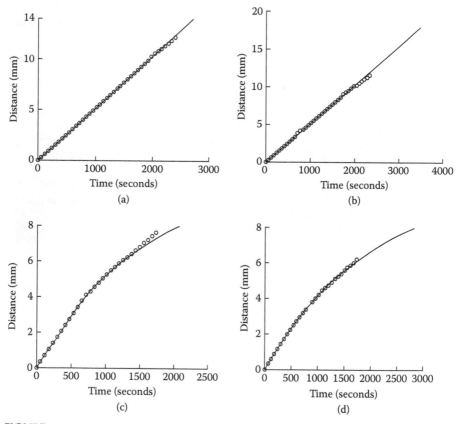

FIGURE 1.10 Comparison of the predicted and experimental position-time curves of the moving boundary in NaOH–IPG–MNB system. The solid lines and "o" denote the theoretically predicted and experimental position-time curves, respectively, of the moving boundary under the same conditions except NaOH concentrations, for four different NaOH concentrations (C_{alk}) in strong alkali rehydration buffers, (a) 2 mmol·L^{-1}, (b) 5 mmol·L^{-1}, (c) 10 mmol·L^{-1}, (d) 12 mmol·L^{-1}.

position-time curves with the recursion equation, Equation 1.16, satisfactorily agreed with experimental points in Figure 1.10c and Figure 1.10d. We easily found that the local velocity, μ_k, of α–β phase boundary of MNB has different values in different time intervals. And it also validates that the recursion equation approach with Equation 1.16 can accurately predict the moving boundary and their migration positions in this MNB system with the time-varying velocity of phase boundary.

1.3.6.2 Recursion Equations for IPG–MNB Migration in Weak Base Buffer

For IPG–MNB migration in weak base buffer, C_{H^+,x_k}^{α} and C_{OH^-,x_k}^{β} in Equation 1.16 are determined by the local dissociation equilibrium on MNB [17]. The concentration of conveyable OH^- on pH 3–6 IPG strip in the ammonia-rehydration buffer is principally determined by the ammonia dissociation equilibrium in the mobile β phase at x_k [17]. However, the ammonia dissociation equilibrium of in the mobile β phase is affected by the pH_{x_k} environments in the neighboring region of a confined surface of the grafted CAs in the immobilized β phase at x_k. Thus, according to Henderson–Hasselbalch equation, the ammonia dissociation equilibrium at x_k in the mobile β phase on the IPG strip can be given by

$$14 - pH_{x_k} - pK_b = \log \frac{[NH_4^+]_{x_k}}{[NH_3 \cdot H_2O]_{x_k}} \tag{1.17}$$

where pK_b is the negative logarithm of dissociation constant of ammonia. Considering the mass conservation law ($[NH_3 \cdot H_2O]_{x_k} + [NH_4^+]_{x_k} = C_{NH_3 \cdot H_2O}$) for the ammonia dissociation equilibrium in the mobile β phase, the concentration of OH^- ions in the mobile β phase of the ammonia–IPG–MNB system at x_k, C_{OH^-,x_k}^{β}, can be written as

$$C_{OH^-,x_k}^{\beta} = [NH_4^+]_{x_k} = \left[\frac{10^{14-pH_{x_k}-pK_b}}{1+10^{14-pH_{x_k}-pK_b}} \right] \cdot C_{NH_3 \cdot H_2O} \tag{1.18}$$

By putting Equations 1.15 and 1.18 into Equation 1.16, an integrated equation of position recursion is obtained for the time-varying process of ammonia–IPG–MNB migration. The position-time curve or the velocity-time curve of the time-varying migration process of the ammonia–IPG–MNB system can be calculated using the integrated recursion equation.

Direct observation of the MNB images in Figure 1.11 reveals that the boundary migrations along the time coordinates are slightly curved. It indicates that the weak base IPG–MNB goes through the nonuniform velocity migration due to the pH gradient along the separation axis of the IPG strip. The nonuniform velocity migration of the MNB primarily resulted from the direct relation of pH_{x_k} and C_{OH^-,x_k}^{β} to x_k, which is shown by Equations 1.15 and 1.18. In addition, different ammonia concentrations, $C_{NH_3 \cdot H_2O}$, in the rehydration solutions resulted in slight difference among the position-time curves by direct observation of the MNB images, as shown in Figure 1.11a through e.

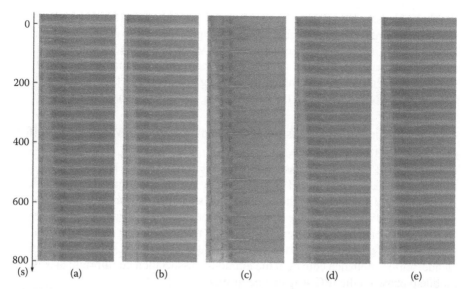

FIGURE 1.11 **(See color insert.)** Images of the moving boundaries in ammonia–IPG–MNB systems on IPG strips (pH 3–6). $C_{\mathrm{NH_3 \cdot H_2O}}$: (a) 0.20, (b) 0.25, (c) 0.30, (d) 0.35, (e) 0.40 mmol·L^{-1} in the 125 µL weak base rehydration solution confected with ammonia of different concentrations, 0.1 mol·L^{-1} KCl and 0.1% (w/v ethanol) bromophenol blue. The IEF experiments were performed using a PROTEAN IEF Cell and IPG strip (pH 3–6). (From Liang et al., *Journal of Separation Science*, 34, 1212–1219, 2011. With permission.)

The weak base (ammonia) rehydration buffer–IPG–MNB system involves the time-varying movements of MNB on IPG strips. In this system, $C^{\beta}_{\mathrm{OH^-},x_k}$ is determined by the local dissociation equilibrium of the weak base (e.g., ammonia) in the local pH environment (pH$_{x_k}$) of grafted CAs at a given position x_k. Equation 1.18 was used to calculate $C^{\beta}_{\mathrm{OH^-},x_k}$ for the ammonia-rehydration buffer–IPG–MNB system. The time-varying characteristics of the boundary movement process in the weak base–IPG–MNB system were discussed on the basis of the position-time curves and the velocity-time curves obtained from the experiments and predictions. The position-time varying local concentrations, $C^{\alpha}_{\mathrm{H^+},x_k}$ and $C^{\beta}_{\mathrm{OH^-},x_k}$, obtained by Equations 1.15 and 1.18, respectively, were introduced into the recursion equation-based calculation program. By comparing the experimental and theoretically predicted position-time curves (as shown in Figure 1.12) of the moving boundaries, the application of the recursion approach in the weak base (ammonia)–IPG–MNRB system was verified. In Figure 1.12a, the predicted position-time curve satisfactorily coincides with the experimental one. Except for the curve in Figure 1.12a, all the other predicted position-time curves in Figure 1.12 are located below the experimental ones, although the predicted and experimental curves bend toward the same directions. The discrepancy between the predicted and experimental position-time curves is possibly attributed to the measurement error of the void strip length x_s of the MNB system. Figure 1.12 shows that the MNB displacement decreased with the increase in ammonia concentration. The nonuniform velocity of migration of the MNB determined by the local dissociation equilibria of ammonia and the grafted CAs, which

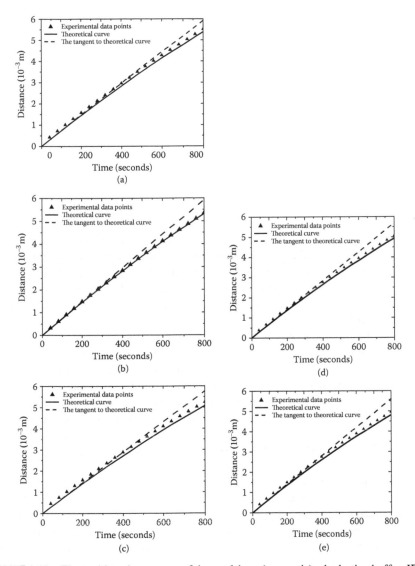

FIGURE 1.12   The position-time curves of the weak base (ammonia) rehydration buffer–IPG–MNB system. Figures (a) through (e) denote $C_{NH_3 \cdot H_2O}$ 0.20, 0.25, 0.30, 0.35, and 0.40 mmol·L^{-1} in the weak base rehydration solutions, respectively. The black triangles, ▲, denote the experimental measurement results. The solid curves denote theoretical prediction curves theoretically derived computer simulation results. with Equations. 1.15, 1.18, and 1.16. The tangent lines were artificially introduced to the solid curves at the origin of the time coordinates.

is related to the position or time, was calculated by the position recursion approach. Thus, the recursion approach is fully qualified to calculate the MNB migration with the nonuniform velocity for the pH IPG strips with the weak base rehydration buffer. The recursion approach may also provide a deeper understanding of IEF such as the instability of pH gradient.

1.4   RECURSION EQUATIONS IN PREDICTING BAND WIDTH UNDER GRADIENT ELUTION

With the mesoscopic description approach [92], we can consider that a separation process is a set of the combination of solutes' volume elements that simply stretch and move on separation path in external force fields. In separation science, space- and time-dependent fields, $u_{i+1,k}$, generally include the chemical potential between mobile and stationary phase due to the space- and time dependent solvent strength under gradient elution [5,63], or due to the space- and time-dependent solute concentration in nonlinear chromatography [68], or due to time-varying electric field in CE [69], and so on. With the local equilibrium assumption, we can define a set of local thermodynamic variables in given volume element j of the solute zone i at the time t_k in separation processes, such as local molar concentration, $c_{i,j,k}$, local migration velocity, $v_{i,j,k}$, local diffusion coefficient, $D_{i,j,k}$, local thermodynamic distribution constant, $K_{i,j,k}$, local capacity factor, $k'_{i,j,k}$, and so on. Accordingly, we can define local external force fields that act on the given volume element j of the solute zone i at the time t_k, these local external force fields include local strong eluent concentration of mobile phase, $\varphi_{i,j,k}$, local chemical potential, $\mu_{i,j,k}$, electrical field strength, $E_{i,j,k}$, and so on. One can use any thermodynamic laws with local thermodynamic variables in the identifiable volume elements [4,6].

The evolution of solute zone under gradient elution is a typical problem of nonlinear continuity equation since the local diffusion coefficient and local migration velocity of the mass cells of solute zones are the functions of position and time due to space- and time-variable mobile phase composition. Based on the mesoscopic approaches (Lagrangian description [56], the continuity axiom [33], and the local equilibrium assumption [6,48,49]), the evolution of solute zones in space- and time-dependent fields is described by the iterative addition of local probability density of the mass cells of solute zones. Furthermore, on macroscopic levels, the recursion equations have been proposed to simulate zone migration and spreading in reversed-phase high-performance liquid chromatography (RP-HPLC) through directly relating local retention factor and local diffusion coefficient to local mobile phase concentration. Actually, bandwidth recursion equation is the accumulation of local diffusion coefficients of solute zones to a series of discrete-time slices. This approach differs entirely from the traditional theories on plate concept with Eulerian description, since bandwidth recursion equation is actually the accumulation of local diffusion coefficients of solute zones to discrete-time slices. Recursion equations and literature equations were used in dealing with same experimental data in RP-HPLC, and the comparison results show that the recursion equations can accurately predict bandwidth under gradient elution. Although the concept of theory plate is appropriate to indicate column efficiency, it fails to predict bandwidth under gradient elution.

In Figure 1.13, we arbitrarily chose three adjoining volume elements of the two components, i and $i + 1$, which are in the region $x_0 \sim x_0 + \delta_{j,k=0}$ ($j = 1, 2, \cdots, J$), where $\delta_{j,k=0}$ is the length of the volume element j on the x axis along separation path at initial time $t_{k=0} = 0$ in Figure 1.13a. The local solute densities of two components,

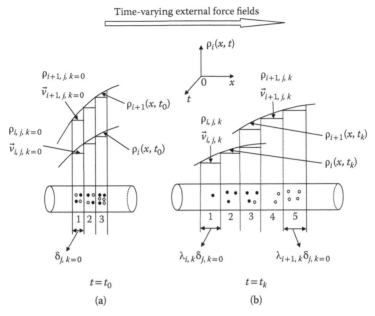

FIGURE 1.13 The two motions of infinitesimal volume elements of solute zones are in both typical motion forms, "deformation" (spreading) and "flow" (relative migration) in time-varying external force fields at the time interval between (a) and (b). (From Liang and Liu, *Journal of Chromatography A*, 1040, 19–31, 2004. With permission.)

i and $i + 1$, are $\rho_{i,j,k=0}$ and $\rho_{i+1,j,k=0}$ in the region. Apparently, these local solute densities determine corresponded macro density distributions, $\rho_i(x,t_0)$ and $\rho_{i+1}(x,t_0)$, of solute zones. Local velocities of the volume elements of the components i and $i + 1$ can be indexed by time-varying velocities, $\vec{v}_{i,j,k}$ and $\vec{v}_{i+1,j,k}$, and so volume elements are in time-varying external fields. For any effective separation process, there is the relationship $\vec{v}_{i,j,k} \neq \vec{v}_{i+1,j,k}$, which just causes the slip motion between the volume elements of two separable components. Along the direction of "the arrow of time" [93] ($t_k < t_{k+1}$), the solute system will evolve a new state that the volume elements would have a new local solute densities in Figure 1.13b, where $\rho_{i,j,k}$ in the region $x_0 \sim x_0 + \lambda_{j,k}\delta_{j,k=0}$ and $\rho_{i+1,j,k}$ in $x_k \sim x_k + \lambda_{i+1,k}\delta_{j,k=0}$ with local velocities, $\vec{v}_{i,j,k}$ and $\vec{v}_{i+1,j,k}$, respectively, where $\lambda_{i+1,k}$ is the stretch ratio of volume elements at time t_k.

We may choose $[-2\sigma, 2\sigma]$ as the research region of a solute zone, where σ is standard deviation of solute zone distribution. Based on the local equilibrium assumption and the continuum principium (as shown in Figure 1.13), the general diffusion recursion equation of the cells of the solute zone was obtained as follows,

$$m_{i,j,k+1} = \sqrt{m_{i,j,k}^2 + \left(\frac{32}{J^2}\right)D_{i,j,k}\Delta t_k} + \left(\vec{v}_{i,j+1,k} - \vec{v}_{i,j,k}\right)\Delta t_k \qquad (1.19)$$

where

$$D_{i,k} = \frac{0.09023L^2}{t_0^3} \frac{\left(a_i + b_i \cdot k_{i,k}'\right)^2}{\left(1 + k_{i,k}'\right)^3} \tag{1.20}$$

where $m_{i,j,k+1}$ and $m_{i,j,k}$ are the volume element width along column axes of the cell j of solute i at the times t_{k+1} and t_k, respectively. $\bar{v}_{i,j+1,k}$ is the linear velocity of the cell $j + 1$ of the solute i in t_k along x coordinates. $D_{i,k}$ is the local diffusion coefficient under gradient elution. a_i and b_i are experimental constants of the component i depending on the solute and on the chromatographic system. t_0 is the column dead time, and L is the column length.

The first term on the right-hand side of Equation 1.19 is the local diffusion term of the cell j, which means the accumulation of the local apparent diffusion coefficient to the migration time, and the increment of this term is always positive since $D_{i,j,k} > 0$. And the second term on the right-hand side of Equation 1.19 is the contribution to the width of the cell due to the overlapping and stretching between adjacent cells, and this term may be negative, zero, or positive, which brings the distortion of band shape. This term plays an important role in the introduction of the vector of the local shock wave in nonlinear chromatography [13]. The half-bandwidth with time unit ($w_{1/2,t,i}$) is given by

$$w_{1/2,t,i} = 2.354\sqrt{\frac{\sigma_0^2 + \sigma_d^2 + \sum_{k=0}^{K-1} 2D_{i,k}\Delta T \left(1 + k_{i,k}'\right)/k_{i,k}'}{\bar{v}_{i,K-1}^2}} \tag{1.21}$$

where σ_0 and σ_d are the standard deviations of solute band that arise from sample injection and gradient dwell time t_d. $D_{i,k=0}$ is the local diffusion coefficient of the solute i in the mobile phase composition $\varphi_{k=0}$. The parameter $\bar{v}_{i,K-1}$ is the local migration velocity of the solute zone along x coordinates in stationary phase in the last time interval, Δt_{K-1}. The parameters $k_{i,k}'$ and $D_{i,k}$ directly relate to predigest local strong eluent φ_k respectively, so $w_{1/2,t,i}$ relates to strong eluent concentration sequence, $\{\varphi_k\}$, through the two local coefficients, $k_{i,k}'$ and $D_{i,k}$. It implies that different $\{\varphi_k\}$ induces different $w_{1/2,t,i}$. Equation 1.21 is the bandwidth equation of iterative addition form corresponding to $\{\varphi_k\}$ in RP-HPLC.

Figure 1.14 shows distinctly the effect of the concentration of the stronger solvent on diffusion coefficients under isocratic elution. The diffusion coefficients increase with the addition of the organic solvent concentration, and the degree of increase changes with different solutes, which means that the local diffusion coefficients change according to the local concentration of the mobile phase in gradient elution processes. The different solutes have also different diffusion coefficients in the same concentration of the mobile phase. Analysts commonly calculate bandwidths through the number of theoretical plate with the assumption that the number of theoretical plates is independent of the mobile phase composition throughout the separation progress, but Figure 1.14 indicates that the assumption is no true in gradient elution processes. In fact, the number of theoretical plate depends not only on the column-self but also on the solute, concentration, and composing of the mobile phase. It implies that it is not appropriate to predict half-bandwidth with the average theoretical plates under gradient elution.

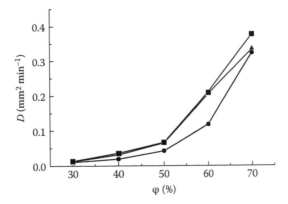

FIGURE 1.14 Dependence of the diffusion coefficient in the isocratic elution on mobile-phase compositions. The D values were calculated with Equation 1.21 through the experimental the retention time and the half-bandwidth of the solute i, $t_{R,i,e}$, and $w_{1/2,i,e}$, on 50-mm column (■) 2-nitrotoluene, (●) 3-nitrotoluene, (▲) 4-nitrotoluene. (From Liang and Liu, *Journal of Chromatography A*, 1040, 19–31, 2004. With permission.)

1.5 LAGRANGIAN DESCRIPTION IN NONLINEAR CHROMATOGRAPHY

1.5.1 LOCAL LAGRANGIAN APPROACH

Under the framework of NTST [8–17], the local Lagrangian approach (LLA) was proposed to deal with the essential issues of the convection and diffusion (shock waves) phenomena in nonlinear chromatography with recursion equations based on the three basic theorems: Lagrangian description, the continuity axiom, and the local equilibrium assumption. This approach is remarkably distinguished from the system of contemporary chromatographic theories (Eulerian description–partial differential equations) [1,2,20,23,46,80] and can felicitously match modern information theory and cybernetics.

Continuity axiom in hydrodynamics is also visually called the "non-interpolating axiom," which does not allow dislocation movements or slippage phenomena among matter cells [94]. The Lagrangian description [33,56] in hydrodynamics adopts the matter cells, which contains the matter with its constant quantity in the space-time-varying flow field, and the matter cells can be regarded as a single particle to study its movement track on a macro-scale, which parallels the Eulerian description. Local equilibrium assumption [6,48,49] is the cornerstone of nonequilibrium thermodynamics, and the thermodynamic relationship in macro-equilibrium states also works in local or mesoscopic matter cells. When Lagrangian description is adopted to describe the evolvement of the position and distribution of solute bands in space-time-depended field of solute distribution itself, it is convenient to use local equilibrium assumption. It is impossible to use traditional Lagrangian description to depict the movement of matter cells in the whole process of nonlinear chromatography. Nevertheless, taking cognizance of the fact that the three basic theorems, Lagrangian description, the continuity axiom, and the local equilibrium assumption, can only be simultaneously valid in the tiny time interval (Δt), in the nonequilibrium processes, then the Lagrangian description can be used to deal with matter cells in Δt, and the position and width of matter cells can be calculated from the time t_k to the time t_{k+1} by introducing a series of localized physical parameters and corresponding local function relationships according to local equilibrium assumption. To ensure the consistency between the continuity axiom and the Lagrangian description, it is necessary that the density and width of "overlapped" or "separated" adjacent solute cells, which stem from the independent calculations based on the formulas of definition of local migration and diffusion of solute cells in Δt, are adjusted according to respective distribution of each solute cell. The operation corresponds to achieving mass transfer among adjacent matter cells, which will make the hydrodynamic calculations simultaneously satisfy the three basic theorems. We name the method LLA, which is used to deal with essential issues of the convection and diffusion phenomena in hydrodynamics based on the three basic theorems.

1.5.2 LAGRANGIAN COORDINATES

The Lagrangian coordinates are visually shown in Figures 1.15 and 1.16 [12]. Let $i = 1, 2, ..., I$ sign each solute in a solute system. $I = 1$ for the single-component system. The time that the solute zones evolve in chromatographic column is divided into a series of time intervals, Δt or $\Delta t_k = t_{k+1} - t_k$, where the subscript k is the serial number of the time grids, $k \in \mathbf{K}_+$, where $\mathbf{K}_+$ is a countable time-set, $\mathbf{K}_+ := \{0, 1, ..., k, ..., K\}$. At appointed time, $t = t_k$, the region, $\kappa \cdot \delta_{i,k}$, of solute zones is chosen as a research region, which always covers the whole bandwidth of solute space distribution in both solid phase and mobile phase along the axial direction of the chromatographic column, where κ is region factor and $\delta_{i,k}$ is the standard deviation of the solute zone i at the time t_k. The defined research region of the solute zone is further divided into a series of solute cells with an equal width along the positive direction of x coordinates. As shown in Figure 1.15, all solute cells are adjacent to each other, but not overlapping or separating each other, marked successively by the serial number, $j = 1, 2, ..., J$, where J is the total number of solute cells of solute zones. This definition of solute cells incarnates the characteristic of continuity axiom in Lagrangian description. The three adjacent cells, the middle cell j (we studied), its previous adjacent cell $j + 1$, and its after adjacent cell $j - 1$ (see Figure 1.17), form a cell unit j, which belongs to a particle-in-cell method [34]. The local equilibrium assumption in linear nonequilibrium thermodynamics ensures that all physicochemical parameters or relationships, which are tenable in classical thermodynamics and kinetics, also hold water in the mesoscopic matter cells in Lagrangian description. Therefore, a number of local physicochemical parameters, such as local solute concentration, local equilibrium isotherms, local capability factor, local migration velocity, and local diffusion coefficient, and corresponding local physicochemical relationships were naturally introduced in the Lagrangian description [33,56].

The width and local solute probability density of cell j at time t_k are indexed as $m_{i,j,k}$ and $p_{i,j,k}$, respectively. They vary with the time t_k and the space x or j. The local probability density, $p_{i,j,k}$, indicates the proportion of quantity of solute i (in stationary and mobile phases) in the solute cell j to the total quantity of the solute band at time t_k.

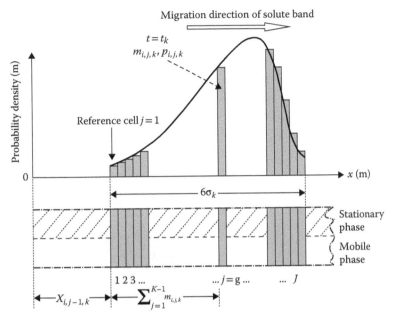

FIGURE 1.15 Spatial distribution of solute band and divided cells at appointed time t_k with Lagrangian coordinates. (From Liang and Liu, *Science in China Series B-Chemistry*, 47, 443–452, 2004. With permission.)

In Figure 1.15, the origin of x coordinates is located at the injecting sample end of the chromatographic column, and the hind edge of the solute cell $j = 1$, which is always appointed as the reference cell in the process of band evolvements, in initial spatial distribution of the solute band is at the origin of x coordinates at time t_0. The cell $j = g$ of the solute i is chosen as research object. Figure 1.16a shows the state of a cell defined by the Lagrangian description at two adjacent times t_k and t_{k+1}. In the cell, there is a local equilibrium of solute distribution between stationary and mobile phases, and we suppose that mass transfer resistances in the two phases can be ignored, and the equilibrium can be instantly reached. According to the Lagrangian description, the solute quantity in the cell and in Δt_k keeps constant. In Δt_k, the solute in the cell does not, by convection and diffusion, exchange with adjacent cells through traversing the cell interfaces. It is just that the local solute migration of the cell changes its position, which indicates the trajectory movement of matter cells in the Lagrangian description, and local solute diffusion of the cell changes its width, which reflects the feature of the deformation of the solute cells allowed by the Lagrangian description. In addition, the overlapping and separating between adjacent solute cells in nonlinear chromatography also contribute to the deformations of the solute cells. Figure 1.16b shows the volume cell with the Eulerian description, which is well known in the science field. The volume cell is fixed at the certain position of the column, and the movements of convection and diffusion make the solute penetrate the interface of the volume cell, and consequently, the solute concentration of the volume cell changes with time. The convection–diffusion partial differential equations are established

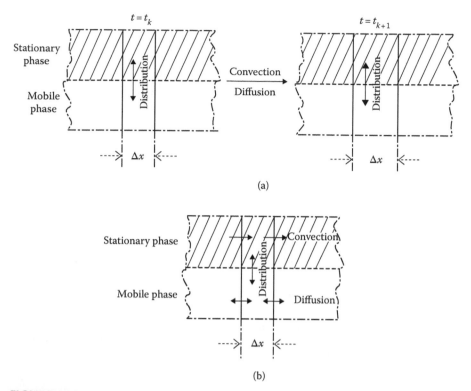

FIGURE 1.16 Two methods of dividing cell based on Lagrangian and Eulerian descriptions in chromatographic theories. (From Liang and Liu, *Science in China Series B-Chemistry*, 47, 443–452, 2004. With permission.) (a) Solute cell with Lagrangian description; (b) Volume cell with Eulerian description.

based on mass conservation in the volume cell. Comparing the two methods of dividing cells in Figure 1.16a and b, the marked differences were found (see Section 1.2.6).

1.5.3 VIRTUAL THERMODYNAMIC PATH OF NONLINEAR NONIDEAL CHROMATOGRAPHY

In Figure 1.17, the spatial distribution conformations of the cell j and its adjacent cells are classified into four types (convex, concave, compressing, and spreading) during Δt_k. The virtual thermodynamic path was designed by applying the property of state functions in thermodynamics that the changes of value of state functions depend not on path but only on the initial and final states [95]. In the virtual thermodynamic path, the cells firstly undergo net diffusion processes from the t_k states to arrive at virtual diffusion states. Then they undergo net migration processes to reach virtual overlapping-separating states at the time t_{k+1}. Lastly, according to the continuity axiom, overlapping-separating cells are adjusted to recover the continuum distributions of the t_{k+1} states, synchronously to realize the mass transfer of the solute between adjacent cells. As an example, Figure 1.17 illustrates the key steps of LLA with the Langmuir isotherm with convex structure ($b > 0$).

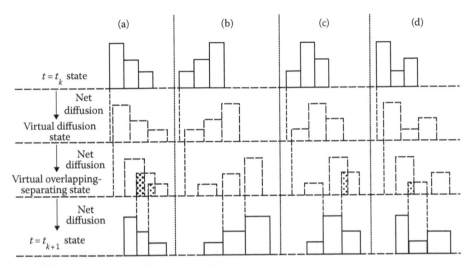

FIGURE 1.17   Net diffusion, net migration, and mass transfer of the four types of solute cells. (From Liang and Liu, *Science in China Series B-Chemistry*, 47, 443–452, 2004. With permission.) (a) Compressing type, (b) spreading type, (c) convex type, (d) concave type. The cell j and its two adjacent cells, the one before it, $j + 1$, and the one after it, $j - 1$ are successively marked along x coordinates. Convex Langmuir isotherm ($b > 0$) works as an example.

Taking into account the relationship between capacity factor (k') and migration velocity of solute band, the local migration velocity ($v_{i,j,k}$) [11] of the cell defined by Lagrangian description along x coordinates of chromatographic column is written as

$$v_{i,j,k} = \frac{u}{1 + k'_{i,j,k}} \tag{1.22}$$

where u is the linear flow velocity of mobile phase $k'_{i,j,k}$ is local capacity factor. k_{ijk} is local capacity factor. In virtual overlapping-separating states, the widths, $\Delta m_{i,(j+1,j),k+1}$, of overlapping or separating between the adjacent cells j and $j + 1$ (or $j - 1$) are defined as

$$\Delta m_{i,(j+1,j),k+1} = \left(v_{i,j+1,k} - v_{i,j,k} \right) \Delta t_k \tag{1.23}$$

where $\Delta m_{i,(j+1,j),k+1}$ is called the local shock wave. The overlapping and separating between adjacent cells are only due to the differences between their local velocity ($v_{i,j,k}$) caused by the differences of their local concentrations ($c_{i,j,k}$) in net migration processes. The magnitude of local shock wave determines the degree of overlapping or separating between the adjacent cells and their degrees for the certain isotherm. For local shock wave, $\Delta m_{i,(j+1,j),k+1}$, with the convex type of Langmuir isotherm ($b > 0$), we have

$$\Delta m_{i,(j+1,j),k+1} \quad \begin{array}{l} > 0, \ \text{cell overlapping, band compressing} \\ = 0, \ \text{linear chromatography} \\ > 0, \ \text{cell separating, band spreading} \end{array} \tag{1.24}$$

We call Equation 1.24 the criterion of generating of local shock waves. The volume element width $(m_{i,j,k+1})$ along column axes of the cell j of solute i at the time t_{k+1} is

$$m_{i,j,k+1} = m'_{i,j,k+1} + \Delta m_{i,(j+1,j),k+1}$$
$$= \sqrt{m^2_{i,j,k} + \left(\frac{\kappa^2}{J^2}\right)D_{i,j,k}\Delta t_k + \left(\vec{v}_{i,j+1,k} - \vec{v}_{i,j,k}\right)\Delta t_k} \quad (1.25)$$

Equation 1.25, called the recursion equations of cell widths [12], is applied to deal with the convection and diffusion (shock waves) phenomena. It distinctly indicates that the cell width at $t = t_{k+1}$ states is composed of two parts, the added value $(m'_{i,j,k+1})$ of cell width due to net diffusion and local shock wave $(\Delta m_{i,(j+1,j),k+1})$. The added value of cell width is codetermined by both local diffusion and local relative migration of the adjacent cells in nonlinear chromatography under the framework of LLA. Moreover, the convex and concave types in Figure 1.17 were regarded as the combination of compressing and spreading types. Thus, the cell widths of the convex and concave types can be calculated from Equation 1.25 while considering their specific cases. Keeping both the continuity axiom and the principle that the mass transfers from high-density cell to low-density one, the mass transfer of solutes among adjacent cells is realized through adjusting the cell widths and corresponding local probability densities at virtual overlapping-separating states to recover their continuum distributions at $t = t_{k+1}$ states in Figure 1.17. The results simulated with LLA are shown in Figure 1.18, which truly agreed with the corresponding experimental results in literature [96] on retention time and peak width, and the shapes of elution profiles between them also agree to a large extent.

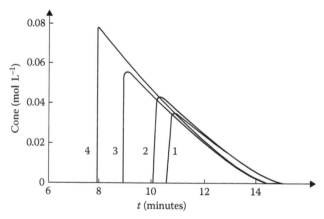

FIGURE 1.18 Simulated results with LLA of nonideal nonlinear chromatography. (From Liang and Liu, *Science in China Series B-Chemistry*, 47, 443–452, 2004. With permission.) $Q = 0.07, 0.10, 0.15, 0.25$ mmol for curves 1–4, respectively; $L = 25$ cm; 4.6 mm i.d.; $\beta = 0.2$; $u = 0.12$ cm s^{-1}; $D = 0.0003$ cm^2 s^{-1}; $\delta_0 = 0.5$ cm, $\kappa = 8$; $\Delta t = 0.05$ seconds; $J = 300$; isotherm: $q = 14.25c/(1+7.8c)$.

1.6 LOCAL LAGRANGIAN APPROACH FOR NONLINEAR EQUILIBRIUM-DISPERSIVE PROCESSES

The matrix forms of LLA were developed based on Lagrangian description for single-component in nonlinear nonideal chromatography [13]. A LTP was designed based on the Lagrangian description [54,55], the local equilibrium assumption [6,48,49], and the thermodynamic state functions [10]. With LTP, the recursion equations of fully thermodynamic states on time sequence in the matrix forms were obtained with the Markov characteristics. The convergence, compatibility, and stability of the LLA based on the LTP were discussed with some theoretical analysis and numerical experiments. The stability condition of the LLA was given. The algorithm of the LLA in the vector form was shown as the computer program to simulate the elution profiles affected by a few factors, space-distribution, axial diffusions, injection samples, and so on. The corresponding relationships were established between the trajectories of discrete-time state and discrete-time control vectors in the ergodic space. The matrix forms of the LLA remove the gap between preparative chromatography theories and optimal control approaches based on discrete-time states with state space representation (see Section 1.2.11).

1.6.1 MARKOV CHAINS OF CHROMATOGRAPHY PROCESSES

On the description of general state spaces, Markov chains of chromatography processes (as shown in Figure 1.19) [13] are defined as a sequence or a collection of determinate local thermodynamic states concerning the space-concentration distributions

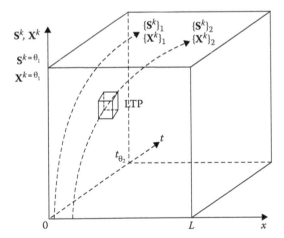

FIGURE 1.19 The ergodic state and control spaces that solute zones evolve in preparative chromatographic processes. (From Liang, *Chin J Chromatography*, 25, 664–680, 2007. With permission.)

of the solute zones (or the solute cells along the axis of chromatographic column) at the pointed time, $t = t_k$, $\mathbf{S}^k = \left\{ \mathbf{S}^{k=0}, \mathbf{S}^{k=1}, \ldots, \right\}$, $k \in \mathbf{K}_+$. The vector $\mathbf{S}^k$ in a state space (S) determinately evolves in time under the action of the vector ($\mathbf{X}^k$) of control variables at the kth stage. The Markov chains of chromatography processes remember their past trajectory only through their most recent values, and their future trajectories just depend on their present, and its past only through the current values, or they depend on that past only through knowledge of the most recent information on their trajectory. In fact, we have made the sense that the $\mathbf{S}^k$ evolving in S is governed by the principles of linear nonequilibrium thermodynamics [4,32] and fluid mechanics [94] in chromatography processes. The optimal process of multiparameters in chromatography processes can be considered as optimal control approaches with state space representation. As shown in Figure 1.19, $\mathbf{S}^k$ ($\mathbf{S}^k \in \mathbf{S} \in \mathbf{R}^{n_s}$) is n_s-dimensional state vector, and $\mathbf{X}^k$ ($\mathbf{X}^k \in \mathbf{X} \in \mathbf{R}^{n_x}$) is the n_x-dimensional control vector. In any objective control system, such as a chromatography system, there is a full reciprocity between $\mathbf{S}^k$ and $\mathbf{X}^k$, which are related to local space-time-varying variables concerning the space-concentration distributions of the solute zones and corresponding external force fields, respectively. The sequences or trajectories of evolution of solute zones, $\left\{ \mathbf{S}^k \right\}$, their corresponding control sequences, $\left\{ \mathbf{X}^k \right\}$, and the state space S and the control space X (look like a cuboid) are shown in Figure 1.19.

1.6.2 THERMODYNAMIC PATHS

Thermodynamic paths can be further divided into LTP and macroscopical thermodynamic path (MTP). The LTP is a virtual thermodynamic path that we design for the convenience of operator splitting [72,73] of the studied physicochemical phenomena in chromatography processes. In LTP, the solute system evolves from $\mathbf{S}^k$ to $\mathbf{S}^{k+1}$ undergoing a few virtual thermodynamic states. For the construction of LTP, the local equilibrium assumption and the thermodynamic state functions are adopted to unify naturally the essential physicochemical phenomena of thermodynamics and kinetics (e.g., equilibria, diffusions, and mass transfer resistances) in nonlinear chromatography processes at the level of nonequilibrium thermodynamics as a close-to-equilibrium thermodynamic system [4,32]. As a method of the operator splitting, for all calculations concerning the studied cell in its cell unit, LTP is constructed to indicate orderly individual physicochemical process, net migration, net diffusion, and net mass transfer among adjacent cells in each cell unit, on the basis of the full localized form of essential definition of equilibrium isotherm, migration velocity, and axial dispersion coefficient as well as the principle of thermodynamic state functions [11]. MTP can be considered as the extension of LTP from a cell unit and thin time interval, Δt, to an integrated solute zone and the whole space-time regions for nonlinear chromatography processes. It indicates all possible trajectories in the ergodic space of state variables. In MTP, the evolvements of distribution states of solute system are space-time-varyingly affected in the space-time-varyingly external force fields or the ergodic spaces of control variables, $\mathbf{X}^k$, in whole chromatography process, as shown in Figure 1.19. Thus, the essential of optimal control of nonlinear chromatography processes is just to find an optimal MTP in the ergodic space of control variables as the problem of the time-varying multiparameters, multistage, and multiple objective decisions with the methods of optimal control approaches based on discrete-time states.

1.6.3   LOCAL THERMODYNAMIC PATH OF NONIDEAL NONLINEAR CHROMATOGRAPHY

The local thermodynamic state $\mathbf{S}^k$ of the solute system at the time t_k is composed by the collection of local thermodynamic states $(\mathbf{S}_i^k)$ of each solute zone i in I components-solute system.

$$\mathbf{S}^k = \left\{ \mathbf{S}_1^k, \mathbf{S}_2^k, \ldots, \mathbf{S}_i^k, \ldots, \mathbf{S}_I^k \right\} \qquad (1.26)$$

$\mathbf{S}^{k=0}$ is the initial state of the MTP for the solute system, and $\mathbf{S}_i^k$ is the initial state of the LTP for the solute zone i. $\mathbf{S}_i^k$ is further defined with three basic vectors of solute zone distributions,

$$\mathbf{S}_i^k = \left[\mathbf{X}_i^k, \mathbf{M}_i^k, \mathbf{P}_i^k \right]^T \qquad (1.27)$$

where the vectors, $\mathbf{X}_i^k$, $\mathbf{M}_i^k$, and $\mathbf{P}_i^k$ are the vectors of position, width, and solute probability density of all cells in the solute zone i at t_k, respectively. In Lagrangian description [54–56], each vector of the solute zone i mentioned above is made up of corresponding variables of a series of cells in the solute zone,

$$\mathbf{X}_i^k = [x_{i,1}^k, x_{i,2}^k, \ldots, x_{i,j}^k, \ldots, x_{i,J}^k]^T \qquad (1.28a)$$

$$\mathbf{M}_i^k = [m_{i,1}^k, m_{i,2}^k, \ldots, m_{i,j}^k, \ldots, m_{i,J}^k]^T \qquad (1.28b)$$

$$\mathbf{P}_i^k = [p_{i,1}^k, p_{i,2}^k, \ldots, p_{i,j}^k, \ldots, p_{i,J}^k]^T \qquad (1.28c)$$

where Equation 1.28 shows the matrix forms of the vectors of $\mathbf{X}_i^k$, $\mathbf{M}_i^k$, and $\mathbf{P}_i^k$. $x_{i,j}^k$, $m_{i,j}^k$, and $p_{i,j}^k$ are the variables of position, width, and local solute probability space density, respectively, of each cell. $p_{i,j}^k$ is further defined as the rate of the solute i (including the solute in the local solid phase and the local mobile phase) in the cell j to its total quantity in the separation process.

In LLA, the velocity vector $(\mathbf{V}_i^k)$ of the solute zone i at t_k is defined by a series of the local velocity, $v_{i,j}^k$, of each cell.

$$\mathbf{V}_i^k = [v_{i,1}^k, v_{i,2}^k, \ldots, v_{i,j}^k, \ldots, v_{i,J}^k]^T \qquad (1.29)$$

According to the local equilibrium assumption and the Lagrangian description, we can further define the local velocity of cells, $v_{i,j}^k$,

$$v_{i,j}^k = \frac{u}{\left\{1 + F\left(\dfrac{\partial q_i}{\partial c_i}\right)\right\}}\Bigg|_{C_n = C_{n,j}^k} \tag{1.30}$$

where F is the phase ratio. C_n is the local concentration of the nth solute in the mobile phases of the cell j of the solute zone i at t_k. Thus, Equation 1.30 holds true for the nonlinear and competitive isotherms in the multicomponent system.

The local forms of equilibrium isotherms under the local equilibrium condition can be given by

$$q_{i,j}^{k*} = f_i\left(c_{1,j}^{k*}, c_{2,j}^{k*}, \ldots, c_{i,j}^{k*}, \ldots, c_{I,j}^{k*}\right) \quad \text{or} \quad q_{i,j}^{k*} = f_i\left(c_{i,j}^{k*}\right) \tag{1.31a}$$

$$q_{i,j}^{k*} = \frac{a_{i,j}^k \cdot c_{i,j}^{k*}}{1 + \sum_{i=1}^{I} b_{i,j}^k \cdot c_{i,j}^{k*}} \quad (\text{for multicomponent system}) \tag{1.31b}$$

$$\begin{cases} q_{1,j}^{k*} = \dfrac{a_{1,j}^k \cdot c_{1,j}^{k*}}{1 + b_{1,j}^k \cdot c_{1,j}^{k*} + b_{2,j}^k \cdot c_{2,j}^{k*}} \\[4mm] q_{2,j}^{k*} = \dfrac{a_{2,j}^k \cdot c_{2,j}^{k*}}{1 + b_{1,j}^k \cdot c_{1,j}^{k*} + b_{2,j}^k \cdot c_{2,j}^{k*}} \end{cases} \quad (\text{for two-component system}) \tag{1.31c}$$

$$q_{i,j}^{k*} = \frac{a \cdot c_{i,j}^{k*}}{1 + b \cdot c_{i,j}^{k*}} \quad (\text{for single-component system}) \tag{1.31d}$$

where $q_{i,j}^{k*}$ and $c_{i,j}^{k*}$ are the local solute concentrations in the jth SCms and jth SCs of the solute zone i at t_k under the local equilibrium condition, respectively. $a_{i,j}^k$ and $b_{i,j}^k$ (or a and b) are the local numerical coefficients in the jth solute cell unit of the solute zone i at t_k under the local equilibrium condition. Actually, $a_{i,j}^k$ and $b_{i,j}^k$ should be space-time-varying variables with the state space representation [76–79], which adapts flow programming, temperature programming, and gradient elution [86,97,98].

Targeting at the studied integral solute zone i at t_k, the vectors of solute concentration, $\mathbf{C}_i^k$ and $\mathbf{Q}_i^k$, in the mobile and solid phases, respectively, should be introduced since the vector, $\mathbf{V}_i^k$, depends on the local space-concentration distributions of the solutes,

$$\mathbf{C}_i^k = \left[c_{i,1}^k, c_{i,2}^k, \ldots, c_{i,j}^k, \ldots, c_{i,J}^k\right]^T \tag{1.32a}$$

$$\mathbf{Q}_i^k = \left[q_{i,1}^k, q_{i,2}^k, \ldots, q_{i,j}^k, \ldots, q_{i,J}^k\right]^T \tag{1.32b}$$

where $c_{i,j}^k$ and $q_{i,j}^k$ are the local solute concentrations in the mobile and solid phases, respectively, in the cell j of the zone i at t_k. According to mass conservation law for solute i in the cell j, we easily calculate the vector, $\mathbf{P}_i^k$, with the vectors $\mathbf{C}_i^k$ and $\mathbf{Q}_i^k$,

$$\mathbf{P}_i^k = \frac{1}{U_i(1+F)} \cdot \left\{ F \cdot \mathbf{C}_i^k + \mathbf{Q}_i^k \right\} \tag{1.33}$$

where U_i is the total quantity of the solute i in nonlinear chromatography processes.

In the first step of an LTP, the solute zone evolves from the t_k state to the virtual diffusion state firstly undergoing the virtual subprocess of net diffusion. In the same way of definition of $\mathbf{S}^k$, we can define the virtual diffusion state, $\mathbf{SD}^{k+1}$, at the time t_{k+1},

$$\mathbf{SD}^{k+1} = \left\{ \mathbf{SD}_1^{k+1}, \mathbf{SD}_2^{k+1}, \ldots, \mathbf{SD}_i^{k+1}, \ldots, \mathbf{SD}_I^{k+1} \right\} \tag{1.34}$$

where $\mathbf{SD}_i^{k+1}$ is further defined as

$$\mathbf{SD}_i^{k+1} = \left[\mathbf{XD}_i^{k+1}, \mathbf{MD}_i^{k+1}, \mathbf{PD}_i^{k+1} \right] \tag{1.35}$$

where $\mathbf{XD}_i^{k+1}$, $\mathbf{MD}_i^{k+1}$, and $\mathbf{PD}_i^{k+1}$ are the vectors of position, width, and solute probability density of the solute zone i at the virtual diffusion state, respectively. In Equation 1.35, each vector concerning the solute zone can be further expressed as

$$\mathbf{XD}_i^{k+1} = \left[xd_{i,1}^{k+1}, xd_{i,2}^{k+1}, \ldots, xd_{i,j}^{k+1}, \ldots, xd_{i,J}^{k+1} \right]^T \tag{1.36a}$$

$$\mathbf{MD}_i^{k+1} = \left[md_{i,1}^{k+1}, md_{i,2}^{k+1}, \ldots, md_{i,j}^{k+1}, \ldots, md_{i,J}^{k+1} \right]^T \tag{1.36b}$$

$$\mathbf{PD}_i^{k+1} = \left[pd_{i,1}^{k+1}, pd_{i,2}^{k+1}, \ldots, pd_{i,j}^{k+1}, \ldots, pd_{i,J}^{k+1} \right]^T \tag{1.36c}$$

where $xd_{i,j}^{k+1}$, $md_{i,j}^{k+1}$, and $pd_{i,j}^{k+1}$ are the vectors of position, width, and local solute probability density, respectively, of the cell j in the zone i at the virtual diffusion state. The diffusion coefficient vector, $\mathbf{D}_i^k$, of the zone i at t_k is defined by

$$\mathbf{D}_i^k = \left[d_{i,1}^k, d_{i,2}^k, \ldots, d_{i,j}^k, \ldots, d_{i,J}^k \right]^T \tag{1.37}$$

where $d_{i,j}^k$ is the local axis diffusion coefficient of the cell j of the zone i at t_k. We introduce an intergradation vector of the square of width of the cell, $\mathbf{WD}_i^{k+1}$, to temporarily express $\mathbf{MD}_i^{k+1}$,

$$\mathbf{WD}_i^{k+1} = \left[wd_{i,1}^{k+1}, wd_{i,2}^{k+1}, \ldots, wd_{i,j}^{k+1}, \ldots, wd_{i,J}^{k+1} \right]^T \tag{1.38}$$

where $wd_{i,j}^{k+1} = \left(md_{i,j}^{k+1}\right)^2$, and $wd_{i,j}^{k+1}$, is the width square of the cell at the virtual diffusion state. $\mathbf{WD}_i^{k+1}$ can be expressed as

$$\mathbf{WD}_i^{k+1} = \mathbf{W}_i^k + \left(\frac{2k^2\Delta t_k}{J^2}\right)\cdot\mathbf{D}_i^k \qquad (1.39)$$

where $\mathbf{W}_i^k$ is the vector of the square of width of the cell at the t_k state, and let $w_{i,j}^k = \left(m_{i,j}^k\right)^2$

$$\mathbf{W}_i^k = \left[w_{i,1}^k,\ w_{i,2}^k,\ldots,w_{i,j}^k,\ldots,w_{i,J}^k\right]^T \qquad (1.40)$$

With Equation 1.40 and Equations 1.37 through 1.39, the real values of $\mathbf{MD}_i^{k+1}$ can be easily obtained from the two vectors $\mathbf{M}_i^k$ and $\mathbf{D}_i^k$ at the t_k state.

In Lagrangian description, the fluid dynamics is formulated in terms of trajectories of fluid elements, and an essential relationship of the solute mass conservation in a solute cell within the virtual subprocess of the net diffusion has been given [12]

$$pd_{i,j}^{k+1} = \frac{m_{i,j}^k \cdot p_{i,j}^k}{md_{i,j}^{k+1}} \qquad (1.41)$$

The real values of $\mathbf{PD}_i^{k+1}$ can be obtained from the real values of $m_{i,j}^k$ and $p_{i,j}^k$ at the t_k state.

In the net diffusion process, the position changes of the solute cells are only derived from the accumulation of axial spread of each solute cell along the positive direction of x coordinates, which is related to the position of the chosen reference cell, such as the first solute cell, $j = 1$. Furthermore, we must emphasize that the migration of the reference cell is only related to net migration process, but not to the net diffusion process. $\mathbf{XD}_i^{k+1}$ can be expressed as

$$\mathbf{XD}_i^{k+1} = \mathbf{XF}_i^k + \mathbf{H}\cdot\mathbf{MD}_i^{k+1} \qquad (1.42)$$

where $\mathbf{XF}_i^k = \left[x_{i,1}^k, x_{i,1}^k,\ldots,x_{i,1}^k\right]^T$; it is the reference vector of position of the solute zone i at the t_k state, for the reference cell $j = 1$. H in Equation 1.42 is a constant lower triangular matrix.

$$\mathbf{H} = \begin{bmatrix} 0 & & & & \\ 1 & 0 & & & \\ 1 & 1 & 0 & & \\ \cdots & \cdots & \cdots & 0 & \\ 1 & \cdots & \cdots & 1 & 0 \end{bmatrix}_{J\times J} \qquad (1.43)$$

Therefore, with Equations 1.42, 1.41, and 1.39, we can calculate the real values of $\mathbf{XD}_i^{k+1}$, $\mathbf{MD}_i^{k+1}$, and $\mathbf{PD}_i^{k+1}$ of $\mathbf{SD}^{k+1}$ with the information of $\mathbf{S}^k$ of the t_k state.

In the second step of an LTP, a solute zone evolves from the virtual diffusion state to the virtual overlapping-separating state undergoing the virtual subprocess of net migration. In the same way, the vector of the virtual overlapping-separating state of the zone i, $\mathbf{SO}_i^{k+1}$, can be defined as

$$\mathbf{SO}_i^{k+1} = \left[\mathbf{XO}_i^{k+1}, \mathbf{MO}_i^{k+1}, \mathbf{PO}_i^{k+1} \right] \tag{1.44}$$

where $\mathbf{XO}_i^{k+1}$, $\mathbf{MO}_i^{k+1}$, and $\mathbf{PO}_i^{k+1}$ are the vectors of position, width, and probability density of the solute zone i at the virtual overlapping-separating state, respectively. They can be further expressed as

$$\mathbf{XO}_i^{k+1} = \left[xo_{i,1}^{k+1}, xo_{i,2}^{k+1}, \ldots, xo_{i,j}^{k+1}, \ldots, xo_{i,J}^{k+1} \right]^T \tag{1.45a}$$

$$\mathbf{MO}_i^{k+1} = \left[mo_{i,1}^{k+1}, mo_{i,2}^{k+1}, \ldots, mo_{i,j}^{k+1}, \ldots, mo_{i,J}^{k+1} \right]^T \tag{1.45b}$$

$$\mathbf{PO}_i^{k+1} = \left[po_{i,1}^{k+1}, po_{i,2}^{k+1}, \ldots, po_{i,j}^{k+1}, \ldots, po_{i,J}^{k+1} \right]^T \tag{1.45c}$$

where $xo_{i,j}^{k+1}$, $mo_{i,j}^{k+1}$, and $po_{i,j}^{k+1}$ are the position, width, and local solute probability density, respectively, of the cell j in the solute zone i at the virtual overlapping-separating state. For the net migration process, we can obtain the vector form of the position of the cell j.

$$\mathbf{XO}_i^{k+1} = \mathbf{XD}_i^{k+1} + \mathbf{V}_i^k \cdot \Delta t_k \tag{1.46}$$

With operator splitting [34,64,72,73], the width and local solute probability density of the cell j in the solute zone i remain invariable in the net migration process; thus we have

$$\mathbf{MO}_i^{k+1} = \mathbf{MD}_i^{k+1}; \mathbf{PO}_i^{k+1} = \mathbf{PD}_i^{k+1} \tag{1.47}$$

Therefore, with Equations 1.46 and 1.47, the real values of $\mathbf{SO}^{k+1}$ with the information of $\mathbf{SD}^{k+1}$ at the virtual diffusion state is obtained.

The third step (a rezoning process) of the LTP evolves from the virtual overlapping-separating state to the t_{k+1} states undergoing the virtual subprocess of net mass transfer among adjacent cells in a cell unit to keep the continuity of space distribution of solute zones. The definitions of all vectors of the t_{k+1} states, such as $\mathbf{S}^{k+1}$, $\mathbf{S}_i^{k+1}$, $\mathbf{C}_i^{k+1}$, $\mathbf{Q}_i^{k+1}$, $\mathbf{X}_i^{k+1}$, $\mathbf{M}_i^{k+1}$, $\mathbf{P}_i^{k+1}$, and $\mathbf{V}_i^{k+1}$, are same as the corresponding ones ($\mathbf{S}^k$, $\mathbf{S}_i^k$, $\mathbf{C}_i^k$, $\mathbf{Q}_i^k$, $\mathbf{X}_i^k$, $\mathbf{M}_i^k$, $\mathbf{P}_i^k$, and $\mathbf{V}_i^k$), at t_k states. The iteration forms from the t_k state to the t_{k+1} state are just a basic demand of the recursion of thermodynamic states of the solute system in Lagrangian description. Figure 1.17 indicated the four conformations (compressing, spreading, convex, and concave structure) in the rezoning processes [12].

With the Lagrangian description, the vector of the local shock wave ($\Delta\mathbf{M}_i^{k+1}$) [12,13] is defined for the virtual overlapping-separating state,

$$\Delta\mathbf{M}_i^{k+1} = \left[\Delta m_{i,1}^{k+1}, \Delta m_{i,2}^{k+1}, \ldots, \Delta m_{i,j}^{k+1}, \ldots, \Delta m_{i,J}^{k+1} \right]^T \tag{1.48}$$

where $\Delta m_{i,j}^{k+1}$ is called the local shock wave of the cell j in the solute zone i, $\Delta m_{i,j}^{k+1}$ is a width of the overlapping or separating between cell $j + 1$ and cell j, as well as cell j and cell $j - 1$ in a cell unit due to the relative movements of two adjacent cells in the net migration process [12]. The real values of $\Delta\mathbf{M}_i^{k+1}$ can be further obtained in the vector form for a given isotherm, such as the convex type of Langmuir isotherm,

$$\Delta\mathbf{M}_i^{k+1} = (\mathbf{A}^{k+1} \cdot \mathbf{V}_i^k - \mathbf{V}_{i,g}) \cdot \Delta t_k \tag{1.49}$$

where $\mathbf{V}_{i,g} = \left[v_{i,g}, 0, 0, \ldots 0 \right]_{1\times J}^T$. $\mathbf{A}^{k+1}$ is a tridiagonal matrix,

$$\mathbf{A}^{k+1} = \begin{bmatrix} a_{1,1}^{k+1} & a_{1,2}^{k+1} & & & & \\ a_{2,1}^{k+1} & a_{2,2}^{k+1} & a_{2,3}^{k+1} & & & \\ & \cdots & \cdots & \cdots & & \\ & & a_{j,j-1}^{k+1} & a_{j,j}^{k+1} & a_{j,j+1}^{k+1} & \\ & & & \cdots & \cdots & \cdots \\ & & & & a_{J,J-1}^{k+1} & a_{J,J}^{k+1} \end{bmatrix}_{J\times J} \tag{1.50}$$

where $a_{j,j}^{k+1}$ is the main diagonal entry, $a_{j,j-1}^{k+1}$ and $a_{j,j+1}^{k+1}$ are the subdiagonal entries. Equation 1.50 indicates the corresponding formulas to calculate the $\Delta\mathbf{M}_i^{k+1}$ according to the four conformations (compressing, spreading, convex, and concave structure) of distribution of local probability densities of a cell unit [12]. The width vector ($\mathbf{M}_i^{k+1}$) is the summation of the cell width at the virtual diffusion state and the local shock wave at virtual overlapping-separating state.

$$\mathbf{M}_i^{k+1} = \mathbf{MD}_i^{k+1} + \Delta\mathbf{M}_i^{k+1} \tag{1.51}$$

It is clear that the unceasing accumulation of local shock wave ($\Delta\mathbf{M}_i^{k+1}$) for time t with Equation 1.51 is just one of reasons of engendering macroscopical shocks or macroscopical sparse waves.

According to the mass conservation in each cell unit, the vector of local solute probability density of the studied cell j in the solute zone i at the t_{k+1} states, $\mathbf{P}_i^{k+1}$, can be obtained by

$$\mathbf{P}_i^{k+1} = \left(\tilde{\mathbf{M}}_i^{k+1} \right)^{-1} \cdot \left[\tilde{\mathbf{P}}\tilde{\mathbf{D}}_i^{k+1} \cdot \left(\mathbf{MD}_i^{k+1} + \mathbf{B}^{k+1} \cdot (\mathbf{G} \cdot \mathbf{V}_i^k) \cdot \Delta t_k \right) \right] \tag{1.52}$$

$\mathbf{G}$ and $\mathbf{B}^{k+1}$ are two bidiagonal matrixes, which have the form with nonzero entries appearing only on two diagonals, respectively:

$$G = \begin{bmatrix} 1 & -1 & & & & \\ & 1 & -1 & & & \\ & & \cdots & \cdots & & \\ & & & 1 & -1 \\ & & & & 0 \end{bmatrix}_{J \times J}$$

and

$$B^{k+1} = \begin{bmatrix} b_{1,1}^{k+1} & & & & \\ b_{2,1}^{k+1} & b_{2,2}^{k+1} & & & \\ & \cdots & & & \\ & & b_{j,j-1}^{k+1} & b_{j,j}^{k+1} & \\ & & & \cdots & \\ & & & & b_{J,J-1}^{k+1} & b_{J,J}^{k+1} \end{bmatrix}_{J \times J}$$

where $\mathbf{B}^{k+1}$ indicates the criterion of adjusting formulas to calculate $\mathbf{P}_i^{k+1}$ for the four conformations at the t_{k+1} states. $\mathbf{X}_i^{k+1}$ is expressed as

$$\mathbf{X}_i^{k+1} = \mathbf{X}\mathbf{F}_i^{k+1} + \mathbf{H} \cdot \mathbf{M}_i^{k+1} \tag{1.53}$$

With Equations 1.51, 1.52, and 1.53, we can calculate the real values of three vectors of the $\mathbf{S}^{k+1}$ with the information of $\mathbf{SO}^{k+1}$ at the virtual overlapping-separating state.

Up to now, an approach of full iteration of thermodynamic state of space distribution of solute zones from $\mathbf{S}^k$ to $\mathbf{S}^{k+1}$ has been founded in matrix forms with the LLA.

1.6.4 BOUNDARY CONDITIONS AND SPACE-TIME DISTRIBUTION TRANSFERS

The terminating condition of chromatography processes that solute zones undergo is defined that the jth cell just arrives at the column end

$$\mathbf{XL}_i^{k_j} \geq \mathbf{L} \tag{1.54}$$

where

$$\mathbf{XL}_i^{k_j} = \left[x_{i,1}^{k_1}, x_{i,2}^{k_2}, \ldots, x_{i,j}^{k_j}, \ldots, x_{i,J}^{k_J} \right]^T; \quad \mathbf{L} = \left[L, L, \ldots, L \right]_{I \times J}^T$$

where $x_{i,j}^{k_j}$ is the position of the cell j of the solute zone i when the cell just has arrived at or passed the column end at the time t_{k_j}, and L is the column length. The information of solute zone space distribution recorded by the detector at the column end is in the time distributions as the elution curves at a pointed position. Thus, the space distributions in LTP at a pointed time should be transferred to the time distributions at the pointed position for single-column chromatography. In describing evolutions of time distribution of solute zones at a pointed position, the collection of local thermodynamic states ($\mathbf{St}_i^{x_p}$) at the pointed position (x_p) can be defined as

$$\mathbf{St}_i^{x_p} = \left[\mathbf{T}_i^{x_p}, \tau_i^j, \mathbf{Pt}_i^{g_i+j} \right]^T \tag{1.55}$$

where $\mathbf{T}_i^{x_p}$, τ_i^j, and $\mathbf{Pt}_i^{g_i+j}$ are the vectors of the time, time width, and local solute probability density, respectively, of the time distribution of all solute cells in the solute zone i at the pointed position, x_p. In the Lagrangian description, the space distribution and the time distribution concerning the solute cell j corresponds to each other due to the local "permanent" identifications. The vectors $\mathbf{T}_i^{x_p}$, $\tau_i^{x_p}$, and $\mathbf{Pt}_i^{x_p}$ are further defined as

$$\mathbf{T}_i^{x_p} = [t_{i,J}^{x_p}, t_{i,J-1}^{x_p}, \ldots, t_{i,j}^{x_p}, \ldots, t_{i,2}^{x_p}, t_{i,1}^{x_p}]^T \tag{1.56}$$

$$\tau_i^{x_p} = [\tau_{i,J}^{x_p}, \tau_{i,J-1}^{x_p}, \ldots, \tau_{i,j}^{x_p}, \ldots, \tau_{i,2}^{x_p}, \tau_{i,1}^{x_p}]^T \tag{1.57}$$

$$\mathbf{Pt}_i^{x_p} = [pt_J^{x_p}, pt_{J-1}^{x_p}, \ldots, pt_{i,j}^{x_p}, \ldots, pt_{i,2}^{x_p}, pt_{i,1}^{x_p}]^T \tag{1.58}$$

where $t_{i,j}^{x_p}$, $\tau_{i,j}^{x_p}$, and $pt_{i,j}^{x_p}$ are the time, time width, and local solute probability density, respectively, of the time distribution of the cell j of the zone i at the pointed position, x_p. We further define the time $\tau_{i,J-(k-\theta_i)}$ that each solute cell spends in passing the detector at the column end.

$$\tau_{i,J-(k-\tau_i)} = \frac{m_{i,J-(k-\theta_i)}^k}{v_{i,J-(k-\theta_i)}^k}, \quad \theta_i \le k \le \theta_i + J - 1 \tag{1.59}$$

Equation 1.59 shows that the operation of space-time transfer of distribution of the solute zone i starts from the $j = J$ cell at the time $t_{k=\theta_i}$ and stops at the $j = 1$ cell at the time $t_{k=\theta_i+J-1}$.

The Lagrangian description ensures that a relationship of mass conservation in one space-time cell is tenable in the process of space-time transfer. Thus, $\mathbf{T}_i^{x_p}$ can be expressed as

$$\mathbf{T}_i^{x_p} = \mathbf{TF}_i^L + \mathbf{H} \cdot \tau_i^{x_p} \tag{1.60}$$

where $\tau_i^{x_p}$ can be obtained by inserting Equation 1.59 in Equation 1.60.

With Equations 1.59 and 1.60, we can finish the operation of the space-time transfer from $\mathbf{S}^k$ to $\mathbf{St}^{x_p}$.

1.7　THE 0-1 MODEL IN NONLINEAR TRANSPORT CHROMATOGRAPHY

The 0-1 model [14] is presented for nonlinear-mass transfer kinetic processes. One solute cell unit can be split into two solute cells, one (SCm) in the mobile phase with the velocity of the mobile phase, and the other (SCs) in the stationary phase with zero velocity with the LED. The Lagrangian–Eulerian coordinates of space distribution of solute zones are visually shown in Figure 1.20. With the local thermodynamic equilibrium assumption [6,48,49], all local physicochemical parameters or

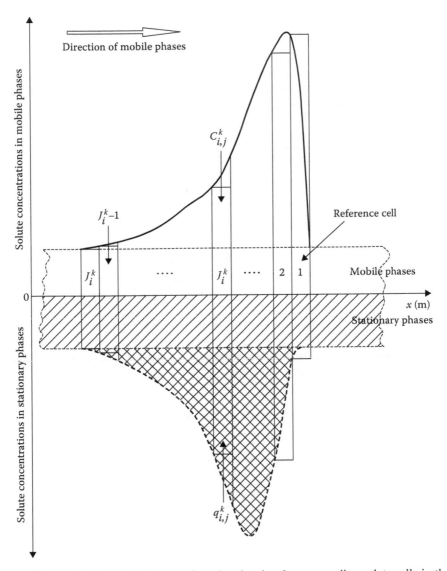

FIGURE 1.20　Spatial distributions of a solute band and corresponding solute cells in the mobile and stationary phases at the given time with the Lagrangian–Eulerian description.

relationships, such as local solute concentrations, local equilibrium isotherms, and local linear driving force kinetic equations, can be naturally introduced in the LED.

The thermodynamic state vector, S^k, which comprises four vector components, that is, the sequence number, the position, and the local solute concentrations in both SCms and SCs, is introduced to describe LTP and MTP. For nonlinear-mass transfer kinetic processes, the LTP is designed for a solute zone to evolve from the state S^k to the virtual migration state S^M undergoing the virtual net migration subprocess, and then to the state S^{k+1} undergoing the virtual net interphase mass transfer subprocess in a short time interval. Complete thermodynamic state iterations with the Markov characteristics [70,73] are derived by using the local equilibrium isotherm and the local lumped mass transfer coefficient. When the local thermodynamic equilibrium is retained, the excellent properties (such as consistency, stability, conservation, and accuracy) of the numerical solution of the 0-1 model are observed in the theoretical analysis and in the numerical experiments of the nonlinear chromatography. It is found that the 0-1 model could properly link up with the optimal control approaches based on discrete-time states and state space.

1.7.1 Physical Descriptions of Whole Nonlinear Chromatographic Processes

The discrete form of concentration curves of process of injection samples is

$$f(c_i^S) = \left\{ c_{i,j=k+1}^k \right\} = \left\{ c_{i,1}^0, c_{i,2}^1, \ldots, c_{i,j=k+1}^k, \ldots, c_{i,J^S}^{k=J^S-1} \right\} 0 \leq k < J^S \qquad (1.61)$$

where c_i^S is the variable of the local solute concentration in SCm at column inlet ($x = 0$) in injection sample process. The superscript "S" in c_i^S and J^S indicates the parameters in the process of injection samples. $c_{i,j=k+1}^k$ in Equation 1.61 is the local solute concentrations in the SCm with $j = k + 1$, and $t = t_k$, at the column inlet during the subprocess of injection samples. The whole nonlinear chromatographic process that includes the three subprocesses (e.g., injection sample, column separation, and sample collection) should be a most general process method close to practical applications with the general initial and boundary conditions with the LED.

1.7.2 Local Thermodynamic Path of Nonlinear Transport Chromatography

LTP is governed by the principles of operator splitting [34,64,72,73] and thermodynamic state functions [95]. Figure 1.21 shows the LTP of the nonlinear transport chromatography, in which the solute zone evolving from the state S^k at t_k to the state S^{k+1} at t_{k+1} is realized by designing an LTP. It is from S^k through virtual net migration subprocess to the virtual migration state, S^M, then through virtual net mass transfer subprocess to S^{k+1}. This is in accordance with the method of the operator splitting. Each subprocess can be directly calculated by an independent operator with a localized form of the quantificational relationship that can define the subprocess.

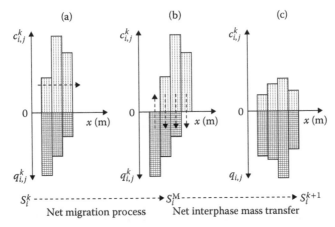

FIGURE 1.21 Local thermodynamic path (LTP) of nonlinear transport chromatography in the 0-1 model with the LED. (From Liang and Jia, *Chin J Chromtogr*, 25, 785–798, 2007. With permission.) The arrows in (a) and (b) indicate the directions of SCm slipping with u_k and the local interphase mass transfer $k_{fi,j}^k$, k_{fij} in a solute cell unit respectively.

The t_k states (S^k): In the LED, the S_i^k of the solute zone i at t_k can be defined by four vectors concerning solute distributions,

$$S_i^k = \left[J_i^k, X_i^k, C_i^k, Q_i^k \right]^T \tag{1.62}$$

where J_i^k, X_i^k, C_i^k, and Q_i^k are the vectors of the sequence number, the position, and the local solute concentrations in the SCms and SCs of the solute zone i at t_k under the actual (including equilibrium or nonequilibrium) conditions, respectively. The four vectors can be further given by

$$J_i^k = [1, 2, \ldots, j_i^k, \ldots, J_i^k]^T \tag{1.63a}$$

$$X_i^k = [x_{i,1}^k, x_{i,2}^k, \ldots, x_{i,j}^k, \ldots, x_{i,J}^k]^T \tag{1.63b}$$

$$C_i^k = [c_{i,1}^k, c_{i,2}^k, \ldots, c_{i,j}^k, \ldots, c_{i,J}^k]^T \tag{1.63c}$$

$$Q_i^k = [q_{i,1}^k, q_{i,2}^k, \ldots, q_{i,j}^k, \ldots, q_{i,J}^k]^T \tag{1.63d}$$

where j_i^k, $x_{i,j}^k$, $c_{i,j}^k$, and $q_{i,j}^k$ are the sequence number, the position, and the local solute concentrations in the jth SCms and jth SCs of the solute zone i at t_k under the actual conditions, respectively.

Net migrations and virtual migration states (S^M): In the first step of the LTP for the nonlinear transport chromatography, a solute zone evolves from the state (S^k) to the virtual migration state (S^M) undergoing the virtual net migration subprocess in Δt (Figure 1.21). In the same way as Equations 1.62 and 1.63, S^M can be defined for the solute zone i.

$$S_i^M = [J_i^M, X_i^M, C_i^M, Q_i^M]^T \tag{1.64}$$

Furthermore

$$J_i^M = [1, 2, \ldots, j_i^M, \ldots, J_i^M]^T \tag{1.65a}$$

$$X_i^M = [x_{i,1}^M, x_{i,2}^M, \ldots, x_{i,j}^M, \ldots, x_{i,J}^M]^T \tag{1.65b}$$

$$C_i^M = [c_{i,1}^M, c_{i,2}^M, \ldots, c_{i,j}^M, \ldots, c_{i,J}^M]^T \tag{1.65c}$$

$$Q_i^M = [q_{i,1}^M, q_{i,2}^M, \ldots, q_{i,j}^M, \ldots, q_{i,J}^M]^T \tag{1.65d}$$

where J_i^M, X_i^M, C_i^M, and Q_i^M are the vectors of the sequence number, the position, and the local solute concentrations for the SCms and the SCs of the solute zone i at S^M under the actual (equilibrium or nonequilibrium) conditions, respectively. j_i^M, $x_{i,j}^M$, $c_{i,j}^M$, and $q_{i,j}^M$ are the sequence number, the position, and the local solute concentrations for the jth SCms and jth SCs of the solute zone i at S^M under the actual conditions, respectively.

With the LED, the virtual net migration subprocess provides a few extraordinarily simple algebraic relationships between S^M and S^k. Since all SCms collectively slip at the u_k along with the mobile phase and all SCs are still fixed in Δt, the total number J_i^M of solute cell units in the S^M is one more than that of S^k. In the net migration subprocess, the LED assumes that the local solute concentrations in all SCms and SCs do not change during the slipping of the SCms with the SCs fixed. We can give the iteration relation of the local solute concentrations in SCs between S^M and S^k by comparing Figure 1.21a and b. The vectors J_i^M, X_i^M, C_i^M, and Q_i^M can be given by

$$J_i^M = [1, 2, \ldots, j_i^k, \ldots, J_i^k, J_i^k + 1]^T \tag{1.66a}$$

$$X_i^M = [(x_{i,1}^k + m), (x_{i,2}^k + m), \ldots, (x_{i,j}^k + m), \ldots, (x_{i,J}^k + m), x_{i,J}^k]^T \tag{1.66b}$$

$$C_i^M = [c_{i,1}^k, c_{i,2}^k, \ldots, c_{i,j}^k, \ldots, c_{i,J^k}^k, 0]^T \tag{1.66c}$$

$$Q_i^M = [0, q_{i,1}^k, q_{i,2}^k, \ldots, q_{i,j}^k, \ldots, q_{i,J^k}^k]^T \tag{1.66d}$$

With Equations 1.62 through 1.66, we can finish the state iteration from S^k to S^M.

Net interphase mass transfers and the t_{k+1} state (S^{k+1}): In the second step of the LTP of the nonlinear transport chromatography, a solute zone evolves from the state S^M to the state S^{k+1} at t_{k+1} undergoing the virtual net interphase mass transfers in Δt, as shown in Figure 1.21. The definitions of all relative vectors of the state S^{k+1}, such as S^{k+1}, S_i^{k+1}, J_i^{k+1}, X_i^{k+1}, C_i^{k+1}, and Q_i^{k+1}, are the same as the corresponding ones, S^k, S_i^k, J_i^k, X_i^k, C_i^k, and Q_i^k, at the state S^k. The same definitions concerning the S^{k+1} and S^k are just an elementary demand of the iteration equations.

With the LED, the discrete and local forms of the solid film linear driving force model [2,99] can be given by

$$q_{i,j}^{k+1} = q_{i,j}^{M}\left(1 - k_{fi,j}^{k} \cdot \Delta t_{k}\right) + k_{fi,j}^{k} \cdot \Delta t_{k} \cdot q_{i,j}^{M*} \tag{1.67}$$

where $k_{fi,j}^{k}$, $q_{i,j}^{M*}$, $q_{i,j}^{M}$, and $q_{i,j}^{k+1}$ are the local lumped mass transfer coefficient, the equilibrium-local and actual-local solute concentrations in stationary phase at S^{M}, and the actual-local solute concentration in stationary phase at S^{k+1}, for the jth SCs of the solute i, respectively. With Equation 1.67, we can find

$$
\begin{aligned}
q_{i,j}^{k+1} > q_{i,j}^{M*}, &\quad \text{if } k_{fi,j}^{k} \cdot \Delta t_{k} < 1, \text{ local solute transports} \\
&\qquad \text{from SCms to SCses} \\
q_{i,j}^{k+1} = q_{i,j}^{M*}, &\quad \text{if } k_{fi,j}^{k} \cdot \Delta t_{k} = 1, \text{ no local solute transports} \\
&\qquad \text{between SCses and SCms} \\
q_{i,j}^{k+1} < q_{i,j}^{M*}, &\quad \text{if } k_{fi,j}^{k} \cdot \Delta t_{k} > 1, \text{ local solute transports} \\
&\qquad \text{from SCses to SCms}
\end{aligned}
\tag{1.68}
$$

Equation 1.68 gives a numerical criterion to distinguish the transport direction of local solute between SCms and SCs. With the Lagrangian characteristic of the LED, we can reveal the relations of mass conservations concerning the total quantity of the solute i in the jth solute cell unit at the state S^{M}, the corresponding equilibrium state (S^{M*}) of S^{M} and the state S^{k+1}. Equation 1.69 can be easily derived from Figure 1.21b and c,

$$c_{i,j}^{M*} + F \cdot q_{i,j}^{M*} = c_{i,j}^{M} + F \cdot q_{i,j}^{M} \tag{1.69a}$$

$$c_{i,j}^{k+1} + F \cdot q_{i,j}^{k+1} = c_{i,j}^{M} + F \cdot q_{i,j}^{M} \tag{1.69b}$$

where $c_{i,j}^{M*}$ and $q_{i,j}^{M*}$ are the local solute concentrations of the solute i in the jth SCms and SCs at S^{M*}, respectively. With Equations 1.31, 1.69, and 1.67, the vectors C_i^{k+1} and Q_i^{k+1} can be obtained. In the net interphase mass transfer process, the sequence number and the position of the solute cell units do not change between S^{M} and S^{k+1}, which means

$$J_i^{k+1} = J_i^{M} \tag{1.70a}$$

$$X_i^{k+1} = X_i^{M} \tag{1.70b}$$

The iteration relationships from S^{M} to S^{k+1} can be obtained with Equations 1.31, 1.69, 1.67, and 1.70. The LTP of the nonlinear transport chromatography has been fully expressed by the iteration forms of the distribution state of solute zones with the concepts of state space and countable space.

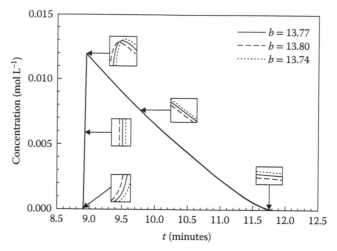

FIGURE 1.22 Influence of small perturbations of the isotherm parameter on the stability of elution profiles in nonlinear ideal chromatography. (From Liang and Jia, *Chin J Chromtogr*, 25, 785–798, 2007. With permission.) $m = 0.01$ mm ($\Delta t = 10^{-4}$ minutes), $q = 29.2 \cdot c / (1 + b \cdot c)$, $b = 13.77 \pm \varepsilon$, $\varepsilon = 0.03$.

1.7.3 STABILITY AND CONVERGENCE

In the iteration equations of the 0-1 model [14], the small perturbation ε of the parameter b in the Langmuir isotherm (Equation 1.31c) should more effectively affect the error propagation of the 0-1 model. The introduction of ε in b will affect each step of iterative computation of the 0-1 model and can be used to verify its error propagation. This equals three kinds of different solutes with small differences (e.g., $b =$ 13.74, 13.77, 13.80) on the Langmuir isotherm. The small perturbation ε contributes to the perturbations of both $c_{i,j}^k$ and $q_{i,j}^k$ during each LTP between S^k and S^{k+1} in the whole iterative process. In Figure 1.22, the left ($b = 13.80$) and right ($b = 13.74$) simulated elution profiles relative to the middle one ($b = 13.77$) with the small perturbation ($\varepsilon = 0.03$) of the parameter (b) do not diverge. This perturbation or error is not enlarged after undergoing a rather large amount ($K = 12 \times 10^4$) of iterative computations although the error of $c_{i,j}^k$ and $q_{i,j}^k$ due to ε is made during each LTP. The left and right elution profiles converge to the middle one, and it just indicates the fact that the solute with bigger parameter b (e.g., $b = 13.80$) should be eluted with relatively smaller time (e.g., the left elution profile), and vice versa. Figure 1.22 shows that the algorithm of the 0-1 model possesses good stability and convergence in nonlinear ideal chromatography.

1.7.4 INTERPHASE MASS TRANSFER COEFFICIENTS

The 0-1 model can be easily used to analyze the influence of the interphase mass transfers on the band profiles due to the nonequilibrium from the solute concentration difference between two phases. Such concentration difference generally results from the relative slippage between corresponding SCms and SCs, as shown

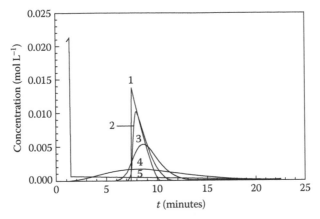

FIGURE 1.23 Influence of the interphase mass transfer coefficient on the band profile of nonlinear chromatography with the 0-1 model. $m = 0.01$ mm ($\Delta t = 10^{-4}$ min); $k_f = 10,000$, 1,000, 100, 10, 1 min^{-1} for the curves 1, 2, 3, 4, and 5, respectively.

in Figure 1.21. Figure 1.23 shows the influences of lumped mass transfer coefficient, k_f, on band profiles in nonlinear chromatography. Again, the simulated elution profiles in Figure 1.23 have the same trend on the change of $k_{fi,j}^{k} \cdot \Delta t_k$ in Equation 1.67 as those in reference [100]. We have noted $k_{fi,j}^{k} \cdot \Delta t_k$ as a new parameter to directly affect the local interphase mass transfers during each LTP and the simulated elution profiles in a MTP. It also existed in the difference scheme found in Equation 4 in the work by Lin et al. [99]. In the difference schemes of the partial differential equations, the small parts of the calculated elution profiles [101,102] with the smallest k_f (e.g., $k_f = 0.1$–1.0 min^{-1}) are in the unreasonable time region, $t < t_0$, which implies that some problems on the boundary or initial conditions possibly exist in the different schemes of PDEs based on Eulerian description.

1.7.5 INJECTION SAMPLE CONCENTRATIONS

The duration (t_p) for the injection sample volume is an important factor on band profiles of nonlinear chromatography. With the 0-1 model [14], it can be shown that the pure t_p with a given c_0 independently affects the band profile of nonlinear chromatography in Figure 1.24. In Figure 1.24a, the concentration flats at c_0 appear to gradually increase with increasing t_p at $t_p = 3.0$ minutes and $t_p = 5.0$ minutes in the curves 4 and 5, respectively, although the concentration flats do not appear in the curves 1–3. These sharp shocks in Figure 1.24a indicate that the artificial dissipations in the 0-1 model with the width of solute cell unit (m), $m = 0.01$ mm, do not propagate. It is to be noted that different injection conditions were used (at constant sample concentration) than the conditions observed by Guiochon et al. in their experiment [103] (at constant sample size), although they deal with an analogical problem. Figure 1.24b shows that the interphase mass transfer as a dispersion factor makes the sharp shocks of corresponding solute bands spread with delaying the peak positions at different t_p and same c_0.

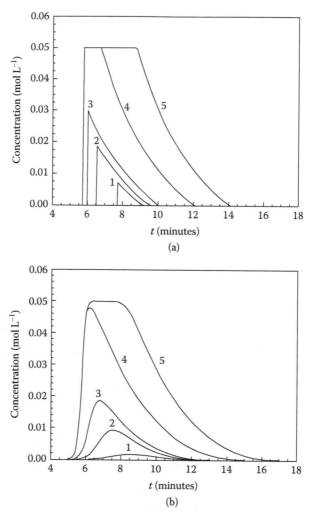

FIGURE 1.24 Influence of the sample volume on the band profile of nonlinear chromatography with the 0-1 model. (From Liang and Jia, *Chin J Chromtogr*, 25, 785–798, 2007. With permission.) (a) $k_f = 10,000$ min^{-1} and $k_{fi,j}^k \cdot \Delta t = 1$ for nonlinear ideal chromatography. (b) $k_f = 100$ min^{-1} and $0 < k_{fi,j}^k \cdot \Delta < 1$ for nonlinear transport chromatography. $c_0 = 0.05$ mol L^{-1} and $m = 0.01$ mm ($\Delta t = 10^{-4}$ minutes) are constant in all cases. $t_p = 0.1, 0.5, 1.0, 3.0, 5.0$ minutes for the curves 1, 2, 3, 4, and 5, respectively.

1.7.6 SPACE-TIME DISTRIBUTION TRANSLATIONS

With the 0-1 model, we can easily obtain the space distribution at the given time, as shown in Figure 1.25. The space distribution of solute zones can be easily obtained from the imaging at given time, such as single-molecule imaging [15]. The Lagrangian characteristic of the LED makes one express the injection sample curves and the elution curves in the discrete space-time form easily. Specially, the Lagrangian characteristics of the 0-1 mode adapt the numerical calculations, optimizing controls, and

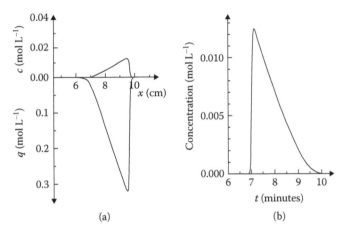

(a)   (b)

FIGURE 1.25 Translation of space distribution to time distribution of solute bands of non-linear ideal chromatography at the column end with the 0-1 mode. (From Liang and Jia, *Chin J Chromtogr*, 25, 785–798, 2007. With permission.) (a) The space distribution of solute band at the time $t = t_{k = 69379} = 6.9379$ minutes; (b) the time distribution of solute band at the column end $x_p = L = 100$ mm. $m = 0.01$ mm ($\Delta t = 10^{-4}$ minutes), $k_f = 10000$ min^{-1}, and $k_{fi,j}^k \cdot \Delta t = 1$.

understanding the processes of the MCC [3,23,104,105] (see Sections 1.12 and 1.13). In single-column chromatography, the results of numerical experiments can be used to reveal the relations between the space distributions and time distributions of solute bands at the column end.

Figure 1.25 is for the space distribution at the given time for the case that $j = 1$ SCm just arrives at the column end. The two space distributions are shown in the positive (in the mobile phase) and negative (in the stationary phase) directions of the vertical (solute concentration) coordinates of Figure 1.25a, respectively. Actually, the local form of Langmuir isotherm, $q_{i,j}^k = 29.2c_{i,j}^{k*}/(1 + 13.77c_{i,j}^{k*})$. Figure 1.25b is for the time distribution at the position $x_p = L$, when $J_i^{x_p}$ SCms pass the column end one by one at different times. In fact, the discrete form of injection sample curve (Equation 1.61) is an initial input that the environment is related to the separation system at the column inlet, and it indicates a transfer process from the solute time distribution in injection samples to the space distribution in column separations. Correspondingly, the discrete form of sample detection (or collection) curve is a final output that the outside contacts the separation system at the column end, and it is a transfer process from the solute space distribution to the time distribution in the process of sample collections.

1.8 LOCALIZED SINGLE-MOLECULE ISOTHERMS ON CONFINED LIQUID–SOLID INTERFACES

The study of dynamics and thermodynamics of single biological molecules at confined liquid–solid interfaces is crucially important, especially in the case of low-copy number molecules in a single cell [15]. Using a high-throughput single-molecule imaging system [37,38] and Lagrangian coordinates [11,12] of single-molecule

images based on the 0-1 model [12,14], we discovered that the local equilibrium isotherms of single λDNA molecules at a confined liquid–solid interface varied from a
stair type for the regions of single or double molecular DNA to a mild "S" type for
the regions of triple molecular DNA spots, which does not agree with the conventional equilibrium isotherms in the literature [2,46]. Single-molecule images in time
sequence for different λDNA concentrations were statistically analyzed by measuring preferential partitioning from shearing effects, which were used to measure
the local velocity of DNA molecules by directly observing the migration of DNA
fluorescence spots for the 12 continuous images. The local equilibrium isotherms
of λDNA at confined liquid–solid surfaces have been obtained from the statistics of
real time, local velocities, and positions of single molecules by measuring preferential partitioning from shearing effects associated with the change in flow velocity
from 0 at the wall to the average value in the center of the capillary. The local linear
velocity of hydrodynamic flow was calculated by the Hagen–Poiseuille equation
[106] in different microregions with LLA [11,12,14]. The local single-molecule isotherms for the tracked molecules in the regions of single, double, or triple molecular DNA layers within the laminar flows were obtained according to the average
local velocities of both the stochastic molecule events and the corresponding local
Poiseuille flows.

1.8.1　Single-Molecule Imaging and Lagrangian Coordinates

The principle of experimental design for single-molecule imaging and its Lagrangian
coordinates are shown in Figure 1.26. Figure 1.26a shows the square capillary and
ICCD camera arrangement for obtaining a series of single-molecule images in time
sequence. The focusing plane with a certain depth in the vertical coordinates (Z)
was sketched as a thin cuboid with green lines in the square capillary. The volume
($V = 2.607 \times 10^4$ μm^3) of the focusing plane was obtained by experimental single-
molecule counting from the linear slope of the observed labeled DNA molecules versus sample concentrations. Its inboard and outboard sides along the radial direction
(r) were the confined surface of polyvinylpyrrolidone (PVP)-coated square fused-
silica capillary, and the upper and lower sides were the free buffer solution. The
labeled DNA molecules in this thin cuboid (the focusing plane) were only observed
in mobile phases, and the λDNA molecules at the above or below boundaries of the
liquid-internal wall of the capillary in the Z direction were not observed, since they
were out of the focusing plane. The effective focal region in the Z direction was
estimated between $Z \sim Z + 2$ μm. Figure 1.26b represents the focusing plane in the
x and r coordinates and divided regions at the near-wall interface depending on the
approximate diameter of single-DNA fluorescent spots. The width ranges of the studied regions (Figure 1.26a and b) were 40.5~47.5 pixels (W_1), 33.5~47.5 pixels (W_2),
27.5~47.5 pixels (W_3), and −27.5~27.5 pixels (W_4), respectively. W_1, W_2, and W_3 just
indicate the regions of fluorescent spots of single, double, or triple molecular DNA
in the radial direction (r), respectively. W_4 indicates the middle region of the focusing
plane in the radial direction (r). Figure 1.26c shows Lagrangian coordinates within
a real single-molecule image; it also shows the local regions of one frame from the
ICCD image of fluorescent DNA molecules with the concentration of 0.5 pmol L^{-1}

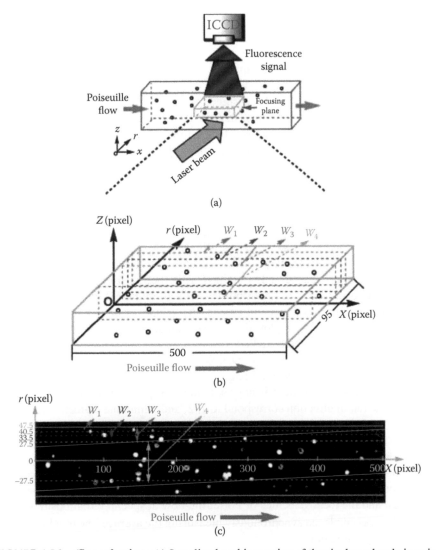

FIGURE 1.26 **(See color insert.)** Localized probing region of the single-molecule imaging system and its Lagrangian coordinates. (From Liang, Cheng, and Ma, *Analytical Chemistry*, 81, 2059–2066, 2009. With permission.) (a) The focusing plane (a thin cuboid). Its volume ($V = 2.607 \times 10^4$ μm³) in the vertical coordinates (Z) was obtained from the linear slope of the observed labeled DNA molecule versus sample concentrations. Its inboard and outboard sides along the radial direction (r) were the confined surface of the PVP-coated square fused-silica capillary (50 × 50 μm), and the upper and lower sides were the free solution. (b) The enlarged focusing cuboid as the ICCD visual field, showing the four regions of molecular layers: single-molecule layer (W_1), double-molecule layer (W_2), triple-molecule layer (W_3), and middle-molecule layer (W_4) in the radial direction (r). (c) Lagrangian coordinates of single-molecule images, showing the local regions of one frame from the CCD image of fluorescent DNA molecules (0.5 pmol L⁻¹). The Lagrangian coordinates and the local regions were fixed even if each identified individual molecule moves along with the hydrodynamic flow (x coordinates) for a countable time-set (**K**) of the continuous images.

DNA sample. Each image has the visible region of 95 × 500 pixels, and each pixel represents 0.53 × 0.53 μm of real space. The distance from the center to the wall of the capillary is 47.5 pixels. The origin of x coordinates in Lagrangian coordinates is at the axial midline of the square capillary. The coordinate is always unchangeable when the DNA imaging spots move along in consecutive images.

In LLA [12,13], $i = 1, 2, ..., I$ is assigned to identify individual molecules in the kth single-molecule image. The time that the molecules evolve at confined liquid–solid surfaces is divided into a series of time intervals, Δt or $\Delta t_k = t_{k+1} - t_k$, where the subscript k is the serial number of the time grids or images. Δt is 110 ms in our experiments. Thus, a given sequence of images is marked by $\mathbf{K} = 1,..., k, ..., K, k \in \mathbf{K}$, where $\mathbf{K}$ is a countable time-set. Through the continuous observation of single-molecule trajectories (fluorescence spots) in the studied regions, each single molecule is identified and tracked. The front spot in these images is not permitted to overstep the visible scale of the last (Kth) image (Figure 1.26c). Thus, all studied molecules are ensured to satisfy the Lagrangian coordinates in LLA. In our studies, the time-set K equals 13 at current technological conditions. The local position (x_i^k, r_i^k) of the individual molecule i is accurately obtained from the kth frame (image), where x_i^k and r_i^k are the local axial and radial positions for the ith DNA molecule in the kth single-molecule image, as shown in Figure 1.26c. All observed individual molecules in the studied regions were employed as our statistical objects. Since none of coordinates of DNA molecules in the studied regions overlap the boundary of two regions, the molecule i with its position x_i^k, r_i^k was assigned clearly to a special region of the kth image. In our single-molecule images, the diameter of the fluorescence spot of each labeled λDNA molecule was approximately 7.0 pixels. It is clear that W_1, W_2, and W_3 in the radial direction correspond to 1, 2, and 3 multiple diameters of a DNA fluorescent spot. It must be emphasized that W_3 includes W_1 and W_2 from the near-wall interface. W_2 includes W_1, and W_1 is at the near-wall interface. W_4 indicates the middle region of the focusing plane of square capillary, which is far from the near-wall interface. The regions, W_1, W_2, and W_3, are considered as the divided local regions for counting the local molecule numbers, their local velocities of single-DNA molecules, and the local velocities of Poiseuille flow at special radial positions. The region (W_4) was employed synchronously measuring the average linear velocity ($\bar{U}_{av}^c$) of the hydrodynamic flow.

1.8.2 LOCALIZED SINGLE-MOLECULE ISOTHERMS AT CONFINED LIQUID–SOLID INTERFACES

In Figure 1.27, the vertical coordinates were defined as q_j (pmol μm^{-2}), which represents the picomole number of λDNA molecules per unit area (μm^2) of the confined inner surface of the square fused-silica capillary. This confined inner surface serves as the confined stationary phase. The horizontal coordinates, c_j (pmol μm^{-3}), represent the picomole number of λDNA molecules per unit volume (μm^3) in the three different regions, $W_j, j = 1, 2, 3$. These three regions serve as the confined mobile phases. With the use of the single-molecule imaging system (as shown as Figure 1.26), q_j and c_j were obtained by

$$q_j = \frac{N_j^s}{(N_A S_c)} = \frac{N_j^s d}{(N_A V)} \tag{1.71a}$$

$$c_j = \frac{N_j^m}{(N_A V)} \tag{1.71b}$$

where N_j^s is the number of λDNA molecules in each studied region (W_j) in the stationary phases, and N_j^m is the number of λDNA molecules in each studied region (W_j) in the mobile phases. S_c and d are the area of the confined inner surface and the diameter observed of the thin cuboid (the focusing plane) in the square fused-silica capillary, respectively. N_A is the Avogadro constant. In our case, $V = 2.607 \times 10^4$ μm^3 through fitting the data of counting local molecule numbers. Also, $d = 47.5$ μm from the single-molecule detection measurements. It is worth mentioning that the number of molecules shown in the isotherm curves is not the total number of DNA molecules actually observed from single-molecule imaging. Instead, they are obtained from the probability of DNA molecules in the mobile phase or in the stationary phase, which depends on both local molecular number and its local velocity ratio. To compare the localized single-molecule isotherms with the bulk measurement-equilibrium isotherms, experimental points from the measurements (q_j, c_j) of single-molecule images were fitted with the equations of sigmoidal dose-response (variable slope) for three different microregions (W_j, $j = 1, 2, 3$).

This trend of change of lumbar highs and slopes were easily discovered by comparing the b_1, b_2, and b_3 in sigmoidal curves in Figure 1.27. It shows that the local equilibrium isotherms of single λDNA molecules at the confined liquid–solid interface varied from the stair types for the regions of single or double molecular DNA to the mild "S" type for the region of triple molecular DNA. The microspace scale effects on the isotherm types were not reported in the conventional description of

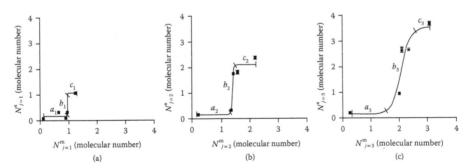

FIGURE 1.27 Local single molecular isotherms at confined liquid–solid interfaces. (From Liang, Cheng, and Ma, *Analytical Chemistry*, 81, 2059–2066, 2009. With permission.) (a) Correlations between the picomole number of λDNA molecules per μm^2 area of the confined inner surface of the square fused-silica capillary, q_j (pmol μm^{-2}), and the picomole number of λDNA molecules per μm^3 volume of the confined mobile phases, c_j (pmol μm^{-3}), at the region of the single molecular DNA layer ($W_{j=1}$). (b) Correlations between q_j and c_j in the region of the double molecular DNA layer ($W_{j=2}$). (c) Correlations between q_j and c_j in the region of the triple molecular DNA layer ($W_{j=3}$).

equilibrium isotherms in the literature. In addition, the data to obtain thermodynamic information from the different microregions close to the wall could not be fitted by Langmuir or Freundlich isotherms, which may be due to the microspace scale effects or the low concentrations of DNA sample (0.1~3 pmol) or other unknown reasons. We can find from Figure 1.27 that the isotherm shapes change as the microregions increase from $W_{j=1}$, $W_{j=2}$ to $W_{j=3}$ although the confined liquid–solid interfaces and other conditions are fixed. It needs to be kept in mind that the intraparticle pores of the chromatography packing materials and the dimensions of microchannels will have great effects on the equilibrium or nonequilibrium adsorptions on the confined liquid–solid interface. The hydrodynamic flow velocity was quite slow in this experiment of single-molecule imaging, since faster flows would likely produce streaks instead of dots in the images. However, as long as the laminar flow with relatively low Reynolds number is ensured, the Hagen–Poiseuille equation will be tenable; the approach of the single molecular isotherm should also be tenable, although some typical chromatography, electrophoresis separations, or other bioseparations are carried out with faster flows. This millisecond and microvolume approach will have significant impact on preparations of low-copy number proteins in the single-cell, membrane separations, and other bioseparation studies.

The local equilibrium isotherms, which were fitted based on the statistical results of the confined surface molecule densities and the local concentrations in the confined volume, demonstrated that they varied from the stair type to the mild "S" type, depending on the defined microradial regions at the near-wall interface for the same labeled λDNA molecules at a given liquid–solid interface. These single-molecule isotherms will have significant impact in many areas, such as micropreparative chromatography or electrophoresis of low-copy number of protein molecules in single cells [107].

1.9 0-1 MODEL-INVERSE METHOD IN REVERSED-PHASE LIQUID CHROMATOGRAPHY

A novel inverse method [108] based on the 0-1 model [14], which is referred to as the 0-1 model-inverse method, of nonequilibrium nonlinear chromatography was developed to simultaneously determine the isotherm parameters and the lumped mass transfer coefficients of the one- and two-component systems in RP-HPLC on a C18-silica gel column. After obtaining the linear portion of the Langmuir isotherm at low solute concentrations, complete chromatographic processes were simulated to fit the experimental elution curves of the one- and two-component systems at relatively high sample concentrations (in the range of 10–80 mg mL^{-1}). By comparing the simulated elution curves with experimental curves with regard to profiles and areas, the suitable isotherm parameters and the lumped mass transfer coefficients were obtained. With a solute cell unit width of 1×10^{-3} cm, the average errors of the peak areas were 0.11% for one component and −0.39% for two components, and the numerical diffusions of the 0-1 model for the contribution to band broadening may be negligible. In addition, the results showed that the lumped mass transfer coefficients decrease as the solute concentration increases. This conclusion is contrary to the conclusion reached by Miyabe and Guiochon [109,110], where the lumped mass transfer coefficients increase with increasing solute concentration, but it is in accordance with Hao et al. [111], when the solute concentration is higher than approximately 1.0 mg mL^{-1}. This study shows that the 0-1 model-inverse method has not only the advantages of high calculation speed and high accuracy in simultaneously obtaining thermodynamic parameters (isotherm parameters) and kinetic parameters (lumped mass transfer coefficients) of two-component systems. This method possesses the potential to optimally design and control the time-variant preparative chromatographic system due to the thermodynamic state recursion and the Lagrangian–Eulerian presentation of the 0-1 model.

1.9.1 ESTIMATION OF THERMODYNAMIC AND DYNAMIC PARAMETERS FOR SINGLE COMPONENT

Figure 1.28 shows the convergence process by which the simulated elution curves fit the experimental curves by changing k_f of benzyl alcohol in RP-HPLC when the isotherm parameters are the optimal values. k_f gradually increases from simulated curves 1 through 5. When k_f increases, the retention time of the band decreases, and as the shock increases, the band profile becomes narrower. This observation agreed with the results obtained by Lin et al. [99]. The simulated peak area does not depend on k_f in the 0-1 model. We can thus adjust the value of k_f under the conditions of the same simulated peak area. The value of k_f ($k_f = 1200$ min^{-1}) of benzyl alcohol in RP-HPLC was obtained from the value of k_f of the simulated curve 3, which is the best curve for fitting experimental data points. Thus, Figure 1.28 shows that the 0-1 model possesses good convergence for changing k_f.

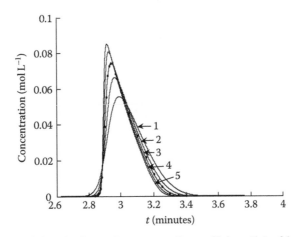

FIGURE 1.28 Obtaining the lumped mass transfer coefficient (k_f) of benzyl alcohol in RPLC through the 0-1 model-inverse method. Black points (·) denote the experimental data points ($n = 21$). $C_0 = 80$ mg mL^{-1}; sample injection volume: 20 μL; Langmuir isotherm, $q = \dfrac{1.592c}{1+1.28c}$. The values of k_f, 1, 500; 2, 800; 3, 1200; 4, 1700; 5, 2300 min^{-1}. Simulated conditions: $m = 10^{-3}$ cm. The other simulated conditions are the same as the experimental conditions in Section 1.3.

In Figure 1.29, from down to up, the sample concentrations of benzyl alcohol vary from 10 mg mL^{-1} to 80 mg mL^{-1}, corresponding to the simulated curves from 1 to 4, respectively. With Figure 1.29, we can find suitable values of b and k_f to fit experimental elution curves through the simulations of the chromatographic processes with the 0-1 model. It indicates that the 0-1 model possesses good convergence and mass conservation characteristics in changing k_f. In the simulated process with four injection concentrations of benzyl alcohol, we also adjust the parameter b to better fit experimental elution curves. Although the simulated optimum values of b vary slightly for each injection concentration, the simulated elution curve with the best values of b surely fits the experimental curve best. Here, the simulated b for each injection concentration is different from the regressive value for b based on all experimental points p $(q_{i,p}, c_{i,p})$. In the isotherm curve, $q_i \sim c_i$, for the p sample concentrations (i.e., $p = 1,\dots,$ 4 in Figure 1.29). The slight variance of the simulated b for each sample concentration is reasonable. Figure 1.29 shows that k_f decreases (from $k_f = 2300$ min^{-1} to $k_f = 1200$ min^{-1}) with increasing C_0 (from 10 to 80 mg mL^{-1}). These data are contrary to the data provided by Miyabe and Guiochon [109,110], in which k_f increases with increasing solute concentration. However, we obtained the result in accordance with Hao et. al [111]. Hao et al. drew their conclusion with the solute concentration higher than 1.0 mg mL^{-1} [111]. The different models can clearly lead to the visible difference for k_f, depending on the solute concentration. In addition, a few parameters (e.g., the flow rate of the mobile phase, the column temperature, and the properties of the column packing materials) may also influence k_f in nonlinear chromatography.

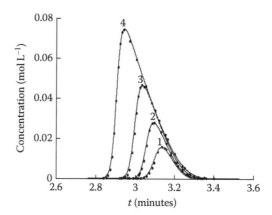

FIGURE 1.29 Simulated elution curves for four injection concentrations of benzyl alcohol through the simulations of chromatographic processes with the 0-1 model. Black points (·) denote the experimental data points. $a = 1.592$ for all the simulated elution curves. 1, $C_0 = 10$ mg mL^{-1}, $b = 1.1$ L mol^{-1}, $k_f = 2300$ min^{-1}; 2, $C_0 = 20$ mg mL^{-1} k_f, $b = 1.2$ L mol^{-1}, $k_f = 2000$ min^{-1}; 3, $C_0 = 40$ mg mL^{-1}, $b = 1.21$ L mol^{-1}, $k_f = 1550$ min^{-1}; 4, $C_0 = 80$ mg mL^{-1}, $b = 1.28$ L mol^{-1}, $k_f = 1200$ min^{-1}.

1.9.2 Estimation of the Parameters of Bi-Langmuir Isotherm

By substituting the Bi-Langmuir isotherm (Equation 1.31c) into Equations 1.67 and 1.69, the 0-1 model-inverse method for the two-component competitive isotherm was established through thermodynamic state recursion in a time series. With the 0-1 model-inverse method, the values of isotherm parameters b_1 and b_2 can be varied simultaneously to simulate the chromatographic processes of two components to obtain the simulated elution curves of two peaks. In Figure 1.30 from curves 1 through 5, the isotherm parameters b_1 and b_2 of the simulated curves gradually decrease in value, their retention times gradually increase, and the shock effects and peak tailings of the simulated curves become less serious. Because the simulated curve 3 fit the experimental points well, the isotherm parameters b_1 of benzyl alcohol and b_2 of phenylethyl alcohol in RP-HPLC were obtained under the reported conditions. A series of the simulated curves in Figure 1.30 indicates the good convergence of the 0-1 model in obtaining the isotherm parameters b_1 and b_2 for two-component systems.

1.9.3 Estimation of the Lumped Mass Transfer Coefficients of Two Components

For a two-component system ($i = 1, 2$), Equation 1.67 as the local form of the lumped kinetic model does not make the thermodynamic state recursion method more

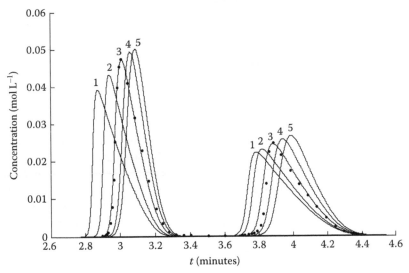

FIGURE 1.30 Obtaining the Bi-Langmuir isotherm parameters (b_1 and b_2) with the 0-1 model-inverse method. Solutes: Benzyl alcohol (left peaks) and phenylethyl alcohol (right peaks). Sample injection concentrations, 40:40 mg mL^{-1}. Black points ($\cdot$) $n = 42$ denote the experimental data points. $m = 1 \times 10^{-3}$ cm. $k_{f1} = 1500$ min^{-1}, $k_{f2} = 1350$ min^{-1}. 1, $b_1 = 3$ L mol^{-1}, $b_2 = 4$ L mol^{-1}; 2, $b_1 = 2$ L mol^{-1}, $b_2 = 3.5$ L mol^{-1}; 3, $b_1 = 1.21$ L mol^{-1}, $b_2 = 2.66$ L mol^{-1}; 4, $b_1 = 0.8$ L mol^{-1}, $b_2 = 2.0$ L mol^{-1}; 5, $b_1 = 0.6$ L mol^{-1}, $b_2 = 1.5$ L mol^{-1}. Bi-Langmuir isotherms: $q_1 = \dfrac{1.592 c_1}{1 + b_1 \cdot c_1 + b_2 \cdot c_2}$, $q_2 = \dfrac{2.671 c_2}{1 + b_1 \cdot c_1 + b_2 \cdot c_2}$.

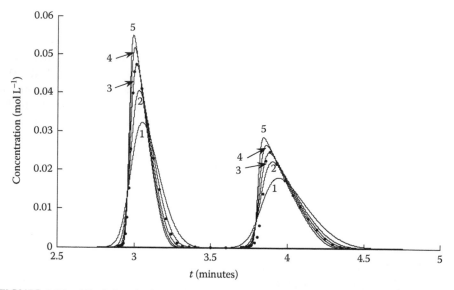

FIGURE 1.31 Obtaining the lumped mass transfer coefficients k_{f1} and k_{f2} through the 0-1 model-inverse method using the Bi-Langmuir adsorption isotherm. Solutes: Benzyl alcohol (left peaks) and phenylethyl alcohol (right peaks). Sample injection concentrations, 40:40 mg mL^{-1}. Black points (·) $n = 42$ denote the experimental data points. $m = 1 \times 10^{-3}$ cm. Bi-Langmuir isotherms: $q_1 = \dfrac{1.592\,c_1}{1+1.21c_1+2.66c_2}$, $q_2 = \dfrac{2.671c_2}{1+1.21c_1+2.66c_2}$. 1, $k_{f1} = k_{f2} = 600$ min^{-1}; 2, $k_{f1} = k_{f2} = 1000$ min^{-1}; 3, $k_{f1} = 1500$ min^{-1}, $k_{f2} = 1350$ min^{-1}; 4, $k_{f1} = 2000$ min^{-1}, $k_{f2} = 1800$ min^{-1}; 5, $k_{f1} = k_{f2} = 2500$ min^{-1}.

complex than it in the case of the one-component system. With a series of k_{f1} and k_{f2}, the transport chromatographic processes are simulated through the state recursion to obtain the simulated elution curves of the two-component system. Figure 1.31 shows that k_{f1} and k_{f2} gradually increase from the simulated curves 1 through 5, their retention times slightly decrease, their shock effects become more intense, and the peak tailing becomes less intense. The values of k_{f1} and k_{f2} of the simulated curve 3 are considered as the lumped mass transfer coefficients of benzyl alcohol and phenylethyl alcohol in RP-HPLC under the given conditions because the profile of the curve 3 is closest to the experimental curve. Figure 1.31 also indicates that the 0-1 model has good convergence for the parameters k_{f1} and k_{f2}.

1.10   THERMODYNAMIC STATE RECURSION METHOD FOR FRONTAL ANALYSIS CHROMATOGRAPHIC PROCESSES

A thermodynamic state recursion method [112], which is based on nonequilibrium thermodynamic path (as shown in Figure 1.32) described by LED, is presented to simulate the whole chromatographic process of FA using the spatial distribution of solute bands in time series. LED of FA chromatography progress is shown in Figure 1.33. All definitions of functions of thermodynamic states; LTP, MTP, and

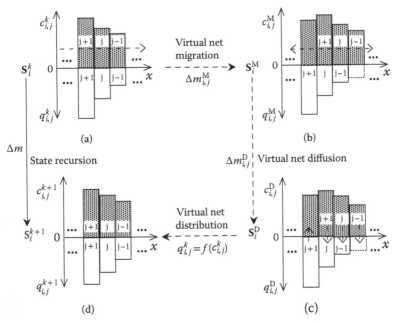

FIGURE 1.32 The local thermodynamic path of the nonideal nonlinear chromatography process. Dash arrows in (a), (b) and (c) show the directions of SCm slipping at u_k, the axial dispersions of solute between a SCm and its adjacent two SCms, and the local interphase mass transfer in a solute cell unit with $k_f = \infty$, respectively. Both the virtual thermodynamic path, $\mathbf{S}_i^k \rightarrow \mathbf{S}_i^M$ (the virtual migration state) $\rightarrow \mathbf{S}_i^D$ (the virtual diffusion state) $\rightarrow \mathbf{S}_i^{k+1}$ (d), and the local thermodynamic path, $\mathbf{S}_i^k \rightarrow \mathbf{S}_i^{k+1}$, can be used to quantitatively describe the state recursion from $\mathbf{S}_i^k$ to $\mathbf{S}_i^{k+1}$.

boundary conditions; and space-time distribution transfers are same as described in Sections 1.6 and 1.7. This method is used to simulate the nonideal nonlinear hydrophobic interaction chromatography (HIC) processes of lysozyme and myoglobin under the discrete complex boundary conditions. The results show that the simulated breakthrough curves agree well with the experimental ones. The Langmuir isotherm parameters of the two proteins in HIC are obtained by the state recursion-inverse method. Due to the time domain and Markov characteristics of thermodynamic state recursion method, it is applicable to the design and online control of the nonlinear multicolumn chromatographic systems.

For the nonideal linear chromatography, the state recursion equation of the local solute concentration ($c_{i,j}^{k+1}$) in the SCm can be obtained by

$$
\begin{aligned}
c_{i,j}^{k+1} = & \left[\frac{u_k \Delta t_k}{\Delta x \cdot \left(1 + F_k \cdot H_k\right)} \right] \left(c_{i,j+1}^k - c_{i,j}^k\right) \\
& + \left[\frac{D_k \Delta t_k}{(\Delta x)^2 \cdot \left(1 + F_k \cdot H_k\right)} \right] \left(c_{i,j+1}^k - 2c_{i,j}^k + c_{i,j-1}^k\right) \\
& + c_{i,j}^k
\end{aligned}
\tag{1.72}
$$

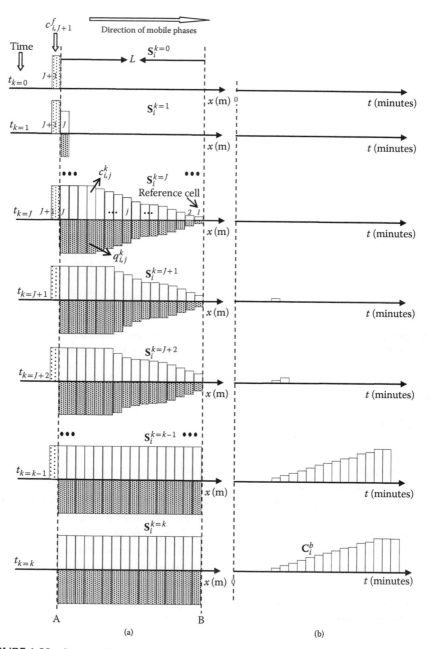

FIGURE 1.33 Lagrangian–Eulerian representation of FA chromatography progress. (a) The horizontal axes and "time" are the space-axis and time sequences of solute space distribution of FA chromatography progress, $\mathbf{S}_i^k$, respectively. The parts above and below the horizontal axis (e.g., $t = t_{k=j}$) depict solute concentrations in SCms and SCs, respectively. The lines A and B denote the positions of the inlets and outlets of columns with the column length (L). (b) The horizontal axis is the time-axis and the vector $\mathbf{C}_{i,j=1}^b$ is a breakthrough curve in discrete time sequence.

where D_k and H_k the time-variant diffusion coefficient and the time-variant Henry constant, respectively. For the nonideal nonlinear chromatography, the state recursion equation of the solute concentration ($c_{i,j}^{k+1}$) in the SCm can be obtained by

$$c_{i,j}^{k+1} + F_{k+1} \cdot \frac{a_{k+1} \cdot c_{i,j}^{k+1}}{1 + b_{k+1} \cdot c_{i,j}^{k+1}} = g^k \tag{1.73}$$

where

$$
\begin{aligned}
g^k = {}& \left(\frac{u_k \Delta t_k}{\Delta x} \right) \left(c_{i,j+1}^k - c_{i,j}^k \right) \\
& + \left(\frac{D_k \Delta t_k}{(\Delta x)^2} \right) \left(c_{i,j+1}^k - 2c_{i,j}^k + c_{i,j-1}^k \right) \\
& + \left(c_{i,j}^k + F_k \cdot \frac{a_k \cdot c_{i,j}^k}{1 + b_k \cdot c_{i,j}^k} \right)
\end{aligned}
\tag{1.73a}
$$

All the terms on the right-hand side of Equation 1.73a are determined by the local parameters of the state $\mathbf{S}_i^k$. Thus, g^k is known in the state recursion process and Equation 1.73 can be solved. The positive solution of physical significance is retained and the concentration of solute i in the jth SCm at t_{k+1} is expressed as

$$
\begin{aligned}
c_{i,j}^{k+1} = {}& \frac{g^k}{2} - \frac{F_{k+1} \cdot a_{k+1} + 1}{2 \cdot b_{k+1}} \\
& + \frac{\sqrt{(F_{k+1} \cdot a_{k+1} - b_{k+1} \cdot g^k + 1)^2 + 4 \cdot b_{k+1} \cdot g^k}}{2 \cdot b_{k+1}}
\end{aligned}
\tag{1.74}
$$

When the functional relation of the time-variant operational parameters, time-variant sample injection, and chromatographic conditions are known, $c_{i,j}^{k+1}$ can be directly calculated with Equation 1.72 or Equation 1.74 according to $c_{i,j}^k$. Thus, we can achieve the recursion from the states $\mathbf{S}_i^k$ to $\mathbf{S}_i^{k+1}$ in the linear or nonlinear nonideal chromatographic process under the time-variant chromatographic conditions. The above recursion method may lay a foundation for the design and optimal control of the time-variant chromatographic process. Figure 1.34 shows the effects of isotherms parameters a and b and the apparent diffusion coefficient D on the positions and shapes of FA breakthrough curves based on the thermodynamic state recursion method.

1.11 FEEDBACK CONTROL OF THE TIME-VARIANT NONIDEAL NONLINEAR CHROMATOGRAPHIC SYSTEM

Figure 1.35 shows the conceptual framework of the potential application of thermodynamic state recursion method to the feedback control of the time-variant nonideal nonlinear chromatographic system. In the system, the column temperature, the

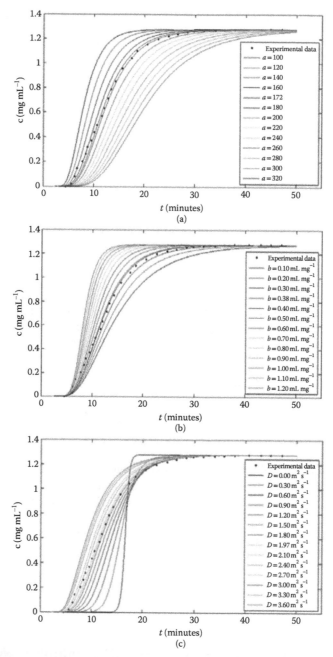

FIGURE 1.34 (**See color insert.**) The effects of isotherms parameters a and b and the apparent diffusion coefficient D on the positions and shapes of FA breakthrough curves. (a) Effects of a on positions and shapes of breakthrough curves with $b = 0.38\,\text{mL mg}^{-1}$ and $D = 1.97 \times 10^{-5}\,\text{m}^2\,\text{s}^{-1}$. (b) Effects of b on positions and shapes of breakthrough curves with $a = 172.57$ and $D = 1.97 \times 10^{-5}\,\text{m}^2\,\text{s}^{-1}$. (c) Effects of D on shapes of breakthrough curves with $a = 172.57$ and $b = 0.38\,\text{mL mg}^{-1}$.

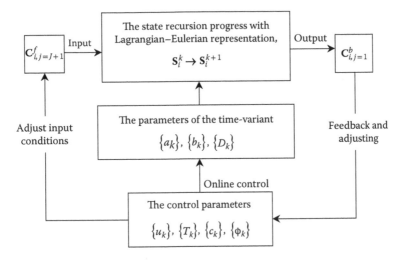

FIGURE 1.35 The feedback control of the time-variant nonideal nonlinear chromatography system.

elution strength, and the flow rate of the mobile phase may vary or fluctuate with time. Thus, the system parameters, H_k, a_k, b_k, u_k, D_k, and F_k, are functions of time (k). This system is thus referred to as a time-variant chromatographic system. This time-variant chromatographic system involves the gradient elution, temperature programming, or flow rate programming of the multicolumn preparative chromatography or the online optimization control of the packing aging of SMB chromatography. Of course, the time-invariant systems are only the special cases of time-variant nonlinear chromatographic systems.

The state recursion of the spatial distribution of the solute concentrations in the mobile phase during a given time interval can be implemented by Equation 1.74, when the sequence of the time-variant local concentrations of the injection sampling solute ($\mathbf{C}_{i,J+1}^{f}$) is known. The breakthrough curves ($\mathbf{C}_{i,j=1}^{b}$) as the output of the chromatographic system can be obtained by Equation 1.74, when MTP ends. As shown in Figure 1.35, the injection process ($\mathbf{C}_{i,J+1}^{f}$), the local isotherm parameters (a_k and b_k), and the local diffusion coefficient (D_k) during Δt_k of the time-variant chromatographic process directly affect the evolution process, $\left\{\mathbf{S}_i^{k}\right\}$, of the solute band in the column and the output solute concentration ($\mathbf{C}_{i,j=1}^{b}$) in the mobile phase. Under given chromatographic conditions, the whole time-variant chromatographic process can be simulated by the $\mathbf{S}_i^{k} \rightarrow \mathbf{S}_i^{k+1}$ thermodynamic state recursions. Accordingly, the simulated breakthrough curves can be obtained. In the design and online optimization control of multicolumn chromatographic systems, the control parameters of the chromatographic system, such as the temperature (T_k), injection concentration ($\mathbf{C}_{i,J+1}^{f}$), flow rate (u_k), and solvent elution strength (ϕ_k) need to be adjusted according to the real-time output solute concentration. This time-variant chromatographic model based on thermodynamic state recursion method may be essential for the optimal design and online control of the time-variant preparative chromatography.

1.12   0-1 MODEL FOR VARICOL PROCESSES OF MULTICOLUMN CONTINUOUS CHROMATOGRAPHY

With the 0-1 model, the numerical description of the MCC becomes more easy compared with the elution single-column chromatography, since the 0-1 model of MCC does not need the space-time distribution transfer step required in the elution single-column chromatography.

1.12.1   THERMODYNAMIC STATE VARIABLES IN THE VARICOL PROCESSES

For the periodic Varicol process of MCC, the discrete time that needs three levels is shown in Figure 1.36.

In Figure 1.36, $\{t_n\}$ indicates a time sequence on the time points of period of big cycles

$$\{t_n\} = \{t_0, t_1, \ldots t_n, \ldots, t_N\} \quad n \in \mathbb{N}_+ \tag{1.75}$$

where t_n is the time point of n th big cycles. In $t_s = t_{n+1} - t_n$, t_s is the time period. We can further divide t_s of a big cycle as G subcycle periods,

$$G = t_s/\tau \quad G_+ := \{0,1,2\ldots g,\ldots G-1\}, \quad g \in G_+ \tag{1.76}$$

where τ is a period of a subcycle in Varicol processes, and g is the serial number of subcycle periods. Furthermore, a series of time intervals, $\Delta t_k = t_{k+1} - t_k$, were used to implement the numerical simulation of Varicol processes as in the 0-1 model [14]. Thus, a sequence of time cells, $\{t_{n,g_k}\}$, of Varicol processes can be expressed

$$\{t_{n,g_k}\} = \{t_{n,g_0}, t_{n,g_1}, \ldots, t_{n,g_k}, \ldots, t_{n,g_K}\}, \mathbf{K}_+ := \{0,1,\ldots,k,\ldots,K\}, k \in \mathbf{K}_+ \tag{1.77}$$

where t_{n,g_k} is the k th time point of the gth subcycle of the nth big cycles. For the discrete space based on the LED, the local position $(x_{r,j}^{n,g_k})$ of the jth solute cell in the rth column at t_{n,g_k} is indicated as

$$x_{r,j}^{n,g_k} = (j-1) \cdot m, \quad (r = 1,2,\ldots,8; \; j = 1,2,\ldots J) \tag{1.78}$$

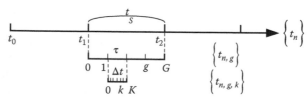

FIGURE 1.36   The discrete times in the periodic Varicol process of MCC. (From Yin, Master's thesis, Xi'an Jiaotong University, Xi'an, 2011. With permission.)

In this section, we will describe the thermodynamic states to build the 0-1 model of MCC. In Varicol processes of MCC, let $i = A, B, \ldots$ for different solutes, the integration thermodynamic state vector $(\mathbf{S}^{n,g_k})$ of space distribution of solutes in MCC systems at t_{n,g_k} can be indicated as

$$\mathbf{S}^{n,g_k} = \left\{ \mathbf{S}_A^{n,g_k}, \mathbf{S}_B^{n,g_k} \right\} \tag{1.79}$$

where $\mathbf{S}_A^{n,g_k}$ and $\mathbf{S}_B^{n,g_k}$ are thermodynamic state vectors of solute A and B at time t_{n,g_k}. The thermodynamic state vector $(\mathbf{S}_i^{n,g_k})$ of solute i can shown as

$$\mathbf{S}_i^{n,g_k} = \left\{ \mathbf{C}_i^{n,g_k}, \mathbf{Q}_i^{n,g_k}, \mathbf{X}^{n,g_k} \right\} \tag{1.80}$$

where $\mathbf{C}_i^{n,g_k}$ and $\mathbf{Q}_i^{n,g_k}$ are the local solute concentration vectors in the SCms and the SCs of the solute zone i at the time t_{n,g_k}, respectively. $\mathbf{X}^{n,g_k}$ is the vector of position of the SCms and SCs of the solute zone i at the time t_{n,g_k}. $\mathbf{C}_i^{n,g_k}$, $\mathbf{Q}_i^{n,g_k}$, and $\mathbf{X}^{n,g_k}$ can be further indicated by

$$\mathbf{C}_i^{n,g_k} = [\mathbf{C}_{i,1}^{n,g_k}, \mathbf{C}_{i,2}^{n,g_k}, s \ldots, \mathbf{C}_{i,r}^{n,g_k}, \ldots, \mathbf{C}_{i,r'}^{n,g_k}]^T$$

$$\mathbf{Q}_i^{n,g_k} = [\mathbf{Q}_{i,1}^{n,g_k}, \mathbf{Q}_{i,2}^{n,g_k}, \ldots, \mathbf{Q}_{i,r}^{n,g_k}, \ldots, \mathbf{Q}_{i,r'}^{n,g_k}]^T \tag{1.81}$$

$$\mathbf{X}^{n,g_k} = \left[\mathbf{X}_1^{n,g_k}, \mathbf{X}_2^{n,g_k}, \ldots, \mathbf{X}_r^{n,g_k}, \ldots, \mathbf{X}_{r'}^{n,g_k} \right]^T$$

$\mathbf{C}_{i,r}^{n,g_k}$, $\mathbf{Q}_{i,r}^{n,g_k}$, and $\mathbf{X}_r^{n,g_k}$ are further expressed by

$$\mathbf{C}_{i,r}^{n,g_k} = [c_{i,r,1}^{n,g_k}, c_{i,r,2}^{n,g_k}, \ldots, c_{i,r,j}^{n,g_k}, \ldots, c_{i,r,J}^{n,g_k}]^T$$

$$\mathbf{Q}_{i,r}^{n,g_k} = [q_{i,r,1}^{n,g_k}, q_{i,r,2}^{n,g_k}, \ldots, q_{i,r,j}^{n,g_k}, \ldots, q_{i,r,J}^{n,g_k}]^T \tag{1.82}$$

$$\mathbf{X}_r^{n,g_k} = [x_{r,1}^{n,g_k}, x_{r,2}^{n,g_k}, \ldots, x_{r,j}^{n,g_k}, \ldots, x_{r,J}^{n,g_k}]^T$$

where $c_{i,r,j}^{n,g_k}$, $q_{i,r,j}^{n,g_k}$, and $x_{r,j}^{n,g_k}$ are the local solute concentrations in the jth SCms, the local solute concentrations in the jth SCs, and the local position of the jth SCms and the jth SCs in the rth column at the time t_{n,g_k}, respectively.

1.12.2 0-1 MODEL FOR VARICOL PROCESSES OF MCC

The 0-1 model includes the LTP and the MTP (see Section 1.6.2). In the Varicol processes of MCC, the LTP can be expressed as the state recursion $\mathbf{S}_i^{n,g_k} \rightarrow \mathbf{S}_i^{M_1,g_k} \rightarrow \mathbf{S}_i^{M,g_k} \rightarrow \mathbf{S}_i^{n,g_{k+1}}$, where $\mathbf{S}_i^{M,g_k}$ and $\mathbf{S}_i^{M,g_{k+1}}$ are the thermodynamic state vectors at the time t_{n,g_k} and $t_{n,g_{k+1}}$, respectively. $\mathbf{S}_i^{M_1,g_k}$ is the virtual migration state, and $\mathbf{S}_i^{M,g_k}$ is cell adjustment state in the LTP of the MCC systems. The LTP only involves the simple recombination of the matrixes and the movements of their lines and rows. We synthetically considered the initial conditions, boundary conditions, period steady-state conditions, and column-switching conditions. These conditions affect the chromatographic process of the Varicol operations.

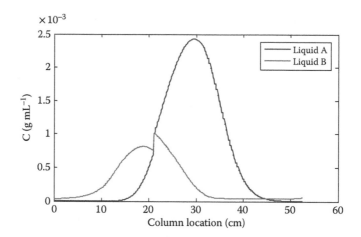

FIGURE 1.37 **(See color insert.)** The solute spatial distribution at the 30th switch time in columns with the 5-column Varicol chromatography system.

$$g = 4, Z(0) = [1,1,1,2], Z(1) = [1,1,2,1], Z(2) = [1,2,1,1], Z(3)$$
$$= [2,1,1,1], \overline{Z}(t) = [1.25,1.25,1.25,1.25]$$

The state recursion of the 0-1 model considering the dead-volumes between the two adjacent columns was presented [113] for the Varicol processes of MCC. This numerical model is more intuitional and suitable for the process control and optimization. The spatial distributions of solutes in the column system (as shown in Figure 1.37) and the outflow curves of the extract and raffinate streams (as shown in Figure 1.38) in Varicol systems were simulated successfully. These can be implemented by the homemade programs based on the 0-1 model with the MATLAB® conveniently and fastly (see Section 1.13). The simulation results showed that the purities and productivities of the Varicol were better than ones of the SMB under the same separation conditions [113]. By comparing the separation results of MCC with different number of chromatographic columns, the advantages of separations in Varicol were demonstrated more obviously in the MCC with the 5-column systems. The function of 36-ported rotary valve was tested and its mechanical structure was promoted to realize the tightness and accuracy of this valve in SMB [114].

1.13 SIMULATION SOFTWARE FOR MCC BASED ON THE 0-1 MODEL

We programmed the simulation software for the optimization of MCC based on the 0-1 model. The optimized algorithm of the 0-1 model promoted the calculation speed in the simulation of the MCC processes, which made it beneficial to design the scheme and to optimize MCC system. We transformed the nonlinear equations with Langmuir adsorption isotherm into linear equations without solving Jacobian matrix. It reduced greatly the computing time from about 10 minutes to 10 seconds

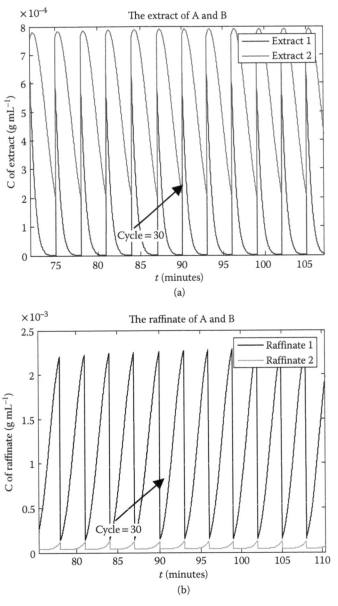

FIGURE 1.38 **(See color insert.)** Outflow curves of extract (a) and raffinate (b) with the 5-column Varicol chromatography system.

for one simulation of a typical MCC process. This simulation software was designed by hybrid programming with VC++ and MATLAB based on the promoted algorithms [115]. Figure 1.39 shows the sketch configuration of user functions of the MCC simulation software based on the 0-1 model. It includes four parts, animation demonstration of the principle of 0-1 model, the obtaining adsorption isotherm

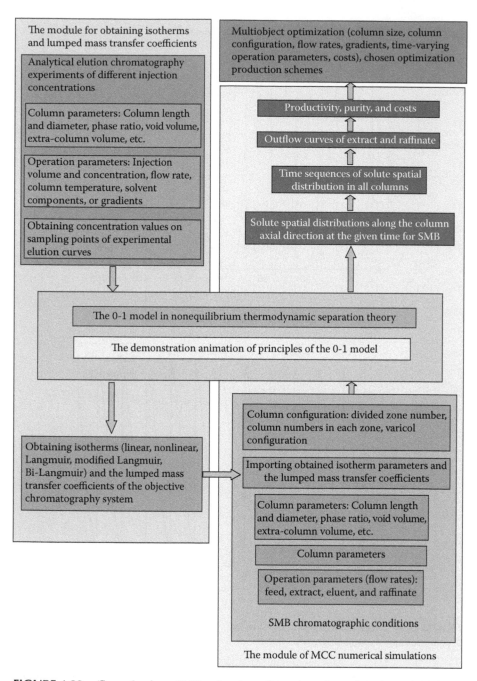

FIGURE 1.39 (**See color insert.**) The sketch configuration of user functions of the MCC simulation software based on the 0-1 model. Note that the different colors with different explanations indicate their subordinate relationships.

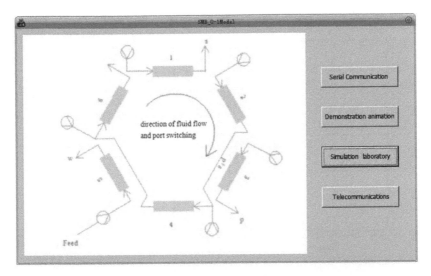

FIGURE 1.40 Start-up interface of the MCC simulations.

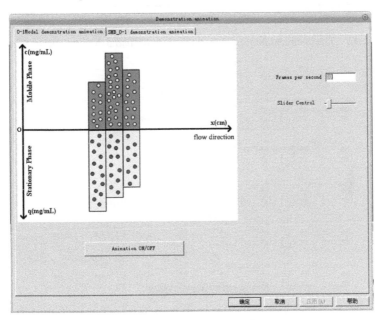

FIGURE 1.41 The interface of demonstration animation of the 0-1 model.

parameters and mass transfer coefficients, the simulation of MCC process with linear/nonlinear adsorption isotherm, and the optimization of productivity, purity, and consumption. By graphical user interface of this software the simulation of MCC process and the analysis of performance index can be executed expediently in evaluating the related production scheme more intuitively and effectively. The start-up interface (as shown in Figure 1.40) includes two push buttons, demonstration animation and simulation laboratory. Figure 1.41 shows an interface of demonstration

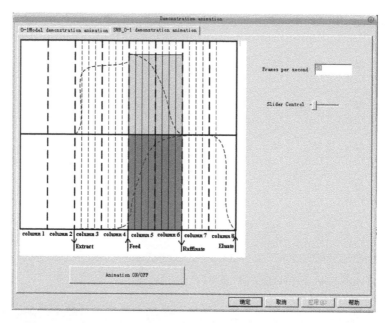

FIGURE 1.42   The demonstration animation interface of spatial distributions of SCm in SMB.

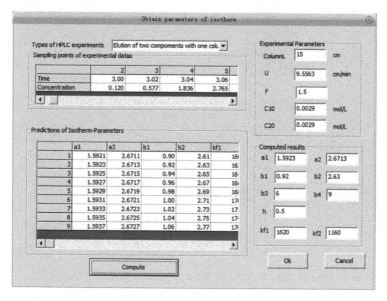

FIGURE 1.43   The calculation results on the parameters of Langmuir isotherm.

animation of the 0-1 model, which demonstrates virtual net migration processes and virtual net mass processes intuitively and dynamically. Figure 1.42 is a demonstration animation interface of all spatial distributions of SCm in all columns of SMB. It can show a time sequence of the varying spatial distributions of SCms. Figure 1.43 is the interface for obtaining parameters of Langmuir isotherm of two components.

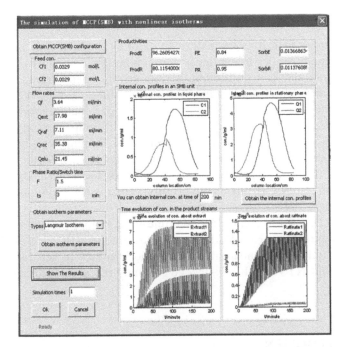

FIGURE 1.44 **(See color insert.)** The interface of simulation results of SMB with Langmuir isotherm of two components.

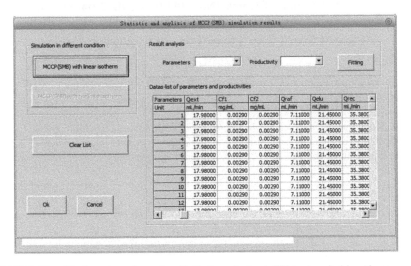

FIGURE 1.45 The host interface of statistical analysis at different switching times.

Figure 1.44 shows the spatial distributions of SCms and SCs at the given times. It also shows the evolution curves of extract and raffinate streams. After many times of the SMB simulations under different conditions, the productivity, purity, and corresponding simulation experiment parameters are listed in the host interface of statistical analysis (as shown in Figure 1.45).

1.14   REFLECTION

It has taken more than 10 years to develop theories and applications of nonequilibrium thermodynamics in nonlinear chromatography and electrophoresis. We have understood that nonequilibrium thermodynamics can unify the phenomena of equilibrium thermodynamics and dynamics in the separation processes truly. Therefore, we developed the NTST in nonlinear chromatography and electrophoresis based on the point of view of nonequilibrium thermodynamics, but not on two separate worlds, the equilibrium thermodynamics or dynamics. The NTST includes some novel concepts and principles (e.g., LTP, thermodynamic state recursions, local thermodynamic state vectors, local thermodynamic equilibrium assumption, Lagrangian description, time-variant system, Markovian representations, and state space representation). However, there is potential for the NTST to be used in controlling the time-variant nonlinear chromatography and electrophoresis system. While these future studies will certainly bring further understanding of nonequilibrium thermodynamics in separation science, we believe that NTST will be extended in many research fields of separation science.

ACKNOWLEDGMENT

This work was financially funded by the funding of the National Science & Technology Pillar Program of China (No. 2009BAK59B02-04) and the National Natural Science Foundation of China (Nos. 20299030, 20175015, and 90209016).

REFERENCES

1. J.C. Giddings, *Unified separation science*, J. Wiley, New York; Toronto, 1991.
2. G. Guiochon, B. Lin, *Modeling for preparative chromatography*, Academic Press, Amsterdam; Boston, 2003.
3. A. Rajendran, G. Paredes, M. Mazzotti, Simulated moving bed chromatography for the separation of enantiomers, *Journal of Chromatography A* 1216 (2009) 709–738.
4. I. Prigogine, *Introduction to thermodynamics of irreversible processes*, Interscience Publishers, New York, 1968.
5. S.R. de Groot, P. Mazur, *Non-equilibrium thermodynamics*, North-Holland Pub. Co.; Interscience Publishers, Amsterdam; New York, 1962.
6. I. Gyarmati, *Non-equilibrium thermodynamics. Field theory and variational principles*, Springer, Berlin; New York, 1970.
7. R.M. Velasco, L.S. Garcia-Colin, F.J. Uribe, Entropy production: Its role in nonequilibrium thermodynamics, *Entropy* 13 (2011) 82–116.
8. H. Liang, B.C. Lin, Frameworks of separation theories from two separate worlds: Dynamics and thermodynamics, *Journal of Chromatography A* 828 (1998) 3–17.
9. H. Liang, Z.G. Wang, B.C. Lin, C.G. Xu, R.N. Fu, Nonequilibrium thermodynamic separation model in capillary electrophoresis, *Journal of Chromatography A* 763 (1997) 237–251.
10. H. Liang, B.C. Lin, Integral optimizing functional of separation efficiency, *Journal of Chromatography A* 841 (1999) 133–146.

11. H. Liang, Y. Liu, Recursion equations in predicting band width under gradient elution, *Journal of Chromatography A* 1040 (2004) 19–31.
12. H. Liang, X.L. Liu, Lagrangian description of nonlinear chromatography, *Science in China Series B-Chemistry* 47 (2004) 443–452.
13. H. Liang, Non-equilibrium thermodynamic separation theory of nonlinear chromatograph I. Local Lagrangian approach for nonlinear equilibrium-dispersive processes, *Chinese Journal of Chromatography* 25 (2007) 664–680.
14. H. Liang, Z. Jia, Non-equilibrium thermodynamic separation theory of nonlinear chromatography. II. The 0-1 model for nonlinear-mass transfer kinetic processes, Chinese *Journal of Chromatography* 25 (2007) 785–798.
15. H. Liang, X.L. Cheng, Y.F. Ma, Localized single molecule isotherms of DNA molecules at confined liquid-solid interfaces, *Analytical Chemistry* 81 (2009) 2059–2066.
16. H. Liang, Y. Chen, L.J. Tian, L. Zhang, Recursion approach for moving neutralization boundary formed on IPG strips Part I: With strong alkali rehydration buffer, *Electrophoresis* 30 (2009) 3134–3143.
17. H. Liang, L.F. OuYang, Q. Liu, L. Zhang, L.J. Tian, Y. Chen, Time-varying migration process of moving neutralization boundary on the immobilized pH gradient strip in the weak-base rehydration buffer, *Journal of Separation Science* 34 (2011) 1212–1219.
18. J.N. Wilson, A theory of chromatography, *Journal of the American Chemical Society* 62 (1940) 1583–1591.
19. N. Wiener, *Cybernetics: Control and communication in the animal and the machine*, Wiley Hermann et Cie, New York; Paris, 1948.
20. M. Mazzotti, G. Storti, M. Morbidelli, Optimal operation of simulated moving bed units for nonlinear chromatographic separations, *Journal of Chromatography A* 769 (1997) 3–24.
21. F. Gritti, G. Guiochon, Systematic errors in the measurement of adsorption isotherms by frontal analysis: Impact of the choice of column hold-up volume, range and density of the data points, *Journal of Chromatography A* 1097 (2005) 98–115.
22. O. Lisec, P. Hugo, A. Seidl-Morgenstern, Frontal analysis method to determine competitive adsorption isotherms, *Journal of Chromatography A* 908 (2001) 19–34.
23. L. Aumann, M. Morbidelli, A continuous multicolumn countercurrent solvent gradient purification (MCSGP) process, *Biotechnology and Bioengineering* 98 (2007) 1043–1055.
24. R. Kuhn, S. Hofstetter-Kuhn, *Capillary electrophoresis: Principles and practice*, Springer-Verlag, Berlin; New York, 1993.
25. P.G. Righetti, The alpher, bethe, gamow of isoelectric focusing, the alpha-centaury of electrokinetic methodologies. Part I, *Electrophoresis* 27 (2006) 923–938.
26. P.G. Righetti, The alpher, bethe and gamow of IEF, the alpha-centaury of electrokinetic methodologies. Part II: Immobilized pH gradients, *Electrophoresis* 28 (2007) 545–555.
27. C.X. Cao, L.Y. Fan, W. Zhang, Review on the theory of moving reaction boundary, electromigration reaction methods and applications in isoelectric focusing and sample pre-concentration, *Analyst* 133 (2008) 1139–1157.
28. H. Svensson, *Electrophoresis by the moving boundary method: A theoretical and experimental study*, Almqvist & Wiksell, Stockholm, 1946.
29. H. Svensson, Isoelectric fractionation, analysis, and characterization of ampholytes in natural ph gradients. I. Differential equation of solute concentrations at a steady state and its solution for simple cases, *Acta Chemica Scandinavica* 15 (1961) 325–341.
30. H. Svensson, Isoelectric fractionation, analysis and characterization of ampholytes in natural ph gradients. II. Buffering capacity and conductance of isoionic ampholytes, *Acta Chemica Scandinavica* 16 (1962) 456–466.

31. P.G. Righetti, *Isoelectric focusing: Theory, methodology, and applications*, Elsevier Biomedical Press; Sole distributors for the U.S.A. and Canada, Elsevier-North Holland, Amsterdam; New York, 1983.

32. H.J. Kreuzer, *Non-equilibrium thermodynamics and its statistical foundations*, Oxford University Press, New York, 1981.

33. I.G. Currie, *Fundamental mechanics of fluids*, McGraw, New York, 1993.

34. I.N. Grigor'ev, V.A. Vshivkov, M.P. Fedoruk, *Numerical "particle-in-cell" methods: theory and applications*, VSP, Utrecht; Boston, 2002.

35. I. Prigogine, I. Stengers, *Order out of chaos: Man's new dialogue with nature*, Bantam Books, Toronto; New York, 1984.

36. I. Prigogine, *From being to becoming: Time and complexity in the physical sciences*, W.H. Freeman, San Francisco, 1980.

37. X.H.N. Xu, E.S. Yeung, Long-range electrostatic trapping of single-protein molecules at a liquid-solid interface, *Science* 281 (1998) 1650–1653.

38. E.S. Yeung, Dynamics of single biomolecules in free solution, *Annual Review of Physical Chemistry* 55 (2004) 97–126.

39. E. Schrödinger, *What is life?: The physical aspect of the living cell with mind and matter & autobiographical sketches*, Cambridge University Press, Cambridge; New York, 1992.

40. H.S. Leff, *Maxwell's demon: Entropy, information, computing*, Hilger, Bristol, 1990.

41. K.G. Denbigh, J.S. Denbigh, *Entropy in relation to incomplete knowledge*, Cambridge University Press, Cambridge [Cambridgeshire]; New York, 1985.

42. I. Procaccia, J. Ross, 1977 Nobel-Prize in Chemistry, *Science* 198 (1977) 716–717.

43. L. Onsager, Reciprocal relations in irreversible processes. I., *Physical Review* 37 (1931) 405–426.

44. L. Onsager, Reciprocal relations in irreversible processes. II., *Physical Review* 38 (1931) 2265–2279.

45. D.K. Kondepudi, I. Prigogine, *Modern thermodynamics: From heat engines to dissipative structures*, John Wiley & Sons, Chichester; New York, 1998.

46. G. Guiochon, S.G. Shirazi, A.M. Katti, *Fundamentals of preparative and nonlinear chromatography*, Academic Press, Boston, 1994.

47. A. Kleidon, R. Lorenz, *Non-equilibrium thermodynamics and the production of entropy life, earth, and beyond*, Springer, Berlin; New York, 2005.

48. P. Glansdorff, I. Prigogine, *Thermodynamic theory of structure, stability and fluctuations*, Wiley-Interscience, London; New York, 1971.

49. J. Keizer, *Statistical thermodynamics of nonequilibrium processes*, Springer-Verlag, New York, 1987.

50. F. Schlögl, *Probability and heat: Fundamentals of thermostatistics*, F. Vieweg, Braunschweig, 1989.

51. E. Grushka, R.M. Mccormick, J.J. Kirkland, Effect of temperature-gradients on the efficiency of capillary zone electrophoresis separations, *Analytical Chemistry* 61 (1989) 241–246.

52. T. Yun, P. Sajonz, Z. Bensetiti, G. Guiochon, Influence of the heat of adsorption on elution band profiles in nonlinear liquid chromatography, *Journal of Chromatography A* 760 (1997) 3–16.

53. J.H. Knox, K.A. Mccormack, Capillary electroseparation methods looking into the crystal ball, *Journal of Liquid Chromatography* 12 (1989) 2435–2470.

54. G.K. Batchelor, *An introduction to fluid dynamics*, Cambridge University Press, Cambridge; New York, 1973.

55. H. Lamb, *Hydrodynamics*, Cambridge University Press, Cambridge, 1932.

56. H. Aref, S.W. Jones, G. Tryggvason, On lagrangian aspects of flow simulation, *Complex Systems* 1 (1987) 545–558.
57. D.H. Wagner, Equivalence of the Euler and Lagrangian equations of gas-dynamics for weak solutions, *Journal of Differential Equations* 68 (1987) 118–136.
58. C.-W. Shu, S. Osher, Efficient implementation of essentially non-oscillatory shock-capturing schemes, II, *Journal of Computational Physics* 83 (1989) 32–78.
59. W.H. Hui, P.Y. Li, Z.W. Li, A unified coordinate system for solving the two-dimensional Euler equations, *Journal of Computational Physics* 153 (1999) 596–637.
60. L. Xiao-Biao, Generalized Rankine–Hugoniot condition and shock solutions for quasilinear hyperbolic systems, *Journal of Differential Equations* 168 (2000) 321–354.
61. I. Malcevic, O. Ghattas, Dynamic-mesh finite element method for Lagrangian computational fluid dynamics, *Finite Elements in Analysis and Design* 38 (2002) 965–982.
62. C.W. Hirt, J.L. Cook, T.D. Butler, A Lagrangian method for calculating the dynamics of an incompressible fluid with free surface, *Journal of Computational Physics* 5 (1970) 103–124.
63. C.-Y. Loh, M.-S. Liou, A new Lagrangian method for three-dimensional steady supersonic flows, *Journal of Computational Physics* 113 (1994) 224–248.
64. W.H. Hui, Y.P. He, Hyperbolicity and optimal coordinates for the three-dimensional supersonic Euler equations, *SIAM Journal on Applied Mathematics* 57 (1997) 893–928.
65. G. Ludyk, *Time-variant discrete-time systems*, Vieweg, Braunschweig [u.a.], 1981.
66. P. Jandera, J. Churácek, *Gradient elution in column liquid chromatography: Theory and practice*, Elsevier, Amsterdam; New York, 1985.
67. P. Jandera, Z. Kucerova, J. Urban, Retention times and bandwidths in reversed-phase gradient liquid chromatography of peptides and proteins, *Journal of Chromatography A* 1218 (2011) 8874–8889.
68. F. Dondi, G. Guiochon, *Theoretical advancement in chromatography and related separation techniques*, Kluwer Academic Publishers, Dordrecht; Boston,1992.
69. P.D. Grossman, J.C. Colburn, *Capillary electrophoresis: Theory & practice*, Academic Press, San Diego, 1992.
70. A. Lasota, M.C. Mackey, *Chaos, fractals, and noise: Stochastic aspects of dynamics*, Springer-Verlag, New York, 1994.
71. S. Meyn, R.L. Tweedie, *Markov chains and stochastic stability*, Cambridge University Press, Cambridge, 2009.
72. K.H. Karlsen, K.A. Lie, J.R. Natvig, H.F. Nordhaug, H.K. Dahle, Operator splitting methods for systems of convection-diffusion equations: Nonlinear error mechanisms and correction strategies, *Journal of Computational Physics* 173 (2001) 636–663.
73. K.H. Karlsen, N.H. Risebro, Corrected operator splitting for nonlinear parabolic equations, *SIAM Journal on Numerical Analysis* 37 (2000) 980–1003.
74. A. Taflove, Application of the finite-difference time-domain method to sinusoidal steady-state electromagnetic-penetration problems, *IEEE Transactions on Electromagnetic Compatibility* 22 (1980) 191–202.
75. A. Taflove, S.C. Hagness, *Computational electrodynamics: The finite-difference time-domain method*, Artech House, Boston, 2005.
76. N.S. Nise, *Control systems engineering*, Wiley, Hoboken, 2004.
77. D. Hinrichsen, A.J. Pritchard, *Mathematical systems theory I: Modelling, state space analysis, stability and robustness*, Springer-Verlag, Berlin, Heidelberg, 2005.
78. B. Friedland, *Control system design: An introduction to state-space methods*, Dover Publications, Mineola, 2005.
79. J. Durbin, S.J. Koopman, *Time series analysis by state space methods*, Oxford University Press, Oxford; New York, 2001.
80. G. Guiochon, Preparative liquid chromatography, *Journal of Chromatography A* 965 (2002) 129–161.

81. S. Katsuo, M. Mazzotti, Intermittent simulated moving bed chromatography: 1. Design criteria and cyclic steady-state, *Journal of Chromatography A* 1217 (2010) 1354–1361.

82. H. Liang, Theoretical studies on solute migration, separation and application in capillary electrophoresis, Ph.D. Thesis, Beijing Institute of Technology, Beijing (1995) 1–126.

83. H. Liang, Z. Wang, R. Fu, B. Lin, Entropy increase and nature of separation, *Journal of Beijing Institute of Technology* 5 (1996) 137–143.

84. J.J. Vandeemter, F.J. Zuiderweg, A. Klinkenberg, Longitudinal diffusion and resistance to mass transfer as causes of nonideality in chromatography, *Chemical Engineering Science* 5 (1956) 271–289.

85. A.J.P. Martin, R.L.M. Synge, A new form of chromatogram employing two liquid phases I. A theory of chromatography 2. Application to the micro-determination of the higher monoamino-acids in proteins, *Biochemical Journal* 35 (1941) 1358–1368.

86. J. Cazes, R.P.W. Scott, *Chromatography theory*, Marcel Dekker, New York, 2002.

87. P.V. Coveney, The 2nd law of thermodynamics: Entropy, irreversibility and dynamics, *Nature* 333 (1988) 409–415.

88. K. De Clerk, C.E. Cloete, Entropy as a general separation criterion, *Separation Science* 6 (1971) 627–635.

89. G.H. Stewart, Measurement and evaluation of separation, *Separation Science and Technology* 13 (1978) 201–213.

90. H. Liang, Q. Yu, J. Gu, Z. Wang, R. Fu, Migration equation of solute in high performance capillary zone electrophoresis, *Chinese Science Bulletin* 41 (1996) 175–176.

91. T. Tsuda, G. Nakagawa, M. Sato, K. Yagi, Separation of nucleotides by high-voltage capillary electrophoresis, *Journal of Applied Biochemistry* 5 (1983) 330–336.

92. R. Serra, *Introduction to the physics of complex systems: The mesoscopic approach to fluctuations, non linearity, and self-organization*, Pergamon, Oxford [Oxfordshire]; New York, 1986.

93. P. Coveney, R. Highfield, *The arrow of time: A voyage through science to solve time's greatest mystery*, W.H. Allen, London, 1990.

94. S.C. Hunter, *Mechanics of continuous media*, E. Horwood; Halsted Press, Chichester [West] Sussex; New York, 1983.

95. H.-G. Lee, *Chemical thermodynamics for metals and materials*, Imperial College Press, London, 2000.

96. S. Golshanshirazi, S. Ghodbane, G. Guiochon, Comparison between experimental and theoretical band profiles in nonlinear liquid-chromatography with a pure mobile phase, *Analytical Chemistry* 60 (1988) 2630–2634.

97. H.L. Liu, C. Sabus, G.T. Carter, M. Tischler, Use of a linear gradient flow program for liquid chromatography-mass spectrometry protein-binding studies, *Journal of Chromatography A* 955 (2002) 237–243.

98. M.H. Moon, P.S. Williams, D.J. Kang, I. Hwang, Field and flow programming in frit-inlet asymmetrical flow field-flow fractionation, *Journal of Chromatography A* 955 (2002) 263–272.

99. B.C. Lin, S. Golshanshirazi, G. Guiochon, Effect of mass-transfer coefficient on the elution profile in nonlinear chromatography, *Journal of Physical Chemistry* 93 (1989) 3363–3368.

100. B.C. Lin, S. Golshanshirazi, Z. Ma, G. Guiochon, Shock effects in nonlinear chromatography, *Analytical Chemistry* 60 (1988) 2647–2653.

101. G. Gotmar, T. Fornstedt, G. Guiochon, Peak tailing and mass transfer kinetics in linear chromatography: Dependence on the column length and the linear velocity of the mobile phase, *Journal of Chromatography A* 831 (1999) 17–35.

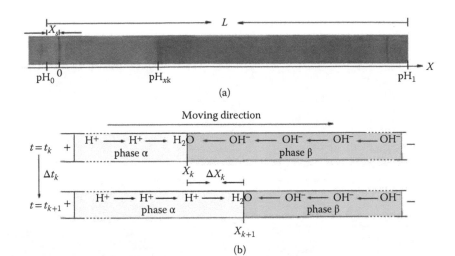

(a)

(b)

FIGURE 1.8 Diagrams of Lagrangian coordinates and the boundary migrations at two adjacent local times in NaOH–IPG–MNB system. (From Liang et al., *Electrophoresis*, 30, 3134–3143, 2009. With permission.)

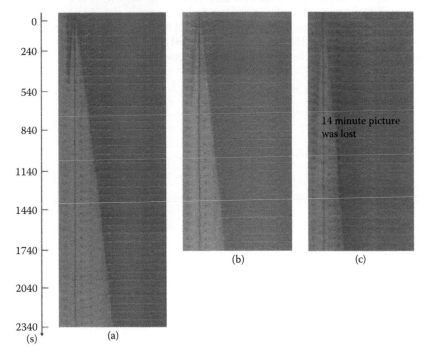

14 minute picture was lost

(b)   (c)

(a)

FIGURE 1.9 Images of the moving boundaries in the NaOH–IPG–MNB systems with 4–7 pH IPG strips. (a) 5 mmol·L^{-1}, (b) 10 mmol·L^{-1}, (c) 12 mmol·L^{-1} with strong alkali rehydration buffers (C_{alk} NaOH + 0.1 mol L^{-1} NaCl + 0.1% (w/v ethanol) bromophenol blue). The IEF experiments were carried on a PROTEAN IEF Cell with limitative current ($I_1 = 5 \times 10^{-5}$ A) and temperature 25°C.

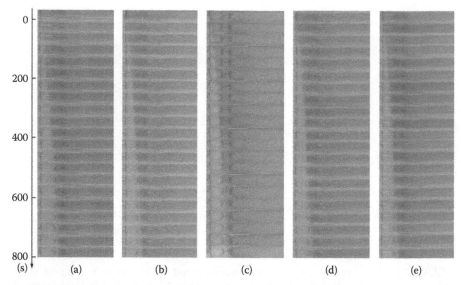

FIGURE 1.11 Images of the moving boundaries in ammonia–IPG–MNB systems on IPG strips (pH 3–6). $C_{NH_3 \cdot H_2O}$: (a) 0.20, (b) 0.25, (c) 0.30, (d) 0.35, (e) 0.40 mmol·L^{-1} in the 125 μL weak base rehydration solution confected with ammonia of different concentrations, 0.1 mol·L^{-1} KCl and 0.1% (w/v ethanol) bromophenol blue. The IEF experiments were performed using a PROTEAN IEF Cell and IPG strip (pH 3–6). (From Liang et al., *Journal of Separation Science*, 34, 1212–1219, 2011. With permission.)

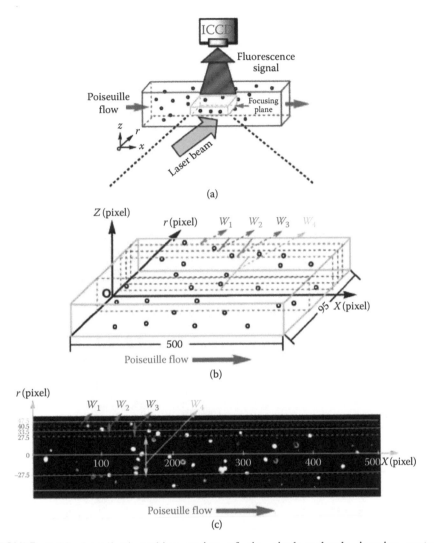

FIGURE 1.26 Localized probing region of the single-molecule imaging system and its Lagrangian coordinates. (From Liang, Cheng, and Ma, *Analytical Chemistry*, 81, 2059–2066, 2009. With permission.) (a) The focusing plane (a thin cuboid). Its volume ($V = 2.607 \times 10^4$ μm³) in the vertical coordinates (Z) was obtained from the linear slope of the observed labeled DNA molecule versus sample concentrations. Its inboard and outboard sides along the radial direction (r) were the confined surface of the PVP-coated square fused-silica capillary (50×50 μm), and the upper and lower sides were the free solution. (b) The enlarged focusing cuboid as the ICCD visual field, showing the four regions of molecular layers: single-molecule layer (W_1), double-molecule layer (W_2), triple-molecule layer (W_3), and middle-molecule layer (W_4) in the radial direction (r). (c) Lagrangian coordinates of single-molecule images, showing the local regions of one frame from the CCD image of fluorescent DNA molecules (0.5 pmol L⁻¹). The Lagrangian coordinates and the local regions were fixed even if each identified individual molecule moves along with the hydrodynamic flow (x coordinates) for a countable time-set (**K**) of the continuous images.

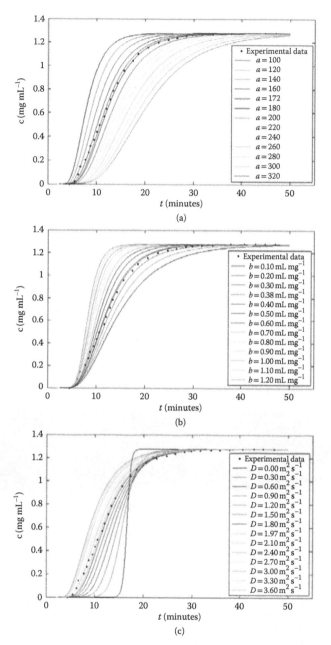

FIGURE 1.34 The effects of isotherms parameters a and b and the apparent diffusion coefficient D on the positions and shapes of FA breakthrough curves. (a) Effects of a on positions and shapes of breakthrough curves with $b = 0.38\,\mathrm{mL\,mg^{-1}}$ and $D = 1.97 \times 10^{-5}\,\mathrm{m^2\,s^{-1}}$. (b) Effects of b on positions and shapes of breakthrough curves with $a = 172.57$ and $D = 1.97 \times 10^{-5}\,\mathrm{m^2\,s^{-1}}$. (c) Effects of D on shapes of breakthrough curves with $a = 172.57$ and $b = 0.38\,\mathrm{mL\,mg^{-1}}$.

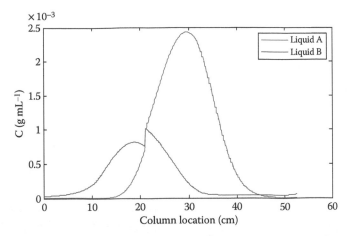

FIGURE 1.37 The solute spatial distribution at the 30th switch time in columns with the 5-column Varicol chromatography system.

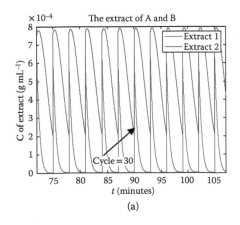

(a)

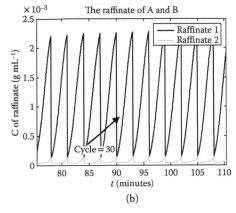

(b)

FIGURE 1.38 Outflow curves of extract (a) and raffinate (b) with the 5-column Varicol chromatography system.

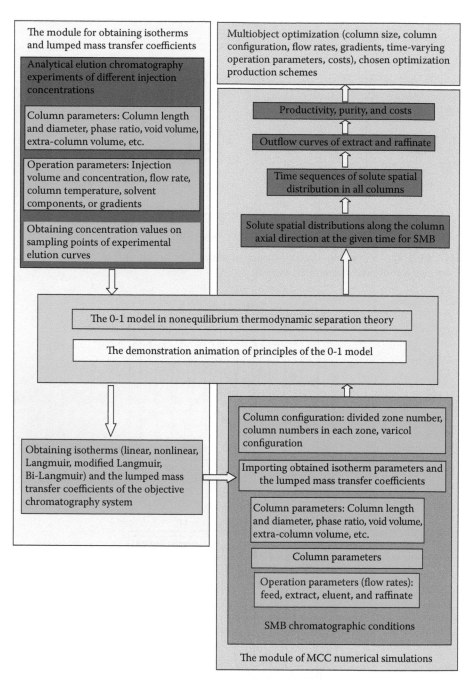

FIGURE 1.39 The sketch configuration of user functions of the MCC simulation software based on the 0-1 model. Note that the different colors with different explanations indicate their subordinate relationships.

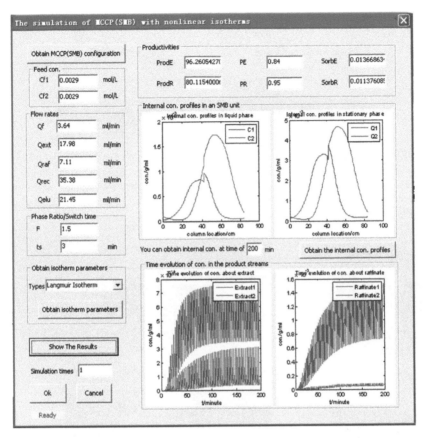

FIGURE 1.44 The interface of simulation results of SMB with Langmuir isotherm of two components.

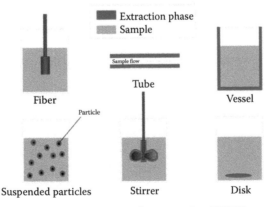

FIGURE 3.1 Configurations of solid-phase microextraction (SPME).

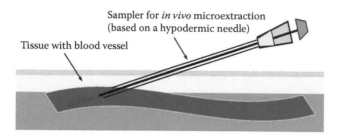

FIGURE 3.4 Schematic representation of the SPME device for *in vivo* monitoring of drug concentrations in the veins of large laboratory animals (Beagles): total length of the device is 8 cm. (Reprinted from *Journal of Biochemical and Biophysical Methods*, 70, Musteata, F. M., and J. Pawliszyn, *In vivo* sampling with solid-phase microextraction, 181–193, Copyright (2006), with permission from Elsevier.)

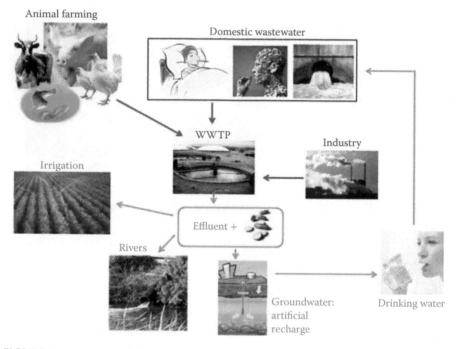

FIGURE 4.1 Sources of pharmaceutical contaminants and their pathways into the aqueous environment.

FIGURE 4.8 Structure of AMX-*S*-oxide.

102. T. Fornstedt, G.M. Zhong, G. Guiochon, Peak tailing and slow mass transfer kinetics in nonlinear chromatography, *Journal of Chromatography A* 742 (1996) 55–68.

103. G. Guiochon, S. Golshanshirazi, A. Jaulmes, Computer-simulation of the propagation of a large-concentration band in liquid-chromatography, *Analytical Chemistry* 60 (1988) 1856–1866.

104. A. Seidel-Morgenstern, L.C. Kessler, M. Kaspereit, New developments in simulated moving bed chromatography, *Chemical Engineering & Technology* 31 (2008) 826–837.

105. M. Mazzotti, Local equilibrium theory for the binary chromatography of species subject to a generalized Langmuir isotherm, *Industrial & Engineering Chemistry Research* 45 (2006) 5332–5350.

106. U. Tallarek, E. Rapp, T. Scheenen, E. Bayer, H. Van As, Electroosmotic and pressure-driven flow in open and packed capillaries: Velocity distributions and fluid dispersion, *Analytical Chemistry* 72 (2000) 2292–2301.

107. B. Huang, H.K. Wu, D. Bhaya, A. Grossman, S. Granier, B.K. Kobilka, R.N. Zare, Counting low-copy number proteins in a single cell, *Science* 315 (2007) 81–84.

108. L. OuYang, Q. Liu, H. Liang, Y. Shi, J. Shi, X. Yin, Simultaneous determination of two-component isotherm parameters and lumped mass transfer coefficients in RPLC with the 0-1 model-inverse method, *Journal of Separation Science* 36 (2013) 645–657.

109. K. Miyabe, G. Guiochon, A kinetic study of mass transfer in reversed-phase liquid chromatography on a C18-silica gel, *Analytical Chemistry* 72 (2000) 5162–5171.

110. K. Miyabe, G. Guiochon, Determination of the lumped mass transfer rate coefficient by frontal analysis, *Journal of Chromatography A* 890 (2000) 211–223.

111. W.Q. Hao, J.D. Wang, X.M. Zhang, Investigation of the concentration dependence of mass transfer coefficients in reversed-phase liquid chromatography, *Journal of Separation Science* 29 (2006) 2745–2758.

112. Q. Liu, L. OuYang, H. Liang, N. Li, X. Geng, Thermodynamic state recursion method for description of non-ideal non-linear chromatographic process of frontal analysis, *Journal of Separation Science* 35 (2012) 1411–1423.

113. X.G. Yin, Master Thesis: Theory Research and Experimental Design of Multicolum Continuous Chromatography Based on 0-1 Model, Xi'an Jiaotong University, Xi'an, 2011.

114. H. Liang, Chinese Invention: A 36 Gateway-Valve Used in High Performance Liquid Preparative Chromatography-Simulated Moving Bed (Patent No. ZL 2007 1 0017257.8), Chinese, January 18, 2007.

115. J.W. Shi, Master Thesis: The Simulation Software for Production Optimization of Simualted Moving Bed Preparative Chromatography System Based on 0-1 Model, Xi'an Jiaotong University, Xi'an, 2011.

2 Biomimetic Chromatography
A Useful Tool in the Drug Discovery Process

Marios Chrysanthakopoulos, Fotios Tsopelas, and Anna Tsantili-Kakoulidou

CONTENTS

2.1　INTRODUCTION

In the past decades, drug discovery faces the challenges of improving compound quality and reducing the attrition rate of drug candidates during clinical phases [1]. It is well known that there are two components that determine drug efficiency, influencing each other: (1) the pharmacodynamic component, which concerns the interaction of drugs with the biological system to elicit its pharmacological and/or toxic response, and (2) the pharmacokinetic component, which concerns the action of the biological system on the drug by absorbing, distributing, metabolizing, and excreting it, events that are termed as ADME properties [2]. Poor absorption and related poor bioavailability has been in the past one of the main reasons for the failure of drug candidates in the drug development process. Nowadays it is recognized that ADME properties should be addressed in parallel with the efforts to enhance affinity and selectivity to the desired targets [3–5]. This strategy necessitates a relevant rapid screening of large compound libraries at an early discovery phase. To this point, it should be noted that the traditional methods to determine pharmacokinetic events are laborious and time consuming. Considerable research efforts are therefore oriented toward the development of alternative techniques with higher throughput, which would permit an estimation and/or ranking of the compounds according to their pharmacokinetic behavior [6].

As a first approximation, physicochemical properties like lipophilicity, size, and hydrogen bond capability can be considered to estimate membrane permeability [7–9]. In fact, upper limit values for these molecular properties have been formulated in the well-known rule of 5, related with oral bioavailability [10]. In particular, lipophilicity, expressed by the logarithm of octanol–water partition coefficient ($\log P$), has a prominent role in various physicochemical models that describe oral absorption [8], permeability [11], central nervous system (CNS) penetration [12], solubility [13], and plasma protein binding [14]. To cope with the high demand for $\log P$ estimation to be used in the various models, a large arsenal for in silico calculation methods implemented in relevant software packages has been developed [15–19].

As a further step, in silico estimates need to be combined with rapid *in vitro* measurements leading to the concept of in combo screening in drug discovery [20]. In this aspect, the development of medium to high-throughput *in vitro* determination of ADME properties is a hot topic in the field of drug discovery.

In regard to membrane permeability, different cell-culture models or cell-free protocols have been developed for the simulation of membrane crossing. Among the cell-based techniques, the Caco-2 cell monolayer and the Madin–Darby Canine Kidney cell lines are the most widely used models for *in vitro* absorption screening [21–23]. Such models have several disadvantages, such as the long culture periods and extensive cost, while results are influenced by interlaboratory differences [24]. As a cell-free technique, parallel artificial membrane permeability assay has been introduced, which, however, necessitates suitable instrumentation [25].

Next to absorption, the distribution profile, the "D" in ADME, constitutes an important issue for drug efficiency. Since the location of the target requires different drug distribution *in vivo*, consideration of the distribution profile and the underlying physicochemical properties before compound selection would greatly contribute to the decrease in attrition rate [26]. Among other factors, distribution is considerably affected by binding to plasma and tissue proteins [27]. Extended plasma protein binding may be associated with drug safety issues and several adverse effects, like low clearance, low brain penetration, drug–drug interaction, loss of efficacy, and so on [28–30]. Traditional experimental protocols to measure plasma protein binding, like ultrafiltration, ultracentrifugation, and equilibrium dialysis, are time-consuming with low throughput and reliability issues in the case of highly bound drugs, while in silico predictions are still not straightforward [31,32]. Lipophilicity is a crucial factor also in this case, since hydrophobic interactions play an important role in protein binding, although other types of interactions, like ionic, dipole, or hydrogen bonding, are also involved [33]. Considering that the binding of some drugs to proteins can be stereoselective, if chiral centers are present in the molecule, such interactions may further influence the fate of enantiomers and diastereoisomers within the body [34–37].

The potential of chromatographic techniques to offer an easy, reliable, and compound sparing way to estimate the ADME properties has been demonstrated several decades ago in regard to the determination of lipophilicity indices. Reversed-phase HPLC has been widely applied as an alternative technique to the tedious and time-consuming direct partitioning experiments, and retention factors can substitute octanol–water partition coefficients or can be used for their calculation by means of relevant calibration equations [38–43]. More details on reversed-phase chromatographic conditions for lipophilicity assessment can be found in the work of Giaginis and Tsantili-Kakoulidou [44,45].

Progress in HPLC column technology succeeded in immobilizing phospholipids as well as different types of proteins on a silica gel skeleton to produce immobilized artificial membrane (IAM) and protein-based stationary phases and gave rise to biochromatography. Biochromatography or biomimetic chromatography confers new perspectives for the simulation of membrane partitioning or protein interactions. It constitutes a liquid chromatographic technique that incorporates a "biologically related agent" in the stationary phase [46]. More specifically, the use of protein-based stationary phases is associated with the term "high-performance affinity chromatography" (HPAC) [47]. In the next paragraphs the latest achievements in IAM chromatography and HPAC are reviewed and their performance to provide useful data for the evaluation of crucial biological drug properties is discussed. For better understanding of the use of protein-based stationary phases, a brief overview of the most important serum proteins, human serum albumin (HSA) and α_1-acid glycoprotein (AGP), is given.

2.2 IMMOBILIZED ARTIFICIAL MEMBRANE CHROMATOGRAPHY

Octanol–water solvent system, the reference system for measuring petition coefficients to be used in correlations with membrane permeability and drug action, although successfully applied in relevant quantitative structure-property relationships (QSPR) and quantitative structure-analysis relationships (QSAR) studies [48–50], has received a lot of criticism since it constitutes an isotropic medium with only a superficial similarity to biomembranes [51]. Efforts to use liposomes as more representative models for membranes did not succeed in a wide-scale application due to the associated difficulties in their preparation and their stability, while liposome/water partitioning studies are more susceptible to experimental conditions [52–55].

The development of IAM stationary phases has unfolded new perspectives in the application of HPLC as a useful tool to combine simulation of membrane permeation with rapid measurements [56–59].

2.2.1 IAM STATIONARY PHASES AND THEIR PROPERTIES

IAM chromatography columns, first developed and described in the end of 1980s, utilize a solid-phase membrane model in which phospholipids (e.g., phosphatidylcholine [PC]) are covalently bound to solid support at membrane densities. The molecular basis for the use of IAM stationary phases for the investigation of chemical membrane interactions is the similarity of IAMs to liposome membranes [56–59].

IAM stationary phases are monolayers of phospholipids covalently bonded to a propylamino-silica skeleton with 300-Å pores as a mechanically stable support. Most commercially available IAM columns contain the prevalent membrane phospholipid, PC. They are prepared by linking diacylphosphatidylcholine molecules covalently to silica-propylamine groups through their ω-carboxylic group on the C2 fatty acid chain. However, some residual free propylamino groups still remain on the silica surface, resulting in a basic IAM.PC subsurface. It is estimated that about 2 mol residual amine groups per mol of immobilized PC exist at a depth of 15 Å below the immobilized PC groups [60]. These residual amine groups can lead to undesired function on the silica backbone. In particular, they decrease chemical stability of the IAM bonded phase (due to lipid leaching) [61], they increase the retention of acidic compounds, and they decrease the retention in the case of basic compounds [62].

To overcome these problems, the remaining propylamine residues are treated using small molecules like glycidol or methyl glycolate (MG). In particular, the reaction of MG with residual amines converts them into chemically neutral amides, but one "new" hydroxyl group is inserted in the subsurface. These hydroxyl groups are located about 10–12 Å below the immobilized PC headgroups. The IAM.PC surface end-capped with MG is commercially available known as IAM.PC.MG stationary phase [63]. It is plausible that the presence of the hydroxyl group influences the interaction of chemical species with IAM surface. Since natural biological membranes do not contain free hydroxyl groups near their center, newer IAM stationary phases free of −OH groups were developed. In this case, this stationary phase is end-capped with decanoic and propionic anhydrite to avoid the presence of hydroxyl groups in the lipophilic core of the immobilized membrane. End-capping procedure is carried out in two steps, firstly using decanoic anhydrite (converting about 85% of accessible residual amine to amide groups) and secondly using propionic anhydrite [64]. Some authors stated the presence of residual amine groups (about 30%) in small crevices of the silica surface even after the second step of deactivation [58]. On the other hand, the attachment of a single-chain PC lacking the glycerol backbone (⁶IAM.PC, 11-carboxylundecyl-phosphocholine) to the silica-propylamine matrix [56] led to the increase of the surface density of phospholipids as well as the stability under acidic conditions. The described stationary phase, known as IAM.PC.DD column, is more hydrophilic than other IAM.PC surfaces due to the presence of monoacylated-PC. Therefore, it was suggested especially for chemical species with short elution times [63]. However, some authors experienced the premature IAM.PC.DD column failure, expressed with significant peak broadening and increased retention times, compared to initial ones, perhaps as a result of removal of immobilized phospholipids by the aqueous mobile phase [65]. To increase both the stability and the hydrophobicity of the IAM stationary phase, a third IAM.PC surface was developed as a combination of IAM.PC.MG and IAM.PC.DD. In this way, a double-chain stationary phase was prepared by immobilization of diacylated-PC on silica-propylamine, while end-capping was performed with decanoic and propionic anhydrite. This new generation IAM column is marketed as IAM.PC.DD2 and it is nowadays the most widely used stationary phase for membrane simulation [63]. IAM.PC.DD2 stationary phases have been reported to simulate better natural phospholipids and the resulting chromatographic indices are more suitable to correlate permeability data [63,66,67]. It should be mentioned that the free silanol sites, which exist on silica-based chromatographic surfaces, elicit silanophilic interactions with basic and polar compounds. To reduce this effect on IAM chromatographic columns, acylation of the propylamine-silica matrix is performed as a standard procedure to introduce an amido moiety, which provides electrostatic shielding and hinders the accessibility to the eventually remaining silanol groups [67]. Up to date, IAM stationary phases are commercially available by Regis Technologies in columns of various lengths (30–150 mm for IAM. PC.MG and 10–150 mm for IAM.PC.DD2). IAM.PC.DD stationary phase is no longer commercially available. [68]. In Figure 2.1, the three IAM stationary phases are presented. Encircled are the charged centers on the PC polar head, an essential feature for the discrimination of IAM columns in regard to the traditional reversed phase columns, used for lipophilicity assessment. Despite end-capping, stability issues of IAM columns should be taken into consideration since there possess several sites that are susceptible to hydrolytic cleavage [61]. Such sensitive sites are the siloxane bonds (Si-O-Si) linking the silylating moiety to the silica surface, the amide bonds formed between the propylamino groups and the immobilized phospholipids, and in the case

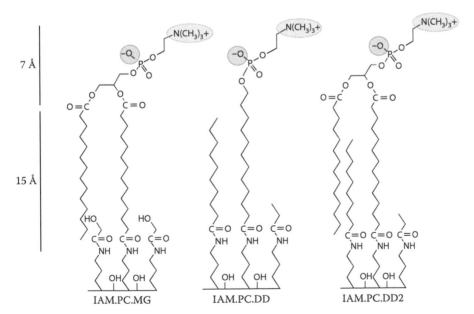

FIGURE 2.1 IAM stationary phases, containing phosphatidylcholine. Encircled are the negatively and positively charged centers.

of double-chain IAM stationary phases the ester bonds between fatty acids and the glycerol backbone of PC.

Although IAM surfaces share many common features with liposomes, the two systems show also important differences. IAM are solid-phase system, where a phospholipid monolayer is covalently linked to the silica skeleton. In contrast, liposomes constitute a fluid phase and each vesicle is formed by a phospholipid bilayer, which separates the external from the inner aqueous phase. Consequently, in IAM surface, the hydrophobic core is half as large as in liposomes and immobilized phospholipids are more ordered and deprived from molecular dynamics. In addition, the density of the polar headgroups in IAM surface is smaller than in liposomes. However, in both systems, the choline groups exhibit larger motional fluctuations than phosphate groups and there is analogous distribution of water near the polar headgroups [69]. Concerning the structural characteristics, the synthetic phospholipids in IAM surface are fully saturated and usually contain acyl chains with 16 carbon atoms (propylamine included), while egg PC, the most widely used phospholipid for liposome preparation, contains longer acyl chains and cis double bonds [57,70].

2.2.2 Chromatographic Conditions and Indices in IAM Chromatography

The application of IAM chromatography to the investigation of drug–membrane interactions involves simply the determination of the retention factor of the solute on the IAM chromatographic column, according to Equation 2.1:

$$k_{IAM} = \frac{t_R - t_0}{t_0} \tag{2.1}$$

where t_R is the retention time of the compound under investigation and t_0 corresponds to void volume.

The determination of void volume can be performed mainly by using an "unretained" chemical although in practice no truly unretained compound exists [71]. According to the manufacturer, for IAM stationary phases, citric acid is recommended as the void volume marker of choice [72]. In a recent publication, a classification of recommended chemicals according to column type and mobile phase conditions was reported. According to this investigation, L-cystine for acidic eluents and potassium bichromate or sodium oxalate in the case of neutral buffers are suggested as best void volume markers for IAM stationary phases [73].

It should be noted that the IAM retention factor, k_{IAM}, is proportional to the equilibrium IAM partition coefficient K_{IAM} according to Equation 2.2:

$$K_{IAM} = K_{IAM} \cdot \frac{V_s}{V_m} \tag{2.2}$$

where V_m is the total volume of solvent within the IAM stationary phase and V_s is the volume of the IAM interphase created by the immobilized phospholipids. The ratio V_s/V_m is constant for a given system.

Generally, retention factors k are used in their logarithmic form, $\log k_{IAM}$, to be linearly related to free energy.

A substantial advantage of IAM stationary phases is the ability to perform measurements using aqueous mobile phases without the addition of organic modifier. Therefore, direct measured $\log k_w$ values in 100% aqueous phase can be obtained without the need of extrapolation. This is important with drugs exhibiting weak affinity for IAM surface, such as cephalosporins or in the case of hydrophilic organometallic compounds [74]. The buffer of choice is phosphate buffered saline (PBS) to mimic physiological conditions, although phosphate buffer and ammonium acetate buffer are also used. Different buffer constitution as well as its ionic strength affects the retention and consequently the correlation between retention factors and octanol–water partition coefficients or biological relevant parameters [75]. A limitation in the use of IAM columns refers to the operation pH of the mobile phase, which ranges from 2.5 to 7.4. Many authors prefer to work using eluents with pH = 7.0, which is close to physiological pH and safe for the column [63,67]. In the case of compounds with strong affinity for the IAM surface, acetonitrile up to 30% can be added and $\log k_w$ values can be obtained by linear extrapolation of the isocratic retention factors according to the equation:

$$\log k = -S \cdot \varphi + \log k_w \qquad (2.3)$$

where φ is the acetonitrile fraction in the mobile phase.

Mobile phases containing higher percentage of acetonitrile should be avoided because their microheterogeneity disrupts the structure of water [76]. Nevertheless small differences may be observed between directly measured and extrapolated $\log k_w$ values [77].

The use of methanol as organic modifier should be avoided, since it provokes instability of the column, due to methanolysis of the phospholipids [63,78]. In any case, the aging of the column should be checked from time to time using standard compounds [72,79].

While standardization of reversed-phase chromatographic conditions for lipophilicity assessment has been suggested [41,42], no such systematic efforts exist for the determination of $\log k_{IAM}$, although some proposed guidelines for technical optimization of retention measurements have been reported [72]. A methodical study for the comparison of 1686 $\log k_{IAM}$ indices of 555 compounds, measured under different conditions, revealed considerable variability as an effect of the stationary phase, the composition and pH of the mobile phase, while the influence of flow rate and temperature over the range studied (22°C–45°C) was rather negligible [75].

In analogy with the chromatographic hydrophobicity index (CHI) [80], Valko et al. have proposed the CHI_{IAM} index. CHI_{IAM} is determined by gradient elution using acetonitrile as organic modifier. Gradient retention times are calibrated with data obtained from isocratic measurements corresponding to the percentage of acetonitrile, which produces equal partitioning of the solute between the stationary and mobile phases, for example, $\log k = 0$. In this sense, CHI can be derived also as the quotient $\log k_w / S$ of Equation 2.3 [81].

2.2.3 Molecular Factors Influencing IAM Retention: Comparison with Octanol–Water and Liposomes Partitioning

According to Ong and Pidgeon [82], solute partitioning seems to be the principal retention mechanism in IAM retention, with similar thermodynamics on both the single-chain and double-chain IAM.PC surfaces. For a set of phenol derivatives, the partitioning into IAM.PC surfaces was found to be both enthalpy and entropy driven, while for beta-blockers, it was entropy driven. Other authors report negative ΔS values for two acidic drugs (warfarin and salicylic acid) and positive ΔS values for three basic drugs (lidocaine, propranolol, and diazepam), while negative ΔH were calculated for all five drugs [83]. Nevertheless, for neutral compounds, the intermolecular forces resemble those underlying partitioning in octanol–water and retention in reversed-phase HPLC. Linear solvation energy relationships (LSER) analysis, based on Abraham's descriptors, has shown that besides the hydrophobic/solvophobic interactions, polar interactions, mainly expressed as H-bond acceptor basicity, are important factors in IAM retention. However, the hydrophobic term seems to have a smaller positive contribution in IAM retention compared to its contribution in octanol–water partitioning [84]. In fact, it is considered that the IAM surface provides a hydrophobic environment that resembles a RP-C3 HPLC column, whereas for lipophilicity assessment, RP-C18 stationary phases are used [61].

The primary feature, however, that distinguishes anisotropic partitioning systems, like IAM chromatography and liposomes, from octanol–water partitioning systems is the presence of the ordered phospholipids, able to elicit ionic interactions. Thus, for ionized species, polar interactions on IAM surface are electrostatic in nature [82]. In particular, protonated basic compounds, or compounds with quaternary nitrogen, are more strongly retained as a result of their interaction with the phosphate anions of the stationary phase. It has been reported that IAM retention of protonated β-blockers is stronger, compared to isolipophilic neutral compounds [85]. In a study concerning structurally diverse basic and neutral compounds, the degree of protonation, expressed as fraction F^+, had to be considered an additional parameter to obtain a good correlation between $\log k_{wIAM}$ and $\log D$ values at pH 7.4 [86]. A positive sign was assigned to the term F^+ denoting the positive contribution of ionic bonds with the phosphate anions of the phospholipids. Otherwise, a better correlation was obtained with $\log P$ values of the neutral form, implying that the decrease in the retention due to ionization was partly compensated by the electrostatic interactions. In the same study, IAM retention was compared to reversed phase chromatographic retention. Very characteristically, the strong base metformin, fully protonated at pH 7.4 and with a negative $\log D_{7.4}$ value, eluted with the dead time in reversed phase HPLC, while it was retained in IAM chromatography due to the ionic bond formation [86]. On the other hand, electrostatic interactions between solute anions and the positively charged choline nitrogen were found to be less pronounced. This finding was attributed to the fact that the choline headgroups, being located at the exterior of the IAM surface, are exposed to solvation and are less inclined to undergo such interactions [87]. Stronger electrostatic interactions of positively charged solutes with the phospholipids on IAM stationary phase, compared to those of negatively charged solutes, has been reported also by Liu et al.

in a study of monofunctional compounds and complex drugs including steroids, nonsteroidal anti-inflammatory drugs, and β-blockers [88]. It should be noted that in $\log k_{wIAM}$ correlations with $\log P$ or $\log D$, the regression coefficient of the lipophilicity term is about 0.6 for structurally diverse compounds [86–89], while a shift toward 1 is observed for congeneric data sets [77,85], indicating a similar balance of forces between IAM retention and octanol–water partitioning in the case of compounds with similar structures. The magnitude of the regression coefficient may be related also to the lipophilicity range of the compounds. For a set of β-blockers, less lipophilic derivatives showed lower slopes in the $\log k_{wIAM}$–$\log P$ regression equation [85]. A curvature in the plot of IAM retention against lipophilicity has also been reported for compounds with $\log P$ close to or smaller than zero [84]. An analogous behavior for hydrophilic compounds has been observed for partitioning into liposomes and may be related to the H-bonding capability of such compounds [90].

Electrostatic forces, however, have been reported to be weaker in IAM chromatography than in liposome partitioning, and the reason may be related to the smaller density of the polar headgroups in IAM surfaces [91]. The contribution of hydrogen bond seems also to be less important for the affinity to IAM stationary phase than to liposomes [92]. Moreover, in the case of basic drugs, silanophilic interactions have been reported to affect the $\log k_{wIAM}$/pH profile as compared to the corresponding pH/partition diagram in liposomes [83,92].

Despite these differences, IAM chromatographic indices have successfully been correlated with liposomes partitioning data; however, such studies include a rather limited number of compounds [83–92].

Less investigated is the effect of conformation in IAM retention. In a comparative study of substituted coumarins, such effects, induced by the extended planar conformation of some derivatives, were found to have a similar impact in IAM retention as in reversed-phase HPLC and TLC, while they did not affect the octanol–water partitioning [77]. They should therefore be attributed rather to the particular features of the chromatographic procedure, in which, contrary to solvent–water partitioning, the solutes approach the stationary phase gradually with the movement of the mobile phase.

Considering the significant contribution of electrostatic interactions in IAM retention mechanism, the question may arise, whether IAM chromatographic indices are suitable to predict passive diffusion of drugs, which according to the well-known pH-partition hypothesis implies reduced absorption of charged species, or they should rather be used to simulate drug–membrane interactions. In fact, IAM chromatography as other anisotropic biomembrane-like systems, such as liposomes, may represent a border case between partitioning and binding. Since there is a poor degree of specificity, partitioning is considered to be the determinant component [93]. The term "phospholipophilicity" is introduced to differentiate the partitioning in such anisotropic systems from octanol–water lipophilicity [93]. Considering the binding component (e.g., binding to the phospholipids), strong IAM retention in the case of basic drugs with poor membrane permeability may be associated with a risk for phospholipidosis [94]. Phospholipidosis has become a significant concern for drug development and safety assessment, although its association with drug toxicity is unclear [95].

2.2.4   CORRELATION OF IAM RETENTION FACTORS WITH PERMEABILITY AND BIOLOGICAL DATA

The potential of IAM chromatographic indices to predict passive transport through various biological barriers as well as to estimate pharmacokinetic properties and certain pharmacological activities has been reviewed by Barbato [96]. More recently, the most significant relationships of membrane-like systems, including IAM chromatography, with passive drug permeation, have been summarized by Liu et al. [97]. Thus, IAM chromatographic indices have been correlated with Caco-2 cells permeability [56,57,98]. For structurally diverse drugs with molecular weight between 200 and 800, the introduction of molecular weight as an additional parameter and exclusion of substrates for p-glycoprotein was necessary to obtain a good relationship [98]. Combination of $\log k_{wIAM}$ with molecular weight described successfully intestinal absorption in rats [99]. Satisfactory correlations between IAM retention indices and skin penetration have also been observed for some steroids and phenol derivatives. [100]. For a small set of structurally diverse drugs with neutral, acidic, and basic functionalities, however, skin penetration did not correlate neither with IAM retention factors nor with $\log P$. On the other hand a satisfactory correlation was obtained using $\Delta \log k_{wIAM}$, which reflects the component of interaction with phospholipids not accounted for by hydrophobicity-based retention [101]. $\Delta \log k_{wIAM}$ is derived in an analogous way as $\Delta \log P$ [102]. The oral absorption of cephalosporin prodrugs in mice could be better predicted by $\log k_{wIAM}$ than by reversed-phase chromatographic indices [56]. Kotecha et al. [103] proposed a sigmoidal model for the oral absorption of a series of 28 structurally diverse drugs based on IAM retention factors in conjunction with polar surface area, expressed by Equation 2.4. To account for the change of pH along the gastrointestinal tract, they measured IAM retention factors in the pH range between 4.5 and 7.4 and introduced the maximum value in the model:

$$\% \, \text{HOA} = \frac{100}{1 + 10^{-(A_1 + A_2 \cdot \log k_{wIAM} + A_3 \cdot \text{PSA})}} \qquad (2.4)$$

where $A_1 = 0.412 \pm 0.171$, $A_2 = 0.407 \pm 0.063$, and $A_3 = -0.491 \pm 0.139$, derived by nonlinear fitting, and $r^2 = 0.900$.

In another study, it was demonstrated that compounds with high oral absorption can be identified using IAM retention factors in combination with the half-life of hepatic microsomal biotransformation [104].

IAM chromatography has also been used in the study of blood–brain barrier penetration. IAM retention factors were found to be superior to octanol–water partition coefficients or reversed phase chromatographic indices for modeling brain uptake when used alone [105] or in combination with other parameters like ionization and size [106]. A rapid screening for the classification of CNS+ and CNS− drugs has been suggested, using IAM retention corrected for the molecular weight in the power of 4. If the expression $(k_{IAM}/Mw^4) \times 10$ is higher or lower than the value 1.01 (Mw) for pH 5.0 (e.g., for compounds administered orally) or 0.85 for pH 7.4, high or low CNS penetration is expected, respectively. Classification is considered uncertain for intermediate values [107]. For a series of nonsteroidal anti-inflammatory drugs, a parabolic relationship was obtained between $\log k_{wIAM}$ and the diffusion across the rat blood–brain barrier, suggesting an optimum lipophilicity of drugs for the transport to brain [108].

It should be noted, however, that most of these studies concern rather limited number of compounds and general models for permeability across the various biological barriers based on IAM retention are still missing. The potential of IAM chromatography to be used in general models for the simulation of particular bio-partitioning systems in comparison with octanol–water partition coefficients and reversed-phase chromatographic indices has been further investigated by means of their characterization with Abraham's solvation parameters [109]. According to this study, IAM stationary phases may not always be the best choice to model biological drug distribution system. Thus, octanol–water partitioning, tadpole narcosis, human skin partition, and blood–brain permeability can be successfully modeled by IAM stationary phases, but human skin permeation is better modeled by retention on reversed-phase C18 columns.

2.3 HIGH-PERFORMANCE AFFINITY CHROMATOGRAPHY

The term "high performance affinity chromatography" is associated with stationary phases containing proteins. Since this review focuses on the use of HPAC for the estimation of protein binding, stationary phases with immobilized serum proteins, for example, HSA and AGP, as well as the characteristics of these proteins in solution will be discussed.

HSA is the most abundant serum protein, constituting more than half of the blood serum protein content; thus, the term "plasma protein binding" is often associated with this protein [110,111]. HSA binds mainly to neutral and acidic drugs and it is regarded as a nonspecific binder and carrier, where hydrophobic interactions are the dominant recognition forces. It should be noted that HSA binding has served in early lipophilicity studies for the establishment of octanol–water as the reference system to measure partition coefficients [48]. Since, however, other factors are also important, poor global correlation with lipophilicity data is obtained if large sets of structurally diverse compounds are considered [32]. Other factors involved in drug binding to albumin are hydrogen bonding and electrostatic interactions. It is well known that acidic drugs have higher affinity for HSA, while there are specific domains in the protein that are able to recognize enantiomers [112,113]. As supported also by crystallographic investigations [114,115], there are mainly two different stereoselective binding sites, while the protein possesses more binding sites for endogenous ligands. The two major drug-binding sites are reported as site I, which binds warfarin and azapropazone, and site II, which binds profens and benzodiazepines. These binding sites are elongated hydrophobic pockets with cationic amino acid residues near their entrances. Dicarboxylic acids and bulky heterocyclic compounds with a negative charge bind mostly to site I, while small aromatic carboxylic acids, such as indole-3-acetic acid, bind preferably to site II [116,117]. HSA, like hemoglobin, may undergo a slow nonenzymatic glycation, mainly by formation of a Schiff base between ε-amino groups of lysine residues mainly and glucose molecules in blood. Glycation has the potential to alter the biological structure and function of HSA. Elevated levels of glycated albumin are observed in patients with diabetes mellitus [118].

The second critical serum protein is AGP. The concentration of AGP in blood is much lower than HSA and its binding capacity is also lower. However, due to its lower isoelectric point (IP), AGP has significant affinity to basic drugs, while neutral and some acidic drugs also bind to AGP. AGP is able to discriminate enantioselectivity if the compounds possess two hydrogen-bonding groups and a rigid or bulky structure close to a chiral center [119,120]. Thus, this protein may considerably affect the pharmacokinetic and pharmacodynamic behavior of both basic and neutral drugs. Not all the properties, however, of the binding sites in AGP are known [121].

Human AGP is a 41–43-kDa glycoprotein with high carbohydrate content (approximately 45% of its mass), which presents high degree of heterogeneity [122]. Different isoforms of AGP can be distinguished in serum depending on the type of glycosylation and multiple amino acid substitutions. Moreover, the protein part shows genetic polymorphism. In most individuals, AGP is a mixture of two or three genetic variants (F1 and/or S and A) encoded by two different genes, but there are large interindividual differences. Pooled commercial AGP contains the three variants in similar proportions. Drug-binding studies have demonstrated that the two main genetic variants (F1/S and A) have different binding properties. Some ligands show high preference for one of the variants, while for other certain ligands, there is no discrimination at all [123–126].

The role of AGP binding in drug development and drug monitoring in clinical therapy is further complicated by other crucial factors such as that its plasma levels may considerably change during disease states, its ligand affinity is influenced by glycosylation status, and the AGP drug-binding profile significantly differs in animal models [127–129]. Moreover, it has been reported that AGP may be associated with immunomodulatory functions [127,128] and may be involved in direct action against pathogens [130].

The chiral recognition ability of proteins is related to the formation of a three-dimensional complex stabilized through hydrophobic, dipole–dipole interactions, hydrogen bonds, as well as electrostatic interactions dependent on their isoelectric points. It was reported already in the early 1950s that the binding of the enantiomers of an anionic azo dye to bovine serum albumin (BSA) or HSA was different [131]. Also, L-tryptophan was found to bind to serum albumin more strongly than the D-isomer. In this aspect, protein-based stationary phases were initially developed and are still used for chiral resolutions. Separation of tryptophan enantiomers on BSA-Sepharose was the first report, dated in 1973, on the application of HPAC for chiral resolution purposes [132]. Since then, protein-based stationary phases have successfully been used in drug analysis for enantiomeric separation. In particular, columns containing immobilized AGP show a broad spectrum for this kind of separations, due to the carbohydrate residues of the protein, which may play a role in determining the stereoselectivity of binding. However, further discussion on the use of HPAC for enantiomeric separation in drug analysis is beyond the scope of this chapter.

Already in the 1990s, several authors [133–135] have demonstrated the application of serum albumin stationary phases to study protein binding, implying that the immobilization of albumin on the silica support does not significantly alter its binding specificity. Hage and Austin have reviewed the theory and practice of HPAC in the study of protein binding [136] and have shown its advantages relatively to the conventional techniques like equilibrium dialysis and ultrafiltration [137–140]. These techniques are tedious and time consuming, they require suitable analytical method for quantitation of the amount of free analyte, and the measurements are very difficult when the compounds show very strong protein binding.

2.3.1 PREPARATION AND PROPERTIES OF PROTEIN-BASED STATIONARY PHASES

Protein-based stationary phases developed so far include albumins such as HSA, BSA, dog serum albumin, and mouse serum albumin enzymes such as trypsin, chymotrypsin, lysozyme, and pepsin and glycoproteins such as AGP from human or bovine serum, cellobiohydrolase I (CBH I), ovoglycoprotein, avidin, ovotransferrin, and Savoprotein (riboSavin-binding protein). Commercially available are stationary phases based on BSA, HSA, pepsin, AGP, CBH I, ovomucoid, and avidin [141,142]. However, most work has focused on HSA, BSA, and AGP columns. Modified proteins can also be used for the preparation of a stationary phase for specific purposes, for example, immobilized glycated HSA is suitable for the investigation of drug protein binding in the case of diabetes [143].

In protein-based stationary phases, high-quality aminopropyl silica gel is used as the support for the immobilization of the protein. Aminopropyl silica gel is activated by N,N-disuccinimidylcarbonate and the protein (HSA, BSA) is bound by an amino group. Or, the protein can be bound via a carboxyl group after its reaction with water-soluble 1-Ethyl-3(3-dimethylaminopropyl)carbodiimie (EDC) and N-hydroxysulfosuccinimide [144–146]. Glyceryl-propylsilica gels (diol-bonded silica) activated with 1,1'-carbonyldiimidazole can also be used to immobilize the protein though a two-step [140] or a three-step [142,147] Schiff reaction-based method. Monolith columns are recently prepared using a bare silica monolith, which is first converted into a diol-bonded form and then again the Schiff-based method is applied to immobilize the protein [148].

Immobilization of AGP is achieved by first oxidizing the carbohydrate residues of AGP with periodate, resulting in the formation of aldehyde groups, which upon increase of pH react with free amine groups on other AGP molecules. The imine bonds formed are then reduced with sodium cyanoborohydride to secondary amine groups. With this procedure, a cross-linked AGP is immobilized on the silica support [144].

Protein-based columns further differ on the protein mass, the isoelectric point (IP), and the carbohydrate content, depending on the immobilized protein. The preceding characteristics for different protein-based columns along with the relevant references are presented in Table 2.1.

HSA columns are more appropriate to measure protein binding since albumin is the most abundant plasma protein. AGP-based columns have broad applicability in the separation of enantiomers in drug analysis. Their applicability in binding studies has not been well established. It seems that cross-linking may produce differentiations in the chromatographic characteristics in comparison with the protein in solution [159,160]. In the next paragraphs, discussion will mainly focus on HSA and AGP columns and their potential to simulate the plasma protein binding process.

TABLE 2.1
Physical Properties of Proteins Immobilized in HPLC Chiral Stationary Phases

Protein	Molecular Mass (Da)	% Carbohydrate	IP (pH)	Origin	Reference
Albumins					
Chiral-HSA	65,000	—	4.7	Human serum	135
Chiral-BSA	66,000	—	4.7	Bovine serum	145
Enzymes					
Cellobiohydrolase I (CBH I)	64,000	6	3.9	Fungus	146
Lysozyme	14,300	—	10.5	Egg white	149
Pepsin	34,600		<1	Porcine stomach	150
Amyloglucosidase	97,000	10–35	4.2	Fungus	151
Glycoproteins					
α_1-acid glycoprotein (AGP)	41,000	45	2.7	Human or bovine serum	152
Ovomucoid (OMCHI)	28,000	30	4.1	Egg white	153
Ovoglycoprotein (OGCHI)	30,000	25	4.1	Egg white	154
Avidin (AVI)	66,000	7	10.0	Egg white	155
Riboflavin-binding protein (RfBP)	32,000–36,000	14	4.0	Egg white	156
Other					
Ovotransferrin (conalbumin)	77,000	2.6	6.1	Egg white	157
β-Lactoglobulin	18,000/36,000		5.2	Bovine milk	158

2.3.2 CHROMATOGRAPHIC CONDITIONS AND INDICES IN AFFINITY CHROMATOGRAPHY

Retention times measured in HPAC are converted to retention factors (k) according to a type (1) equation, using a suitable void volume marker (an unretained compound) to determine the dead time, to L-Cystine can be properly applied throughout the whole operational pH range. Sodium citrate is also a satisfactory void volume marker, especially with neutral buffers as eluents. Attention should be paid in the use of potassium bichromate, which can lead to substantial overestimation of void volume on HSA and AGP columns under acidic conditions [73].

The retention factor, k, equals to the ratio of the number of moles in the stationary and mobile phases and is related to the relevant equilibrium constant by Equation 2.5:

$$k_x = K_x \cdot \frac{V_s}{V_m} \tag{2.5}$$

where V_s and V_m are the volume of stationary and mobile phases, respectively. The subscript x denotes the type of the protein.

Protein-based stationary phases can be used with aqueous mobile phase. PBS and ammonium acetate buffer are commonly used. However, for strongly retained compounds, for example, compounds with high protein binding, an organic modifier should be added in the mobile phase to reduce the retention time. Usually, there is a good linearity between retention factor $\log k$ and volume fraction of organic modifier, and $\log k_w$ values corresponding to 100% aqueous mobile phase can be obtained by linear extrapolation according to a type (3) equation. 2-Propanol or acetonitrile are the organic modifiers of choice, although the use of methanol with HSA columns has also been reported [34]. The manufacturers recommend the use up to 30% organic modifier in AGP columns and up to 10% in HSA. However, it has been found that higher concentrations of organic modifier up to 30% do not induce irreversible conformational change in HSA stationary phases and do not affect its performance [161]. Directly measured $\log k_w$ values do not always coincide with extrapolated $\log k_w$ values [162], so care should be given for relevant calibration before using such values together. Alternatively, gradient retention times (gt_R) may be used. The gradient approach speeds up the analysis and aids in the separation of strongly bound molecules [163,164].

Ionic strength and pH may affect the retention and resolution. There are pH limitations on the use of protein-based columns depending on the protein since in strong acidic or alkaline solution denaturation may occur. According to the manufacturers, for HSA columns, a pH range 5–7 is recommended while for AGP a broader pH range 4–7 is tolerated. Depending on the nature of the compounds, change in pH may exert different effects. In AGP column, decrease of pH leads to reduction of the degree of the net negative charge on the protein. Consequently, for basic compounds, a decrease in retention is expected due to the reduction of ionic bond formation. In contrast, the retention of acidic compounds should increase with decrease in pH due to reduction of the repulsion between the compounds and the stationary phase. In HSA columns, retention of acidic compounds increases with decrease of pH.

2.3.3 Use of HPAC in Protein-Binding Studies

HSA chromatography can be used to determine binding association constants $\log K$ as well as relative binding to albumin. For the determination of binding constants by HPAC, zonal elution and frontal analysis are the techniques widely recommended.

2.3.3.1 Zonal Elution

Zonal elution is the most popular method to study solute protein binding by HPAC. Zonal elution is generally performed by injecting a small amount of the analyte into the column as in most analytical applications and using a mobile phase, usually buffer at a physiological pH, which contains a fixed amount of a competing agent [165,166]. The competing agent can be the same as the analyte (self-competition) or different. Shift in the retention of the analyte is monitored by changing the concentration of the competing agent in the mobile phase. This shift is produced by the change in the strength of binding in the binding sites or the relative distribution of these sites induced by the competing agent. In Figure 2.2, typical zonal elution chromatograms in the presence of different concentrations of competing agent are presented.

In the case of self-competition (the same competing agent as the analyte A is added in the mobile phase), the interaction is represented by the equilibrium

$$A + L \underset{}{\overset{K_A}{\rightleftharpoons}} A - L$$

where A is the analyte, L is the immobilized ligand (protein), and K_A is the association constant. Retention factors depend on the concentration of the same competing agent according to Equation 2.6:

$$k_A = \frac{K_A \cdot m_L}{V_M \cdot \left(1 + K_A \cdot [A]\right)} \tag{2.6}$$

where K_A is association constant of the analyte, m_L the total moles of analyte in the column, and V_M the column void volume.

The plot of the reciprocal $1/k_A$ versus the concentration [A] of the same competing agent results in a linear relationship with positive slope:

$$\frac{1}{k_A} = \frac{V_M \cdot [A]}{m_L} + \frac{V_M}{K_A \cdot m_L} \tag{2.7}$$

The ratio of the slope to the intercept corresponds to the association constant K_A.

In the case of competing agent [I] different than the analyte, the plot of $1/k_A$ versus competing agent concentration [I] provides useful information on the type of competition of the two compounds for the immobilized protein. For noncompetition, a

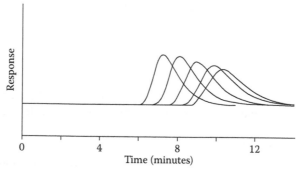

FIGURE 2.2 Typical zonal elution chromatograms of a solute, in the presence of decreasing concentrations of a competing agent in the mobile phase.

random behavior is observed. For systems with direct competition at a single binding site, a linear relationship with a positive slope is obtained analogous to Equation 2.8:

$$k_A = \frac{K_A \cdot m_L}{V_M \cdot (1 + K_I \cdot [I])}$$ (2.8)

where K_I is the association constant of the competing agent I at the site of competition.

In this case, the ratio of the slope to the intercept of Equation 2.9 corresponds to the association constant K_I:

$$\frac{1}{k_A} = \frac{V_M \cdot K_I \cdot [I]}{K_A \cdot m_L} + \frac{V_M}{K_A \cdot m_L}$$ (2.9)

If multisite interactions or allosteric competition are taking place, deviations from linearity are observed. For positive allosteric competition, a negative slope in the $1/k$ versus [I] relationship may be obtained. A related plot of $1/(k$-$X)$ versus [I] can be used, where X is a fitted value that represents the analyte retention due to sites of noncompetition [167].

The advantage of zonal elution is that it requires a small amount of analyte per run and it permits relatively rapid measurements of the binding constants in a good precision and agreement with the values produced by traditional solution phase techniques. In addition, change in activity of the immobilized protein does not affect significantly the results [166].

Zonal elution was initially applied to study the properties of the proteins immobilized on stationary phases. In the case of HSA and BSA columns, there is enough evidence that the proteins retain their binding characteristics, as when they are in solution. Thus, there is an increasing interest in using zonal elution with these columns for quantitative studies of drug–protein interactions [167–169].

Displacement experiments for cross-linked AGP stationary phases, however, have shown differences in the binding properties as compared to the behavior of AGP in solution [159].

2.3.3.2 Frontal Analysis

Frontal analysis is the second commonly used method for the study of drug–protein interactions. Its difference from zonal elution is that the analyte in solution of known concentration is applied continuously into the protein-based stationary phase resulting in a titration of the number of binding sites within the column. As the analyte binds to the immobilized ligand (protein), the column becomes saturated and the amount of the analyte (and the measured signal, e.g, absorbance) increases rapidly. After a breakthrough time, the analyte reaches a plateau with a constant signal depending on the concentration of the analyte in the solution. The breakthrough time corresponds to the time point, which is half way between the baseline and the upper plateau. If the curve is asymmetric, the breakthrough time can be determined by integration below the lower part and above the upper part of the curve to a point

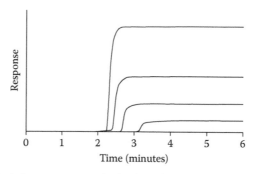

FIGURE 2.3 Typical chromatograms obtained by frontal analysis, at different concentrations of the analyte.

at which the two areas are equal [169,170]. In Figure 2.3, a typical chromatogram obtained by frontal analysis is shown.

Frontal analysis requires higher amount of analyte; however, it provides useful information and permits measurements of both equilibrium constants and number of binding sites within the column. For this purpose, the breakthrough times corresponding to the apparent moles of analyte to saturate the column (m_{Lapp}) at different concentrations of the analyte [A] are measured and data are fitted to various models depending on the number of binding sites. If the analyte has a single site on the immobilized protein, the double reciprocal plot $1/m_{Lapp}$ versus $1/[A]$ should be a linear relationship. In presence of more than one independent binding site, negative deviations from linearity are observed at high analyte concentrations (low values of $1/[A]$). In the first case, the association constant K_A is obtained from the ratio of the slope ($1/K_A m_L$) and the intercept ($1/m_L$) of the linear relationship, while the inverse of the intercept (m_L) corresponds to the total binding capacity of the column. For systems that involve multisite interactions, Scatchard plots [171] or a combination of nonlinear and linear fits can be used [172,173]. The results obtained by this method have very good precision and agree with those obtained by traditional techniques [140,174]. Competition and displacement studies can also be performed by frontal analysis by adding fixed concentrations of a competing agent in the mobile phase and monitor changes in breakthrough times as a function of competing agent concentration. Direct competition between the analyte and the competing agents results in smaller breakthrough times, while positive or negative allosteric effects lead to a shift to higher or lower breakthrough times, respectively [166].

2.3.3.3 Estimation of Relative Binding

Zonal elution and frontal analysis provide reliable measurements for association constants and insight on the type of drug–protein interactions while they proved that HSA immobilized on a chromatographic column retains the binding properties of albumin in solution. Based on this evidence, HSA chromatography can be used for rapid screening of large series of compounds for protein binding by simply correlating percentage of drug binding to albumin to chromatographic retention [161]. In this aspect, retention factors k may be used to calculate the percentage of protein binding according to Equation 2.10:

$$\%PPB = 100 \cdot \frac{k}{k+1} \qquad (2.10)$$

Equation 2.10 is based on the definition of retention factor as the ratio of the moles or fraction of the solute on the stationary phase, for example, the moles b bound to the protein to the moles in the mobile phase, for example, the free moles (f) that remain free ($k = b/f$) and the fact that bound plus free fractions equals to 1.

Equation 2.10 is valid only when compound retention is independent of the injected amount. Moreover, the chromatographic retention factor, k, is equivalent to the protein (albumin) partition coefficient (the binding equilibrium constant K) if the V_s/V_m ratio is unity.

In any case, the expression $k'/(k' + 1)$ can be further used to correlate %PPB.

Yoshida et al. [175] were the first to suggest the application of HPLC for the determination of relative drug protein binding using a physically coated octadecylsilica (ODS) stationary phase. The development of commercially available protein-based stationary phases offered new dynamic perspectives in this field and successful correlations have been obtained in the case for HSA [133,135,167,176,177]. For AGP, there is some skepticism as to whether chromatographic data correlate with biochemical binding data [159]. There are, however, some reports on the good agreement between the behavior of drugs on immobilized AGP and that on soluble AGP [178–180].

In past studies, correlations between percentage of plasma protein binding and HSA or BSA chromatographic retention factors were restricted to structurally related compounds [161,168,169,178]. To improve the performance of biochromatography, in the case of structurally diverse compounds, Beaudry et al. propose the use of normalized peak width as additional chromatographic parameter to improve the correlation of the *in vitro* protein binding versus $k/(k + 1)$. For 40 structurally unrelated pharmaceutical compounds, they obtained a correlation coefficient $r^2 = 0.824$ [181].

In more recent studies, %PPB values are converted to free energy-related apparent protein partition coefficients $\log K$ according to Equation 2.11:

$$\log K = \log\left[\frac{\%PPB}{101 - \%PPB}\right] \qquad (2.11)$$

In Equation 2.11, the number 101 is used instead of 100 to avoid a zero denominator for compounds with %PPB approaching 100%, which are clearly distinguished by the chromatographic system. The $\log K$ values are correlated with the logarithm of retention factors or the logarithm of gradient retention time. In any case, it is recommended to calibrate the chromatographic system using a set of standard drugs to correlate the chromatographic data against published protein-binding data. Apparent $\log K$ values calculated by the calibration equations can be backconverted to %PPB according to Equation 2.12:

$$\%PPB = 101 \cdot \frac{10^{\log K}}{1 + 10^{\log K}} \qquad (2.12)$$

Valko et al. [163] constructed calibration equations using nine standard compounds of compounds known to bind to HSA major binding sites for estimating relative HSA binding using the logarithm of gradient retention time as HSA chromatographic index with 2-propanol as organic modifier. Validation of this equation with a large set of structurally diverse compounds resulted in calculated %HSA values, which showed a good correlation with %HSA data, measured by solution techniques ($r^2 = 0.873$). However, for certain compounds like cephalosporins or ampholytes, predicted values showed large residuals [163].

Reilly et al. comment that HSA-HPLC-based methodology is particularly useful in rank ordering of highly protein-bound compounds, often allowing discrimination in the extent of binding of such compounds that are not differentiated by equilibrium dialysis or ultracentrifugation do not show any differentiation [164]. Protein-binding estimation for a large set of compounds using gradient retention times (1200 compounds training set and 121 rest set) provide satisfactory relationship with data measured by conventional techniques and better fit in the case of highly bound compounds.

Chrysanthakopoulos et al. suggested the isocratic $logk_{10}$ chromatographic retention factor measured at 10% acetonitrile as the best index to simulate apparent protein $logK$ data [162]. A 1:1 correlation was reported with $r^2 = 0.923$ for a set of 43 structurally diverse drugs. The generated calibration equation was successfully validated using another 21 compounds as blind test set. This approach has the advantage that it necessitates only one measurement, while the presence of 10% acetonitrile permits measurements of strongly retained compounds. The same authors report a $logk_{10}$ value equal to 1.85 practically corresponds to 100% PPB. However, deviating behavior was observed for certain compounds [162]. Such routine methods can be very useful to drug discovery chemists, providing a ranking order and quick estimate of albumin binding of candidate molecules prior to be submitted to further development.

2.3.4 MOLECULAR FACTORS INFLUENCING HPAC RETENTION

Although numerous reports support the assumption that immobilized HSA retains its binding properties and behaves like the protein in solution under physiological conditions, it is important to know whether all interactions involved in HSA-HPLC are pharmacologically relevant. In particular, the silica support of immobilized HSA may play its own active role in retention, while the peptide portions of immobilized HSA not involved in the binding sites may also contribute [182]. It has been reported that HSA-HPLC retention measurements of compounds with low affinity for the protein show less accuracy with data obtained with the protein in solution [183]. Moreover, strong retention corresponding to protein binding higher than 100% provides further evidence for additional non-HSA interactions with the stationary phase [162]. In fact, pharmacologically relevant interactions may be distinguished from non-HSA interactions by adding on the mobile phase the same ligand competing agent at excess concentration. Under such conditions, the residual retention time of the analyte corresponds to non-HSA interactions with the stationary phase, nondependent on protein saturation [182,184].

The need for property-based drug design and the establishment of lipophilicity scales for rapid screening of compound libraries [6] has triggered the analysis of HSA and AGP chromatographic indices to more fundamental properties. In this sense, the study of the degree to which hydrophobic interactions determine the retention mechanism on the different affinity stationary phases (mainly on HSA and AGP) has been considered of primary importance. As already mentioned, good correlation between HSA-HPLC retention and lipophilicity is not always the case, depending on the diversity of the data set and on specific characteristics of compounds. Hence, often different subsets should be considered separately [32]. A crucial factor in such correlations is the acid-base functionality, while zwitterionic drugs represent in most cases a separate category [163,176]. Chrysanthakopoulos et al. [162] reported poor and moderate correlation of HSA retention with $\log D_{7.4}$ and $\log P$, respectively, for 59 structurally diverse drugs. However, a considerably improved regression equation ($r = 0.905$) was found if $\log P$ was combined with the molecular fraction of anionic (F^-) and protonated species (F^+). A positive coefficient was assigned to F^- in agreement with the contribution of electrostatic interactions of anions with the opposite charged center on the stationary phase. The term "F^+" had a negative coefficient, an indication that protonation is also an important factor in the case of basic drugs, leading to lower retention, as a result of reduced partitioning of the protonated species. The use of $\log D_{7.4}$ in combination with F^- and F^+ led also to satisfactory correlation. In this case, the regression coefficient of F^- was considerably larger to compensate for the ionization effect, incorporated in $\log D_{7.4}$. F^+ although less important had a positive sign. This result may reflect either suppressed ionization in the biomimetic environment or development of electrostatic interactions between protonated basic analytes and anionic sites on the HSA surface, although to a lesser extent [162].

The contribution of charged species in both HSA in solution and HSA stationary phase was investigated by the same authors by means of LSER analysis. The molecular fractions F^+ and F^- were introduced as additional terms in the regression equations. In both cases, F^- was statistically significant with a positive regression coefficient, while F^+ proved marginally significant only in the analysis of HSA-HPLC retention data with a negative sign [162].

The better performance of the $\log P$ than $\log D$ to explain HSA binding was supported also by the work of Valko et al. [163]. Moreover, a very informative plot of $\log K_{HSA}$ versus $\log D$ values was obtained, showing that most basic compounds are slightly shifted from the trend line for neutral compounds, while this shift is considerably larger for acidic compounds. Most zwitterionic compounds show their own behavior with larger deviations (Figure 2.4).

These observations provide further evidence of the large influence of negatively charged species in HSA retention, compared to the influence of positively charged species.

The lipophilicity magnitude may also influence the correlation with HSA retention factors. Thus, plotting $\log P$ of the neutral form against $\log k_{HSA}$ and $\log k_{AGP}$ a quasi-saturation curve was obtained in both cases with linearity holding for the derivatives with lower lipophilicity [161]. On the other hand, poor correlation with lipophilicity was reported for a series of polar selenium species measured on both HSA and AGP columns [74].

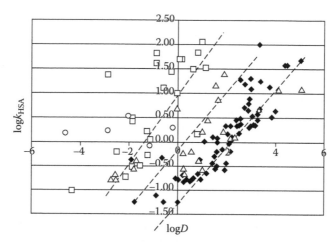

FIGURE 2.4   Plot of logk_{HSA} against structurally diverse neutral (◆), acidic (△), basic (□), and zwitterionic (○) compounds. (Modified from Valko, K. et al., *J. Pharm. Sci.* 92, 2236–2248, 2003.)

2.4 COMPARISON OF THE DIFFERENT BIOMIMETIC STATIONARY PHASES

Despite the different biological material immobilized in the particular biomimetic stationary phases, often good correlation between retention data on IAM, HSA, and AGP columns have been reported [34,163,185]. Among the different IAM columns, retention factors on IAM.DD2 shows better correlation with both HSA and AGP retention factors. Thus, in the case of β-blockers, logk_{HSA} versus logk_{IAMDD2} results in a correlation coefficient $r = 0.933$, whereas $r = 0.920$ and $r = 0.836$ are obtained for IAM.DD and IAM.MG columns. For the same data set, the corresponding correlation between AGP and IAM retention factors yielded $r = 0.986$, $r = 0.937$, and $r = 0.789$ for IAM.DD2, IAM.DD, and IAM.MG, respectively [34]. In all cases, good correlations were obtained also with log$P$ and log$D$. These findings show the importance of nonspecific interactions in the case of β-blockers on IAM.DD2, HSA, and AGP columns. On the other hand, for a set of quinolones, a relatively good relationship was obtained between AGP and HSA retention, while poor correlation was obtained with lipophilicity, unless the data set was separated into two subsets according to the acidic or zwitterionic character of the compounds [176]. This behavior may indicate a significant degree of similarity in the retention mechanism on both columns that cannot be expressed by the nonspecific octanol–water partitioning.

Similarly, very good correlation between HSA and AGP retention factors has been reported also for a set of Se species, with $r^2 = 0.975$, but rather poor correlation of both chromatographic indices with log$D$. Based on this data set, further comparison of the retention on IAM.MG, IAM.DD2, HSA, AGP, three reversed-phase stationary phases, and octanol–water partitioning at two pH values 5 and 7 using principal component analysis showed a high degree of similarity between all systems. Very similar loadings values were obtained on the first principal component, which accounts for most of the variance (80.9%). Differentiation in the loadings on the second principal component was found to correspond to differences in the pH and was attributed to the electrostatic interactions influencing at different extent the retention on biomimetic columns according to pH, but not the octanol–water partitioning [185].

For a large set of 135 drug molecules, the plot of the HSA-HPLC-based $\log K_{HSA}$ values against IAM retention revealed discrimination according to their acid-base or neutral character [186]. Characteristically acidic compounds fit on a trend line with much higher slope reflecting the high affinity for HSA column, while basic compounds with high affinity for IAM stationary phase lie on a trend line with low slope (Figure 2.5).

From the preceding investigations, it is evident that lipophilicity governs to great extent retention on IAM, HSA, and AGP. Positively charged compounds show higher affinity for IAM as well as for AGP column. Negative charge increases retention on HSA. This behavior on HSA and AGP columns is in agreement with the behavior of these proteins in solution.

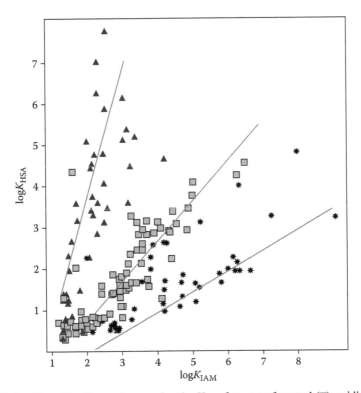

FIGURE 2.5 Plot of $\log K_{HSA}$ values against $\log K_{IAM}$ for a set of neutral (▫), acidic (▲), and basic (✳) compounds. (Modified from Valko, K., 1st World Conference on Physico-Chemical Methods in Drug Discovery and Development, Rovinj, Croatia, 2009.)

2.5   FURTHER APPLICATION OF BIOMIMETIC CHROMATOGRAPHIC DATA

Based on the different features of IAM, HSA, and AGP retention factors and their correspondence with the physiological environment, the application of biomimetic HPLC can be further extended to estimate other biological relevant properties. Thus, the combined use of HSA and AGP retention data may be suitable for estimating total protein binding according to Equation 2.13 [186]:

$$K_{\text{Plasma}} = V_{r(\text{HSA})} \cdot K_{\text{HSA}} + V_{r(\text{AGP})} \cdot K_{\text{AGP}} \tag{2.13}$$

where V_r represents the physological volume ratio of the proteins.

Volume of distribution (VD) is considered as the measure of the extent of distribution of a drug from plasma into tissues and constitutes another important pharmacokinetic property. Its estimation in early drug discovery process would be of great benefit for further development. The VD of a compound depends on the plasma and tissues volumes and the ratio of the bound and unbound drug in plasma and tissue, for example, the plasma and tissue partition coefficients of the compounds [187]. According to Brodie theory [188], the logarithm of VD can be expressed by Equation 2.14:

$$\log V_{\text{D}} = a \cdot \log K_{\text{TISSUE}} - b \cdot \log K_{\text{PLASMA}} \tag{2.14}$$

where $\log K_{\text{TISSUE}}$ and $\log K_{\text{PLASMA}}$ correspond to tissue partition and plasma partition coefficients.

Since membrane/lipid partition and plasma protein binding can be estimated by biomimetic HPLC, the combined use of IAM and HSA chromatography has been proposed for the rapid estimation of V_{D} by replacing the terms of Equation 2.14 accordingly [188]:

$$\log V_{\text{D}} = a \cdot \log K_{\text{IAM}} - b \cdot \log K_{\text{HSA}} + c \tag{2.15}$$

Although Equation 2.14 represents an oversimplification of the process, a satisfactory model was reported for 179 structurally diverse compounds (Equation 2.16) [189]:

$$\log V_{\text{D}} = 0.44 \cdot \log K_{(\text{IAM})} - 0.22 \cdot \log K_{(\text{HSA})} - 0.66$$
$$n = 179, \ r^2 = 0.76, \ s = 0.33, \ F = 272 \tag{2.16}$$

where n is the number of compounds, r^2 is the correlation coefficient, s is the standard deviation, and F is the Fisher test.

2.6   CONCLUSIONS

Biomimetic chromatography is a useful tool for drug discovery and development strategy, which comprises a multiobjective approach. It offers a promising alternative for rapid and friendly measurements that can be used to estimate various biological processes, determining drug efficacy and safety. In particular, retention on IAM stationary phases simulates membranes partitioning and/or drug–membrane interactions.

It may be applied to assess permeability through various biological barriers, such as gastrointestinal, skin, or blood–brain barrier, while strong IAM retention of basic drugs may be associated with the risk of phospholipidosis. Among the protein-based stationary phases, HSA columns attract the highest interest, since the therein immobilized albumin retains the binding characteristics of the protein in solution. There are various ways in which HPAC can be used, providing a wealth of information on drug–HSA interactions. Besides the determination of association constants and of the binding mode, it can be applied for rapid estimation of relative plasma protein binding. Retention on AGP stationary phases, although less investigated and despite initial skepticism in regard to the underlying mechanism, may also lead to successful simulation of drug binding to AGP. There is further investigation for the use of combined biomimetic chromatographic data to estimate other pharmacokinetic properties, like V_D. Most importantly, biomimetic chromatography has initiated an era and an area for facilitating *in vitro* biological protocols. Considering the advances in HPLC column technology, immobilization of other kinds of target proteins or enzymes is feasible, thus offering the possibility for measurements of the biological activity of drugs as well.

REFERENCES

1. I. Kola, J. Landis, "Can the pharmaceutical industry reduce attrition rates?" *Nature Rev. Drug Discovery* 3 (2004) 711–715.
2. G. Gaviraghi, R.J. Barnaby, M. Pellegatti, "Pharmacokinetic challenges in lead optimization," In: B. Testa, H. van de Waterbeemd, G. Folkers, R. Guy (Eds.), *Pharmacokinetic Optimization in Drug Research*, Verlag Helvetica Chimica Acta: Zürich and Wiley-VCH, Weinheim, 2001, pp. 3–14.
3. T. Kennedy, "Managing the drug discovery/development interface," *Drug Discov. Today* 2 (1997) 436–444.
4. B. Testa, G. Vistoli, A. Pedretti, "Musings on ADME predictions and structure—activity relations," *Chem. Biodivers.* 2 (2005) 1411–1427.
5. M.C. Rogge, D.R. Taft, "Preclinical drug development," In: J. Swarbrick (Ed.), *Drugs and the Pharmaceutical Sciences*, 2nd ed., Vol. 187, Informa Healthcare Publications, USA, 2010.
6. D.A. Smith, H. van de Waterbeemd, "Pharmacokinetics and metabolism in early drug discovery," *Curr. Opin. Chem. Biol.* 4 (1999) 373–378.
7. H. van de Waterbeemd, D.A. Smith, K. Beaumont, D.K. Walker, "Property-based design: Optimization of drug absorption and pharmacokinetics," *J. Med. Chem.* 44 (2001) 1–21.
8. H. van de Waterbeemd, B.C. Jones, "Predicting oral absorption and bioavailability," *Prog. Med. Chem.* 41 (2003) 1–59.
9. T.J. Carlson, M.D. Segall, "Predictive, computational models of ADME properties," *Curr. Drug Disc.* (March) (2002) 34–36.
10. C.A. Lipinski, F. Lombardo, B.W. Dominy, P.J. Feeney, "Experimental and computational approaches to estimate solubility and permeability in drug discovery and development settings," *Adv. Drug Deliv.* 23 (1997) 3–25.
11. G. Camenish, G. Folkers, H. van de Waterbeemd, "Review of theoretical passive drug absorption models: Historical background, recent developments and limitations," *Pharm. Acta Helv.* 71 (1996) 309–327.
12. C. Hansch, J.P. Björkroth, A. Leo, "Hydrophobicity and central nervous system agents: On the principle of minimal hydrophobicity in drug design," *J. Pharm. Sci.* 76 (1987) 663–687.

13. S.H. Yalkowski, S.C. Valvani, "Solubility and partitioning I: Solubility of nonelectrolytes in water," *J. Pharm. Sci.* 69 (1980) 912–922.
14. G. Colmenarejo, A. Alvarez-Pedraglio, J.-L. Lavandera, "Chemoinformatic models to predict binding affinities to human serum albumin," *J. Med. Chem.* 44 (2001) 4370–4378.
15. A. Leo, D. Hoekman, "Calculating log P(oct) with no missing fragments; the problem of estimating new interaction parameters," *Perspect. Drug Discov. Design* 18 (2000) 19–38.
16. R.F. Rekker, R. Mannhold (Eds.), *Calculation of Drug Lipophilicity*, Wiley-VCH, Weinheim, 1992.
17. A.K Ghose, G.M Crippen, "Atomic physicochemical parameters for three-dimensional structure-directed quantitative structure-activity relationships. I. Partition coefficients as a measure of hydrophobicity," *J. Comput. Chem.* 7 (1986) 565–577.
18. W.M. Meylan, P.H. Howard, "Atom/fragment contribution method for estimating octanol-water partition coefficients," *J. Pharm. Sci.* 84 (1995) 83–92.
19. R. Mannhold, "Calculation of lipophilicity: A classification of methods," In: B. Testa, S.D. Kramer, H. Wunderli-Allenspach, G. Folkers (Eds.), *Pharmacokinetic Profiling in Drug Research*, Wiley, VCH, Zurich, 2006, pp. 333–352.
20. H. van de Waterbeemd, "Which in vitro screens guide the prediction of oral absorption and volume of distribution?" *Basic Clin. Pharmacol. Toxicol.* 96 (2005) 162–166.
21. S. Yee, "In vitro permeability across Caco-2 cells (colonic) can predict in vivo (small intestinal) absorption in man–fact or myth," *Pharm. Res.* 14 (1997) 763–766.
22. H.H. Usansky, P.J. Sinko, "Estimating human drug oral absorption kinetics from Caco-2 permeability using an absorption-disposition model: Model development and evaluation and derivation of analytical solutions for k_a and F_a," *J. Pharm. Exp. Ther.* 314 (2005) 391–399.
23. V.E. Thiel-Demby, J.E. Humphreys, L.A. St John Williams, H.M. Ellens, N. Shah, A.D. Ayrton, J.W. Polli, "Biopharmaceutics classification system: Validation and learnings of an in vitro permeability assay," *Mol. Pharm.* 6 (2009) 11–18.
24. D.A. Volpe, "Variability in Caco-2 and MDCK cell-based intestinal permeability assays," *J. Pharm. Sci.* 97 (2008) 712–725.
25. B. Faller, "Artificial membrane assays to assess permeability," *Curr. Drug Metab.* 9 (2008) 886–892.
26. H. Wan, M. Rehngren, F. Giodanetto, F. Bergstrom, A. Tunek, "High-throughput screening of drug-brain tissue binding and in silico prediction for assessment of central nervous system drug delivery," *J. Med. Chem.* 50 (2007) 4606–4615.
27. G.L. Trainor, "The importance of plasma protein binding in drug discovery," *Expert Opin. Drug Discov.* 2 (2007) 51–64.
28. K. Ito, T. Iwatsubo, S. Kanamitsu, Y. Nakajima, Y. Sugiyama, "Quantitative prediction of in vivo drug clearance and drug interactions from in vitro data on metabolism together with binding and transport," *Ann. Rev. Pharmacol. Toxicol.* 38 (1998) 461–499.
29. M. Rowley, J.J. Kulagowski, A.P. Watt, D. Rathbone, G.I. Stevenson, R.W. Carling, R. Baker, et al., "Effect of plasma protein binding on in vivo activity and brain penetration of glycine/NMDAreceptor antagonists," *J. Med. Chem.* 40 (1997) 4053–4068.
30. P.E. Rolan, "Plasma protein binding displacement interactions—Why are they regarded as clinically important?" *Br. J. Clin. Pharmacol.* 37 (1994) 125–128.
31. H. Wang, H. Zou, Z. Zhang, "Multi-site binding of fenoprofen to human serum albumin studied by a combined technique of microdialysis with high performance liquid chromatography," *Biomed. Chromatogr.* 12 (1998) 4–7.
32. R.D. Saiakhov, L.R. Stefan, G. Klopman, "Multiple computer-automated structure evaluation model of the plasma protein binding affinity of diverse drugs," *Perspect. Drug Discov. Des.* 19 (2000) 133–155.

33. N.A. Kratochwil, W. Huber, F. Müller, M. Kansy, P.R. Gerber, "Predicting plasma protein binding of drugs: A new approach," *Biochem. Pharmacol.* 64 (2002) 1355–1374.

34. F. Barbato, C. Carpentiero, L. Grumetto, M.I. La Rotonda, "Enantioselective retention of β-blocking agents on human serum albumin and α_1-acid glycoprotein HPLC columns: Relationships with different scales of lipophilicity," *Eur. J. Pharm. Sci.* 38 (2009) 472–478.

35. R.H. McMenamy, J.L. Oncley, "The specific binding of L-tryptophan to serum albumin," *J. Biol. Chem.* 233 (1958) 1436–1447.

36. W.E. Muller, In: I.W. Wainer, D.E. Drayer (Eds.), *Drug Stereochemistry: Analytical Methods and Pharmacology*, Marcel Dekker, New York, 1988, p. 408.

37. E. Domenici, C. Bertucci, P. Salvadori, S. Motellier, I.W. Wainer, "Immobilized serum albumin: Rapid HPLC probe of stereoselective protein-binding interactions," *Chirality* 2 (1990) 263–268.

38. T. Braumann, G. Wber, L.H. Grimme, "Quantitative structure-activity relationships for herbicides. Reversed-phase liquid chromatographic retention parameter, log k(w), versus liquid-liquid partition coefficient as a model of the hydrophobicity of phenylureas, s-triazines and phenoxycarbonic acid derivatives," *J. Chromatogr.* 261 (1983) 329–343.

39. J.G. Dorsey, M.G. Khaledi, "Hydrophobicity estimations by reversed-phase liquid chromatography. Implications for biological partitioning processes," *J. Chromatogr. A* 656 (1993) 485–499.

40. F. Lombardo, M.Y. Shalaeva, K.A. Tupper, F. Gao, "ElogDoct: A tool for lipophilicity determination in drug discovery. 2. Basic and neutral compounds," *J. Med. Chem.* 44 (2001) 2490–2497.

41. X. Liu, H. Tanaka, A. Yamauchi, B. Testa, H. Chuman, "Determination of lipophilicity by reversed-phase high-performance liquid chromatography: Influence of 1-octanol in the mobile phase," *J. Chromatogr. A* 1091 (2005) 51.

42. C. Giaginis, S. Theocharis, A. Tsantili-Kakoulidou, "Contribution to the standardization of the chromatographic conditions for the lipophilicity assessment of neutral and basic drugs," *Anal. Chim. Acta* 573–574 (2006) 311–318.

43. C. Giaginis, S. Theocharis, A. Tsantili-Kakoulidou, "Octanol/water partitioning simulation by reversed phase HPLC for structurally diverse acidic drugs: Effect of octanol as mobile phase additive," *J. Chromatogr. A* 1166 (2007) 116–125.

44. C. Giaginis, A. Tsantili-Kakoulidou, "The current state of the art in HPLC methodology for lipophilicity assessment of basic drugs. A review," *J. Liq. Chromatogr. & Relat. Technol.* 31 (2008) 79–96.

45. C. Giaginis, A. Tsantili-Kakoulidou, "Alternative measures of lipophilicity. From octanol-water partitioning to IAM retention," *J. Pharm. Sci.* 97 (2008) 2984–3004.

46. R.R. Walters, "Affinity chromatography," *Anal. Chem.* 57 (1985) 1099–1114.

47. D.S. Hage, In: E. Katz, R. Eksteen, N. Miller (Eds.), *Handbook of HPLC*, Mercel Dekker, New York, 1998, Chapter 13.

48. A. Leo, C. Hansch, D. Elkins, "Partition coefficients and their uses," *Chem. Rev.* 71 (1971) 525–616.

49. C. Hansch, A. Leo, D. Hoekman D, *Fundamentals and Applications in Chemistry and Biology in Exploring QSAR*, American Chemical Society, Washington, DC, 1995.

50. H. van de Waterbeemd, B. Testa, "The parametrization of lipophilicity and other structural properties in drug design," In: B. Testa (Ed.), *Advances in Drug Research*, Vol. 16, Academic Press, New York, 1987, pp. 85–227.

51. R.P. Mason, D.G. Rhodes, L.G. Herbette, "Reevaluating equilibrium and kinetic binding parameters for lipophilic drugs based on a structural model for drug interaction with biological membranes," *J. Med. Chem.* 34 (1991) 870–877.

52. K. Balon, B.W. Mueller, B.U. Riebesehl, "Drug liposome partitioning as a tool for the prediction of human passive intestinal absorption," *Pharm. Res.* 16 (1999) 882–888.

53. S. Krämer, "Liposomes/water partitioning: Theory techniques and applications," In: B. Testa, H. van de Waterbeemd, G. Folkers, R. Guy (Eds.), *Pharmacokinetic Optimization in Drug Research*, Verlag Helvetica ChimicaActa, Zürich and Wiley-VCH, Weinheim, 2001.
54. G.V. Betageri, J.A. Rogers, "The liposomes as a distribution model in QSAR studies," *Int. J. Pharm.* 46 (1988) 95–102.
55. L. De Young, K.A. Dill, "Solute partitioning into lipid bilayer membranes," *Biochemistry* 27 (1988) 5281–5289.
56. C. Pidgeon, S. Ong, H. Liu, X. Qiu, M. Pidgeon, A.H. Dantzig, J. Munroe, et al., "IAM chromatography: An in vitro screen for predicting drug membrane permeability," *J. Med. Chem.* 38 (1995) 590–594.
57. S. Ong, H. Liu, C. Pidgeon, "Immobilized artificial membrane chromatography: Measurements of membrane partition coefficient and predicting drug permeability," *J. Chromatogr. A* 728 (1996) 113–128.
58. D. Rhee, R. Markovich, W.G. Chae, X. Qiu, C. Pidgeon, "Chromatographic surfaces prepared from lysophosphatidylcholine ligands," *Anal. Chim. Acta* 297 (1994) 377–386.
59. S. Ong, H. Liu, X. Qiu, G. Bhat, C. Pidgeon, "Membrane partition coefficients chromatographically measured using immobilized artificial membrane surfaces," *Anal. Chem.* 67 (1995) 755–762.
60. C. Pidgeon, U.V. Venkataram, "Immobilized artificial membrane chromatography: Supports composed of membrane lipids," *Anal. Biochem.* 176 (1989) 36–47.
61. R.J. Markovich, J.M. Stevens, C. Pidgeon, "Fourier transform infrared assessed of membrane lipids immobilized to silica: Leaching and stability of immobilized artificial membrane –bonded phase," *Anal. Biochem.* 182 (1989) 237–244.
62. R.J. Markovich, X. Qiu, D.E. Nichols, C. Pidgeon, B. Invergo, F.M. Alvarez, "Silica subsurface amine effect on the chemical stability and chromatographic properties of end-capped immobilized artificial membrane surfaces," *Anal. Chem.* 63 (1991) 1851–1860.
63. A. Taillardat-Bertschinger, P.A. Carrupt, F. Barbato, B. Testa, "Immobilized artificial membrane HPLC in drug research," *J. Med. Chem.* 46 (2003) 655–665.
64. C. Pidgeon, C. Marcus, F. Alvarez, "Immobilized artificial membrane chromatography: Surface chemistry and applications," In: J.W. Kelly, T.O. Baldwin (Eds.), *Applications of Enzyme Biotechnology*, Plenum Press, New York, 1991, pp. 201–220.
65. G.W. Caldwell, J.A. Masucci, M. Evangelisto, R. White, "Evaluation of the immobilized artificial membrane phosphatidylcholine. Drug discovery column for high-performance liquid chromatographic screening of drug-membrane interactions," *J. Chromatogr. A* 800 (1998) 161–169.
66. H. Liu, S. Ong, L. Glunz, C. Pidgeon, "Predicting drug-membrane interactions by HPLC: Structural requirements of chromatographic surfaces," *Anal. Chem.* 67 (1995) 3550–3557.
67. F. Barbato, G. di Martino, G. Grunetto, M.I. La Rotonda, "Prediction of drug-membrane interactions by IAM-HPLC: Effects of different phospholipid stationary phases on the partition of bases," *Eur. J. Pharm. Sci.* 22 (2004) 261–269.
68. Regis Technologies, INC., 8210 Austin Avenue, IL 60053, USA, http://www.registech.com.
69. Q. Sheng, K. Schulten, C. Pidgeon, "Molecular dynamics simulation of immobilized artificial membranes," *J. Phys. Chem.* 99 (1995) 11018–11027.
70. D.D. Lasic, *Liposomes: From physics to applications*, Elsevier, Amsterdam, 1993.
71. C.A. Rimmer, C.R. Simmons, J.G. Dorsey, "The measurement and meaning of void volumes in reversed-phase liquid chromatography," *J. Chromatogr. A* 965 (2002) 219–232.
72. A. Taillardat-Bertschinger, A. Galland, P.A. Carrupt, B. Testa, "Immobilized artificial membrane liquid chromatography: Proposed guidelines for technical optimization of retention measurements," *J. Chromatogr. A* 953 (2002) 39–53.

73. F. Tsopelas, M. Ochsenkühn-Petropoulou, A. Tsantili- Kakoulidou, "Void volume markers in reversed-phase and biomimetic liquid chromatography," *J. Chromatogr. A* 1217 (2010) 2847–2854.

74. F. Tsopelas, A. Tsantili-Kakoulidou, M. Ochsenkühn-Petropoulou, "Biomimetic chromatographic analysis of selenium species: Application for the estimation of their pharmacokinetic properties," *Anal. Bioanal. Chem.* 397 (2010) 2171–2180.

75. M.R. Ledbetter, S. Gutsell, G. Hodges, J.C. Madden, S. O'Connor, M.T.D. Cronin, "Database of published retention factors for immobilized artificial membrane HPLC and an assessment of the effect of experimental variability," *Environ. Toxicol. Chem.* 30 (2011) 2701–2708.

76. Y. Marcus, Y. Migron, "Polarity, hydrogen bonding, and structure of mixtures of water and cyanomethane," *J. Phys. Chem.* 95 (1991) 400–406.

77. D. Vrakas, D. Hadjipavlou-Litina, A. Tsantili-Kakoulidou, "Retention of substituted coumarins using immobilized artificial membrane (IAM) chromatography: A comparative study with n-octanol partitioning and reversed-phase HPLC and TLC," *J. Pharm. Biomed. Anal.* 39 (2005) 908–913.

78. K. Morse, C. Pidgeon, "Importance of mobile phase in immobilized artificial membrane chromatography," In: B. Testa, H. van de Waterbeemd, G. Folkers, R. Guy (Eds.), *Pharmacokinetic Optimization in Drug Research*, Verlag Helvetica Chimica Acta, Zürich and Wiley-VCH, Weinheim, 2001, pp. 351–381.

79. M. Hanna, V. de Biasi, B. Bond, P. Camilleri, A. Hutt, "Biomembrane lipids as components of chromatographic phases: Comparative chromatography on coated and bonded phases," *Chromatographia* 52 (2000) 710–720.

80. K. Valko, C. Bevan, D. Reynolds, "Chromatographic hydrophobicity index for fast-gradient RP-HPLC: A high-throughput alternative to logP/logD," *Anal. Chem.* 69 (1997) 2022–2029.

81. K. Valko, C.M. Du, C.D. Bevan, D.P. Reynolds, M.H. Abraham, "Rapid-gradient HPLC method for measuring drug interactions with immobilized artificial membrane: Comparison with other lipophilicity measures," *J. Pharm. Sci.* 89 (2000) 1085–1096.

82. S. Ong, C. Pidgeon, "Thermodynamics of solute partitioning into immobilized artificial membranes," *Anal. Chem.* 67 (1995) 2119–2128.

83. C. Ottiger, H. Wunderli-Allenspach, "Immobilized artificial membrane (IAM)-HPLC for partition studies of neutral and ionized acids and bases in comparison with the liposomal partition system," *Pharm. Res.* 16 (1999) 643–650.

84. A. Taillardat-Bertschinger, F. Barbato, M.T. Quercia, P.A. Carrupt, M. Reist, M.I. La Rotonda, B. Testa, "Structural properties governing retention mechanisms on immobilized artificial membrane (IAM) HPLC columns," *Helv. Chim. Acta* 85 (2002) 519–532.

85. F. Barbato, G. diMartino, L. Grumetto, M.I. La Rotonda, "Can protonated beta-blockers interact with biomembranes stronger than neutral isolipophilic compounds? A chromatographic study on three different phospholipid stationary phases (IAM-HPLC)," *Eur. J. Pharm. Sci.* 25 (2005) 379–386.

86. D. Vrakas, C. Giaginis, A. Tsantili-Kakoulidou, "Different retention behaviour of structurally diverse basic and neutral drugs in immobilized artificial membrane (IAM) and reversed-phase HPLC. Comparison with octanol-water partitioning," *J. Chromatogr. A* 1116 (2006) 158–164.

87. D. Vrakas, C. Giaginis, A. Tsantili-Kakoulidou, "Electrostatic interactions and ionization effect in IAM retention. A comparative study with octanol- water partitioning," *J. Chromatogr. A* 1187 (2008) 67–78.

88. X. Liu, H. Hefesha, G. Scriba, A. Fahr, "Retention behavior of neutral and positively and negatively charged solutes on an immobilized-artificial-membrane (IAM) stationary phase," *Helv. Chim. Acta* 91 (2008) 1505–1512.

89. F. Barbato, I.M. La Rotonda, F. Quaglia, "Chromatogrpahic indexes on immobilized artificial membranes for local anesthetics: Relationships with activity data on closed sodium channels," *Pharm. Res.* 14 (1997) 1699–1705.
90. A. Avdeef, "Physicochemical profiling (solubility, permeability and charge state)," *Curr. Top. Med. Chem.* 1 (2001) 277–351.
91. R. Fruttero, G. Caron, E. Fornatto, D. Boschi, G. Ermondi, A. Gasco, P.A. Carrupt, B. Testa, "Mechanisms of liposomes/water partitioning of (p-methylbenzyl) alkylamines," *Pharm. Res.* 15 (1998) 1407–1413.
92. A. Taillardat-Bertschinger, C.A. Marca-Martinet, P.A. Carrupt, M. Reist, G. Caron, R. Fruttero, B. Testa, "Molecular factors influencing retention on immobilized artificial membranes (IAM) compared to partitioning in liposomes and n-octanol," *Pharm. Res.* 19 (2002) 729–737.
93. G.P. van Balen, C.A.M. Martinet, G. Caron, G. Bouchard, M. Reist, P.A. Carrupt, R. Fruttero, A. Gasco, B. Testa, "Liposome/water lipophilicity: Methods, information content, and pharmaceutical applications," *Med. Res. Rev.* 24 (2004) 299–324.
94. V. Stepanić, D. Žiher, V. Gabelica-Marković, D. Jelić, S. Nunhuck, K. Valko, S. Koštrun, "Physicochemical profile of macrolides and their comparison with small molecules," *Eur. J. Med. Chem.* 47 (2012) 462–472.
95. U.M. Hanumegowda, G. Wenke, A. Regueiro-Ren, R. Yordanova, J.P. Corradi, S. P. Adams, "Phospholipidosis as a function of basicity, lipophilicity, and volume of distribution of compounds," *Chem. Res. Toxicol.* 23 (2010) 749–755.
96. F. Barbato, "The use of immobilized artificial membrane (IAM) chromatography for determination of lipophilicity," *Curr. Comp. Aided Drug Design* 2 (2006) 341–352.
97. X. Liu, B. Testa, A. Fahr, "Lipophilicity and its relationship with passive drug permeation," *Pharm. Res.* 28 (2011) 962–977.
98. E.C.Y. Chan, W.L. Tan, L.J. Fang, "Modeling Caco-2 permeability of drugs using immobilized artificial membrane chromatography and physicochemical descriptors," *J. Chromatogr. A* 1072 (2005) 159–168.
99. M. Genty, G. Gonzalez, C. Clere, V. Desangle-Gouty, J.Y. Legendre, "Determination of passive absorption through the rat intestine using chromatographic indices and molar volume," *Eur. J. Pharm. Sci.* 12 (2001) 223–229.
100. A. Nasal, M. Sznitowska, A. Bucinski, R. Kaliszan, "Hydrophobicity parameter from high-performance liquid chromatography on an immobilized artificial membrane column and its relationship to bioactivity," *J. Chromatogr. A* 692 (1995) 83–89.
101. F. Barbato, B. Cappello, A. Miro, M.I. La Rotonda, F. Quaglia, "Chromatographic indices on immobilized artificial membranes for the prediction of transdermal transport of drugs," *Il Farmaco* 53 (1998) 655–661.
102. N. El Tayar, R.S. Tsai, B. Testa, P.A. Carrupt, A. Leo, "Partitioning of solutes in different solvent systems: The contribution of hydrogenbonding capacity and polarity," *J. Pharm. Sci.* 80 (1991) 590–598.
103. J. Kotecha, S. Shah, I. Rathod, G. Subbaiah, "Prediction of oral absorption in humans by experimental immobilized artificial membrane chromatography indices and physicochemical descriptors," *Int. J. Pharm.* 360 (2008) 96–106.
104. B.S. Shin, C.H. Yoon, J.P. Balthasar, B.Y. Choi, S.H. Hong, H.J. Kim, J.B. Lee, S.W. Hwang, S.D. Yoo, "Prediction of drug bioavailability in humans using immobilized artificial membrane phosphatidylcholine column chromatography and in vitro hepatic metabolic clearance," *Biomed. Chromatogr.* 23 (2009) 764–769.
105. A. Reichel, D.J. Begley, "Potential of immobilized artificial membranes for predicting drug penetration across the blood-brain barrier," *Pharm. Res.* 15 (1998) 1270–1274.
106. T. Salminen, A. Pulli, J. Taskinen, "Relationship between immobilized artificial membrane chromatographic retention and the brain penetration of structurally diverse drugs," *J. Pharm. Biomed. Anal.* 15 (1997) 469–477.

107. C.H Yoon, S.J. Kim, B.S. Shin, K.C. Lee, S.D. Yoo, "Rapid screening of blood-brain barrier penetration of drugs using the immobilized artificial membrane phosphatidylcholine column chromatography," *J. Biomol. Screen.* 11 (2006) 13–20.

108. F. Pehourcq, M. Matoga, B. Bannwarth, "Diffusion of arylpropionatenon-steroidal anti-inflammatory drugs into the cerebrospinal fluid: A quantitative structure-activity relationship approach," *Fundam. Clin. Pharmacol.* 18 (2004) 65–70.

109. E. Lazaro, C. Rafols, M.H. Abraham, M. Roses, "Chromatographic estimation of drug disposition properties by means of immobilized artificial membranes (IAM) and C18 columns," *J. Med. Chem.* 49 (2006) 4861–4870.

110. T. Peters, *All About Albumin: Biochemistry, Genetics and Medical Applications*, Academic Press, San Diego, 1996.

111. W.E. Muller, U. Wollert, "Human serum albumin as a 'silent receptor' for drugs and endogenous substances," *Pharmacology* 19 (1979) 59–67.

112. K. Williams, E. Lee, "Importance of drug enantiomers in clinical pharmacology," *Drugs* 30 (1985) 333–354.

113. F. Jamali, R. Mehvar, F.M. Pasutto, "Enantioselective aspects of drug action and disposition: Therapeutuc pitfalls," *J. Pharm. Sci.* 78 (1989) 695–715.

114. I. Petitpas, A.A. Bhattacharya, S. Twine, M. East, S. Curry, "Crystal structure analysis of warfarin binding to human serum albumin: Anatomy of drug site I," *J. Biol. Chem.* 276 (2001) 22804–22809.

115. S. Curry, P. Brick, N.P. Franks, "Fatty acid binding to human serum albumin: New insights from crystallographic studies," *Biochim. Biophys. Acta* 1441 (1999) 131–140.

116. G. Sudlow, D.J. Birkett, D.N. Wade, "The characterization of two specific drug binding sites on human serum albumin," *Mol. Pharmacol.* 11 (1975) 824–832.

117. I. Sjoholm, B. Ekman, A. Kober, I. Ljungstedt-Pahlman, B. Seiving, T. Sjodin, "Binding of drugs to human serum albumin:XI. The specificity of three binding sites as studied albumin immobilized in microparticles," *Mol. Pharmacol.* 16 (1979) 767–777.

118. N. Iberg, R. Flückiger, "Nonenzymatic glycosylation of albumin in vivo. Identification of multiple glycosylated sites," *J. Biol. Chem.* 261 (1986) 13542–13545.

119. J. Hermansson, M. Eriksson, "Direct liquid chromatographic resolution of acidic drugs using a chiral alpha 1-acid glycoprotein column (Enantiopac)," *J. Liq. Chromatogr.* 9 (1986) 621–639.

120. J. Hermansson, M. Eriksson, O.J. Nyquist, "Determination of (R)- and (S) disopyramide in human plasma using a chiral alpha 1-acid glycoprotein column," *J. Chromatogr.* 336 (1985) 321–328.

121. K. Schmid, In: F.W. Putnam (Ed.), *Plasma Proteins*, Vol. 1, Academic Press, New York, 1975, p. 184.

122. T. Fournier, N.N. Medjoubi, D. Porquet, "Alpha-1-acid glycoprotein," *Biochim. Biophys. Acta* 1482 (2000) 157–171.

123. D.L. Schonfeld, R.B.G. Ravelli, U. Mueller, A. Skerra, "The 1.8-Å crystal structure of α_1-acid glycoprotein (orosomucoid) solved by UV RIP reveals the broad drug-binding activity of this human plasma lipocalin," *J. Mol. Biol.* 384 (2008) 393–405.

124. I. Fitos, J. Visy, F. Zsila, G. Mády, M. Simonyi, "Selective binding of imatinib to the genetic variants of human α_1-acid glycoprotein," *Biochim. Biophys. Acta* 1760 (2006) 1704–1712.

125. S. Ongay, C. Neusüß, "Isoform differentiation of intact AGP from human serum by capillary electrophoresis–mass spectrometry," *Anal. Bioanal. Chem.* 398 (2010) 845–855.

126. F. Herve, G. Caron, J.-C. Duche, P. Gaillard, N.A. Rahman, A. Tsantili-Kakoulidou, P.-A. Carrupt, P. D'Athis, J.-P. Tilleman, B. Testa, "Ligand specificity of the genetic variants of human alpha-1-acid glycoprotein. Generation of a 3D-QSAR model for drug binding to the A variant," *Mol. Pharmacol.* 54 (1998) 129–138.

127. J.K. Lacki, K. Klama, W. Samborski, S.H. Mackiewicz, W. Muller, "Comparison of microheterogeneity of alpha-1-acid-glycoprotein in serum and synovial fluid from rheumatoid arthritis patients," *Clin. Rheumatol.* 13 (1994) 598–604.

128. T. Hochepied, F.G. Berger, H. Baumann, C. Libert, "Alpha(1)-acid glycoprotein: An acute phase protein with inflammatory and immunomodulating properties," *Cytokine Growth Factor Rev.* 14 (2003) 25–34.

129. L. Budai, O. Ozohanics, K. Ludányi, L. Drahos, T. Kremmer, J. Krenyacz, K. Vékey, "Investigation of genetic variants of α-1 acid glycoprotein by ultra-performance liquid chromatography–mass spectrometry," *Anal. Bioanal. Chem.* 393 (2009) 991–998.

130. T. Hochepied, W. Van Molle, F.G. Berger, H. Baumann, C. Libert, "Involvement of the acute phase protein alpha 1-acid glycoprotein in nonspecific resistance to a lethal gram-negative infection," *J. Biol. Chem.* 275 (2000) 14903–14909.

131. F. Karush, "The competitive interaction of organic anions with bovine serum albumin," *J. Am. Chem. Soc.* 72 (1950) 2714–2718.

132. K.K. Stewart, R.F. Doherty, "Resolution of DL-tryptophan by affinity chromatography on bovine-serum albumin-agarose columns," *Proc. Natl. Acad. Sci. USA* 70 (1973) 2850–2852.

133. T.A.G. Noctor, M.J. Diaz-Perez, I.W. Wainer, "Use of a human serum albumin-based stationary phase for high-performance liquid chromatography as a tool for the rapid determination of drug plasma protein binding," *J. Pharm. Sci.* 82 (1993) 675–676.

134. P.R. Tiller, I.M. Mutton, S.J. Lane, C.D. Bevan, "Immobilized human serum albumin: Liquid chromatography/mass spectrometry as a method of determining drug-protein binding," *Rapid Commun. Mass Spectrom.* 9 (1995) 261–263.

135. E. Domenici, C. Bertucci, P. Salvadori, G. Felix, I. Cahagne, S. Motellier, I.W. Wainer, "Immobilized serum albumin: Rapid HPLC probe of stereoselective protein-binding interactions," *Chromatographia* 29 (1990) 170–176.

136. D.S. Hage, J. Austin, "High-performance affinity chromatography and immobilized serum albumin as probes for drug- and hormone-protein binding," *J. Chromatogr. B* 739 (2000) 39–54.

137. W.E. Lindup, In: J.W. Bridges, L.F. Chasseaud, G.G. Gibson (Eds.), *Progress in Drug Metabolism*, Vol. 10, Taylor & Francis, New York, 1987, Chapter 4.

138. T.C. Kwong, "Free drug measurements: Methodology and clinical significance," *Clin. Chim. Acta* 151 (1985) 193–216.

139. J. Barre, F. Didey, F. Delion, J.-P. Tillement, "Problems in therapeutic drug monitoring: Free drug level monitoring," *Ther. Drug Monit.* 10 (1988) 133–143.

140. B. Loun, D.S. Hage, "Characterization of thyroxine-albumin binding using high-performance affinity chromatography. I. Interactions at the warfarin and indole sites of albumin," *J. Chromatogr. B Biomed. Appl.* 779 (1992) 225–235.

141. J. Haginaka, "Protein-based chiral stationary phases for high-performance liquid chromatography enantioseparations," *J. Chromatogr. A* 906 (2001) 253–273.

142. V. Tittelbach, M. Jaroniec, R.K. Gilpin, "Synthesis and characterization of silica-immobilized serum albumin stationary phases for HPLC," *J. Liq. Chromatogr. Rel. Technol.* 19 (1996) 2943–2965.

143. M.J. Yoo, D.S. Hage, "Use of peak decay analysis and affinity microcolumns containing silica monoliths for rapid determination of drug-protein dissociation rates," *J. Chromatogr. A* 1218 (2011) 2072–2078.

144. S. Allenmark, *Chromatographic Enantioseparation: Methods and Applications*, 2nd ed. Ellis Horwood, New York, 1991, Chapter 7.

145. S. Allenmark, B. Bomgren, H. Boren, "Direct liquid chromatographic separation of enantiomers on immobilized protein stationary phases. III. Optical resolution of a series

of N-aroyl D,L-amino acids by high-performance liquid chromatography on bovine serum albumin covalently bound to silica," *J. Chromatogr.* 264 (1983) 63–68.

146. P. Erlandsson, I. Marle, L. Hansson, R. Isaksson, C. Petterson, G. Petterson, "Immobilized cellulase (CBH I) as a chiral stationary phase for direct resolution of enantiomers," *J. Am. Chem. Soc.* 112 (1990) 4573–4574.

147. K. Harada, Q. Yuan, M. Nakayama, A. Sugii, "Effects of organic modifiers on the chiral recognition by different types of silica-immobilized bovine serum albumin," *J. Chromatogr. A* 740 (1996) 207–213.

148. R. Mallik, D.S. Hage, "Development of an affinity silica monolith containing human serum albumin for chiral separations," *J. Pharm. Biomed. Anal.* 46 (2008) 820–830.

149. J. Haginaka, T. Murashima, C. Seyama, "Separation of enantiomers on a lysozyme-bonded silica column," *J. Chromatogr. A* 666 (1994) 203–210.

150. J. Haginaka, Y. Miyano, Y. Saizen, C. Seyama, T. Murashima, "Separation of enantiomers on a pepsin-bonded column," *J. Chromatogr. A* 708 (1995) 161–168.

151. A. Nystrom, A. Strandberg, A. Aspegren, S. Behr, A. Karlsson, "Use of immobilized amyloglucosidase as chiral selector in chromatography. Immobilization and performance in liquid chromatography," *Chromatographia* 50 (1999) 209–214.

152. J. Hermansson, "Direct liquid chromatographic resolution of racemic drugs using α_1-acid glycoprotein as the chiral stationary phase," *J. Chromatogr.* 269 (1983) 71–80.

153. T. Miwa, M. Ichikawa, M. Tsuno, T. Hattori, T. Miyakawa, M. Kayano, Y. Miyake, "Direct liquid chromatographic resolution of racemic compounds. Use of ovomucoid as a column ligand," *Chem. Pharm. Bull.* 35 (1987) 682–686.

154. J. Haginaka, C. Seyama, N. Kanasugi, "The absence of chiral recognition ability in ovomucoid: Ovoglycoprotein-bonded HPLC stationary phases for chiral recognition," *Anal. Chem.* 67 (1995) 2539–2547.

155. T. Miwa, T. Miyakawa, Y. Miyake, "Characteristics of an avidin-conjugated column in direct liquid chromatographic resolution of racemic compounds," *J. Chromatogr. A* 457 (1988) 227–233.

156. N. Mano, Y. Oda, N. Asakawa, Y. Yoshida, T. Sato, "Development of a flavoprotein column for chiral separation by high-performance liquid chromatography," *J. Chromatogr.* 623 (1992) 221–228.

157. N. Mano, Y. Oda, N. Asakawa, Y. Yoshida, T. Sato, "Conalbumin-conjugated silica gel, a new chiral stationary phase for high- performance liquid chromatography," *J. Chromatogr.* 603 (1992) 105–109.

158. G. Massolini, E. De Lorenzi, D.K. Lloyd, A.M. McGann, G. Caccialanza, "Evaluation of β-lactoglobulin as a stationary phase in high-performance liquid chromatography and as a buffer additive in capillary electrophoresis: Observation of a surprising lack of stereoselectivity," *J. Chromatogr. B* 712 (1998) 83–94.

159. R.C. Jewell, K.L.R. Brouwer, P.J. McNamara, "α_1-Acid glycoprotein high-performance liquid chromatography column (AnantioPAC) as a screening tool for protein binding," *J. Chromatogr. B Biomed. Appl.* 487 (1989) 257–264.

160. G. Schill, I.W. Wainer, S.A. Barkan, "Chiral separations of cationic and anionic drugs on an α_1-acid glycoprotein-bonded stationary phase (enantioPac®). II. Influence of mobile phase additives and pH on chiral resolution and retention," *J. Chromatogr.* 365 (1986) 73–88.

161. D.S. Ashton, C.R. Beddell, G.S. Cockerill, K. Gohil, C. Gowrie, J.E. Robinson, M.J. Slater, K. Valko, "Binding measurements of indolocarbazole derivatives to immobilised human serum albumin by high-performance liquid chromatography," *J. Chromatog. B* 677 (1996) 194–198.

162. M. Chrysanthakopoulos, C. Giaginis, A. Tsantili-Kakoulidou, "Retention of structurally diverse drugs in human serum albumin chromatography and its potential to simulate plasma protein binding," *J. Chromatogr. A* 1217 (2010) 5761–5768.

163. K. Valko, S. Nunhuck, C. Bevan, M.H. Abraham, D.P. Reynolds, "Fast gradient HPLC method to determine compounds binding to human serum albumin. Relationships with octanol-water and immobilized artificial membrane lipophilicity," *J. Pharm. Sci.* 92 (2003) 2236–2248.

164. J. Reilly, D. Etheridge, B. Everatt, Z. Jiang, C. Springer, R.A. Fairhurst, "Studies in drug albumin binding using HSA and RSA affinity methods," *J. Liq. Chromatogr. Relat. Technol.* 34 (2011) 317–327.

165. I.M. Chaiken (Ed.), *Analytical Affinity Chromatography*, CRC Press, Boca Raton, FL, 1987.

166. D.S. Hage, "High performance affinity chromatography: A powerful tool for studying serum protein binding," *J. Chromatogr. B* 768 (2002) 3–30.

167. T.A.G. Noctor, D.S. Hage, I.W. Wainer, "Allosteric and competitive displacement of drugs from human serum albumin by octanoic acid, as revealed by high-performance liquid affinity chromatography, on a human serum albumin-based stationary phase," *J. Chromatogr. B Biomed. Appl.* 577 (1992) 305–315.

168. T.A.G. Noctor, C.D. Pham, R. Kaliszan, I.W. Wainer, "Stereochemical aspects of benzodiazepine binding to human serum albumin. I. Enantioselective highperformance liquid affinity chromatographic examination of chiral and achiral binding interactions between 1,4-benzodiazepines and human serum albumin," *Mol. Pharm.* 42 (1992) 506–511.

169. E. Domeneci, C. Bertucci, P. Salvadori, I.W. Wainer, "Use of a human serum albumin-based high-performance liquid chromatography chiral stationary phase for the investigation of protein binding: Detection of the allosteric interaction between warfarin and benzodiazepine binding sites," *J. Pharm. Sci.* 80 (1991) 164–166.

170. D.S. Hage, S.A. Tweed, "Recent advances in chromatographic and electrophoretic methods for the study of drug-protein interactions," *J. Chromatogr. B* 699 (1997) 499–525.

171. S. Tweed, B. Loun, D.S. Hage, "Effects of ligand heterogeneity in the characterization of affinity columns by frontal analysis," *Anal. Chem.* 69 (1997) 4790–4798.

172. N.I. Nakano, Y. Shimamori, S. Yamaguchi, "Binding capacities of human serum albumin monomer and dimer by continuous frontal affinity chromatography," *J. Chromatogr.* 237 (1982) 225–232.

173. C. Lagercrantz, T. Larsson, H. Karlsson, "Binding of some fatty acids and drugs to immobilized bovine serum albumin studied by column affinity chromatography," *Anal. Biochem.* 99 (1979) 352–364.

174. B. Loun, D.S. Hage, "Chiral separation mechanisms in protein-based HPLC columns. 1. Thermodynamic studies of (R)- and (S)-warfarin binding to immobilized human serum albumin," *Anal. Chem.* 66 (1994) 3814–3822.

175. H. Yoshida, I. Morita, G. Tamai, T. Masujima, T. Tsuru, N. Takai, H. Imai, "Some characteristics of a protein-coated ODS column and its use for the determination of drugs by the direct injection analysis of plasma samples," *Chromatographia* 19 (1984) 466–472.

176. F. Barbato, G. di Martino, L. Grumetto, M.I. La Rotonda, "Retention of quinolones on human serum albumin and α_1-acid glycoprotein HPLC columns: Relationships with different scales of lipophilicity," *Eur. J. Pharm. Sci.* 30 (2007) 211–219.

177. N. Lammers, H. De Bree, C.P. Groen, H.M. Ruijten, B.J. Jong, "Determination of drug protein-binding by high-performance liquid chromatography using a chemically bonded bovine albumin stationary phase," *J. Chromatogr. B Biomed. Appl.* 496 (1989) 291–300.

178. A. Nasal, A. Radwanska, K. Osmialowski, A. Bucinski, R. Kaliszan, G.E. Barker, P. Sun, R.A. Hartwick, "Quantitative relationships between the structure of β-adrenolytic and antihistamine drugs and their retention on an α_1-acid glycoprotein HPLC column," *Biomed. Chromatrogr.* 8 (1994) 125–129.

179. R. Kaliszan, A. Nasal, M. Turowski, "Quantitative structure–retention relationships in the examination of the topography of the binding site of antihistamine drugs on alpha 1-acid glycoprotein," *J. Chromatogr. A* 722 (1996) 25–32.

180. H. Xuan, D.S. Hage, "Immobilization of alpha(1)-acid glycoprotein for chromatographic studies of drug-protein binding," *Anal. Biochem.* 346 (2005) 300–310.

181. F. Beaudry, M. Coutu, N.K. Brown, "Determination of drug-plasma protein binding using human serum albumin chromatographic column and multiple linear regression model," *Biomed. Chromatogr.* 13 (1999) 401–406.

182. G.A. Ascoli, E. Domenici, C. Bertucci, "Drug binding to human serum albumin: Abridged review of results obtained with high-performance liquid chromatography and circular dichroism," *Chirality* 18 (2006) 667–679.

183. B. Ravindranath, "In Principles and practice of chromatography," Ellis Horwood, Chichester, UK, 1989, pp. 149–150.

184. G.A. Ascoli, C. Bertucci, P. Salvadori, "Ligand binding to a human serum albumin stationary phase: Use of same-drug competition to discriminate pharmacologically relevant interactions," *Biomed Chromatogr.* 12 (1998) 248–254.

185. F. Tsopelas, M. Ochsenkühn-Petropoulou, A. Tsantili-Kakoulidou, "Exploring the elution mechanism of selenium species on liquid chromatography," *J. Sep. Sci.* 34 (2011) 376–384.

186. K. Valko, "HPLC based property measurements of drug discovery compounds to predict in vivo drug distribution and aid lead optimisation," 1st World Conference on Physico-Chemical Methods in Drug Discovery and Development, 2009, Rovinj, Croatia. http://www.iapchem.org/lectures/Rovinj%202009%20klv%20with%20copyright.pdf.

187. G.R. Wilkinson, "Pharmacokinetics: The dynamics of drug absorption, distribution, and elimination," In: J.G. Hardman, L.E. Limbird, A.G. Gillman (Eds.), *Goodman & Gilman's The Pharmacological Bases of Therapeutics*, 10th ed. McGraw-Hill, New York, 2001, pp. 20–22.

188. B.B. Brodie, H. Kurtz, L.J. Schanker, "The importance of dissociation constants and lipid solubility on influencing the passage of drugs into the CSF," *J. Pharmacol. Exp. Ther.* 130 (1960) 20–25.

189. F. Hollosy, K. Valko, A. Hersey, S. Nunhuck, G. Keri, C. Bevan, "Estimation of volume of distribution in humans from high throughput HPLC-based measurements of human serum albumin binding and immobilized artificial membrane partitioning," *J. Med. Chem.* 49 (2006) 6958–6971.

3 Solid-Phase Microextraction for *In Vivo* Pharmacokinetics and Other Stages of Drug Development

Barbara Bojko and Janusz Pawliszyn

CONTENTS

3.1 INTRODUCTION

The process of drug discovery and development involves many stages including drug design, *in vitro* studies of drug candidate metabolism, evaluation of potential toxicity of drugs and their metabolites, and determination of protein binding and *in vivo* pharmacokinetics (PK) with animal models. Most of these steps are very challenging because *in vitro* drug activity cannot always be extrapolated easily to its *in vivo* activity and because the limited availability of a new compound restricts *in vivo* studies to one or two animal species [1]. Although intraspecies variations in absorption, distribution, metabolism, and elimination (ADME) of drugs may result in different pharmacokinetic profiles in humans than expected from animal data, animal studies remain a critical step in the process of drug development. However, there is a low throughput of *in vivo* PK screening, a limited number of methods feasible for *in vivo* studies, and strict regulations that force pharmaceutical companies to minimize the number of animals used for experimentation. These factors lead to the continuous introduction of new strategies to address these issues [2]. Finally, when a drug is approved for preclinical and clinical uses in human subjects, the pharmacological and pharmacokinetic properties of the drug must be verified. Polymorphism of enzymes, clinical condition of the patient, coexisting diseases, and multidrug therapy may all significantly affect the predicted concentration of the drug; thus, dosing regimen verification and therapeutic drug monitoring are required. Recently, the monitoring of drug-induced metabolome profile changes presented the opportunity to personalize pharmacotherapy and increase its effectiveness.

Continuous improvement of analytical instrumentation provides excellent sensitivity, selectivity, precision, and accuracy of data. Among the various analytical methods used in the field of drug discovery, the most commonly used instruments are liquid chromatography (LC) and gas chromatography (GC) platforms coupled with mass spectrometers or nuclear magnetic resonance (NMR) spectrometers. Whereas the latter method does not require extensive sample preparation, LC(GC)/ mass spectrometry (MS) analysis needs appropriate sample pretreatment to avoid extraction of interferences affecting ionization and contributing to matrix effect. A wide range of extraction methods is available for *in vitro* studies, whereas microdialysis (MD) is currently the main method used for *in vivo* studies in clinical practice [3,4]. New analytical tools such as biosensors or microfluidic devices are also available, but these are used for biomedical research rather than clinical practice [4]. Drug labeling and imaging techniques can be used for screening drug distribution, although these methods are not feasible for precise qualitative analysis of drug metabolism and tissue drug distribution [3] or for metabolomic analysis [5]. Solid-phase microextraction (SPME) is an analytical technique that combines sampling, sample preparation, and extraction in one step. SPME was initially used for environmental and food analysis and became recognized in the bioanalytical area. In this field SPME was primarily used for *in vitro* studies, but over time it emerged as a very promising tool for *in vivo* analyses. SPME has overcome many disadvantages of standard approaches, with no need of sample withdrawal; none or minimal disturbance to the system under study; and the possibility of determining both free and total concentrations of the analyte and, subsequently, binding parameters. In this chapter, we describe the applicability of SPME in drug discovery with a particular focus on *in vivo* PK.

3.2 THEORY OF SOLID PHASE MICROEXTRACTION

SPME was invented by Pawliszyn in the early 1990s. This equilibrium-based nonexhaustive extraction method is feasible for gaseous, liquid, and solid samples. The various geometries of supporting materials holding extraction phases including fiber, blade supporting thin film, in tube or stir bar make SPME a feasible tool for various applications (Figure 3.1).

For biomedical analysis, coated fiber or thin film are the methods of choice. Whereas the fiber is applicable for *in vivo* sampling (direct extraction from blood or tissue), thin film achieves better sensitivity of the assay by its increased surface area. Irrespective of the geometry of the supporting material for the coating, principles of SPME are based on the interaction of the free fraction of the analyte with the extraction sorbent exposed directly to the sample matrix (Figure 3.2). The following discussion is limited to coated fibers; however, analogous equations are valid for any geometry.

Once the analyte is in contact with the polymeric sorbent, it diffuses to the extraction phase until distribution equilibrium between sample matrix and coating is reached. In this case, convection conditions do not affect the amount extracted. In addition, longer exposure of the analyte does not further increase the amount of analyte extracted by the sorbent. The equilibrium conditions and the distribution coefficient K_{fs} are described by Equations 3.1 and 3.2, respectively:

$$C_0 V_s = C_s^\infty V_s + C_f^\infty V_f \tag{3.1}$$

where C_0 is initial concentration of the analyte in the sample; C_f^∞ and C_s^∞ are equilibrium concentrations in the fiber coating and the sample matrix, respectively; and V_f and V_s are the volume of the fiber coating and the sample, respectively.

$$K_{fs} = \frac{C_f^\infty}{C_s^\infty} \tag{3.2}$$

When Equations 3.1 and 3.2 are combined and rearranged, C_f^∞ can be calculated as follows:

$$C_f^\infty = C_0 \frac{K_{fs} V_s}{K_{fs} V_f + V_s} \tag{3.3}$$

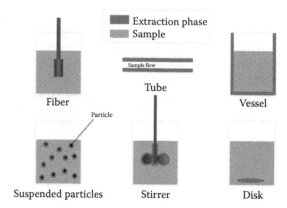

FIGURE 3.1 (**See color insert.**) Configurations of solid-phase microextraction (SPME).

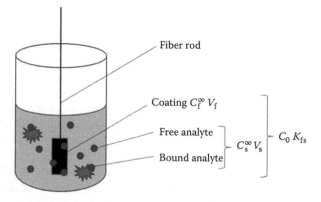

FIGURE 3.2 Schematic illustration of the interaction occurring during sample preparation with SPME: V_f is volume of the fiber coating; C_f^∞ and C_s^∞ are equilibrium concentrations in the fiber coating and the sample, respectively; V_f and V_s are the volumes of the fiber coating and the sample, respectively; C_0 is initial concentration of the analyte in the sample; and K_{fs} is distribution coefficient between fiber coating and sample matrix.

Subsequently, the number of moles of analyte (n) extracted by the coating can be calculated from the following equation:

$$n = C_f^\infty V_f = C_0 \frac{K_{fs} V_s V_f}{K_{fs} V_f + V_s} \tag{3.4}$$

It is evident that the amount of analyte extracted by the sorbent (n) is proportional to the initial concentration of the analyte in the sample (C_0). Moreover, when the sample volume V_s is much larger than the product $K_{fs} V_f$, Equation 3.4 can be further simplified as follows:

$$n = K_{fs} V_f C_0 \tag{3.5}$$

The product $K_{fs}V_f$ can be directly obtained from the regressing slope, indicating the relation between the amount of analyte extracted and the initial concentration of the analyte plotted for calibration standards extracted from the matrix-free solution at equilibrium. Based on Equation 3.5, it is concluded that calculating the amount of analyte can be performed for an unknown sample volume, which in practice gives the opportunity of *in vivo* sampling by introducing a fiber, for example, to the tissue or blood stream. This means that the determination of analyte concentration in living systems does not require sample withdrawal (e.g., biopsy material) and, at the same time, provides much less disturbance to the organism [6].

SPME can be also used for the bioanalysis of liquid or solid samples in the head-space (HS) mode and followed by GC analysis. In such cases, the analyte needs to be released from the matrix and transferred to the extraction phase. Whereas the transfer of compounds during *in vitro* analysis can be enhanced by various agitation methods such as sample heating, *in vivo* studies are limited in this regard. However, the HS approach can still be successfully used for *in vivo* breath and skin sampling of volatile compounds [7–18]. Direct-immersion SPME (DI-SPME) of biological samples has one significant advantage over HS sampling: biological systems (fluids and solid tissues) are an aqueous environment where polar compounds are freely dissolved in water, whereas hydrophobic species form complexes with macromolecules such as proteins, and only a small fraction of the analyte remains free. Most of the extraction phases used in SPME have high affinity toward hydrophobic compounds, and lower recoveries are expected for polar compounds. As explained in Figure 3.3, in the studied system SPME coating strives to achieve equilibrium with free molecules of the analyte; thus, despite the high affinity of hydrophobic species, their extraction efficiency does not provide displacement of lower-affinity

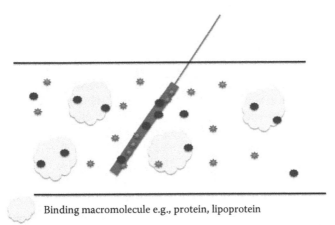

⬭ Binding macromolecule e.g., protein, lipoprotein

● Small hydrophobic molecule, high K_{fs} value, low free concentration in the tissue, high protein binding

✳ Small hydrophobic molecule, low K_{fs} value, high free concentration in the tissue, low protein binding

FIGURE 3.3 The graphic explanation of balanced coverage of DI-SPME; the biological system acts as a "buffer" for SPME coating by binding hydrophobic species and preventing from fiber saturation.

analytes and consequently does not affect the adsorption of polar compounds. As a result, SPME offers a balanced coverage of the metabolites with different physical and chemical properties, as has been shown in studies on the systematic evaluation of biocompatible SPME coatings [19]. Nevertheless, it must be emphasized that an appropriate extraction phase, which protects the adhesion of macromolecules present in complex biological matrices, needs to be carefully selected to avoid the fiber fouling effect.

3.3   EXTRACTION PHASES FOR LC APPLICATIONS

Many different SPME coatings have already been developed and commercialized for GC applications for the analysis of volatile components. A comparison of their properties as well as guidelines for proper fiber selection can be found in the literature [6]. In this chapter, we present fibers applicable for direct immersion from biological matrices, for both *in vitro* and *in vivo* SPME. For many years, direct-immersion sampling with SPME was limited due to the small number of biocompatible sorbents. "Biocompatibility" is one of the main criteria of materials used for manufacturing devices used in bioanalytical applications. In this case, the term biocompatibility not only applies to the properties of a material, which ensures a lack of side effects when the material comes in contact with an organism, but also means that the complex biological matrix will not affect the properties of the material during analysis. For example, a polymeric sorbent is protected by the adhesion of large molecules, such as proteins, what prevents fouling of the coating. Several biocompatible materials including polypyrrole (PPY) [20–25], polyethylene glycol (PEG) [21,26,27], polyacrylonitrile (PAN) [28], polydimethylsiloxane (PDMS) [29], and restricted access materials (RAMs) [30–34] were used for coating preparation. However, many of these lab-made fibers suffered from poor interfiber reproducibility, which limited their applicability for studies requiring precise and accurate measurements such as PK. However, constant attempts to improve interfiber variability have been successful with good results reported [35,36]. In addition, biocompatible fibers characterized by very good interfiber reproducibility are already available commercially. Ideally, probes used for *in vivo* sampling should be disposable, which means that high interfiber reproducibility and cost-effectiveness are key characteristics. Improved robustness of fibers can be achieved by the preparation of thicker coatings, but this significantly increases the extraction time required to reach equilibrium and preequilibrium extractions; therefore, it becomes the method of choice. However, as long as the sensitivity of the preequilibrium extraction method is sufficient for a particular application, a wide range of calibration approaches ensures precise and accurate quantitative data. Devices dedicated to *in vivo* studies should also be miniaturized to dimensions that avoid tissue damages and disturbance to the organism. They must also be very stable mechanically to avoid core breakage or coating stripping. One major concern of *in vivo* animal and human sampling is the necessity of working in a sterile environment. Several *in vivo* SPME studies were conducted successfully with the use of sterile probes. It was reported that autoclaving affects the extraction efficiency of PPY fibers [24]; thus, the changing of coating performance by a sterilization procedure during method development and optimization should be considered.

In such cases, the selection of alternative sterilization procedures, that is, γ irradiation or soaking in alcohol, can be used [26,27].

Another alternative among biocompatible coatings is RAMs. These sorbents consist of two layers: the outer porous surface acts as a barrier for large biomolecules, whereas small molecules can easily transfer through the pores into the inner part, which is the actual extraction phase. It has been demonstrated that RAM coatings can be successfully used for the extraction of exogenous (i.e., drugs) as well as endogenous compounds. Mullet et al. [33,37] and Walles et al. used RAM-SPME for the determination of diazepam and its metabolites from spiked serum [37] and urine [33] and blood samples [32], respectively. Chaves et al. [38] determined interferon alpha$_{2a}$ in plasma samples, and Musteata [30] demonstrated the suitability of RAM coatings for the extraction of angiotensin 1 from whole blood. As mentioned earlier, most of the available extraction phases have a higher affinity toward hydrophobic species; thus, the extraction of polar analytes is often challenging. To overcome this problem, new sorbents suitable for the extraction of polar compounds were presented [19,39]. He et al. [39] used a boronate-based sorbent to extract catecholamine *cis*-diol-containing molecules. The results of evaluating 42 coatings, including commercially available, lab-made, and prototype fibers, were recently published [19]. Three coatings showed good extraction efficiency of polar compounds and balanced coverage of species varying in polarity/hydrophobicity. Absolute recovery allowing reliable detection and quantification was reported even for very polar species with log P about −3.7 (sucrose, glutamic acid). In this study, the feasibility of DI-SPME for global metabolomics studies was presented for the first time.

SPME also offers various selective coatings suitable for the extraction of trace analytes, highly bound drugs (e.g., >99%), or analytes in which a low affinity toward conventional sorbents forecloses their detection. Molecularly imprinted polymers (MIPs) or immunoaffinity sorbents provide an outstanding opportunity to extract these compounds from complex biological matrices. Immunoaffinity sorbents are selective antibodies or antibody-binding mimics immobilized on solid supports. These sorbents were used for the extraction of several drugs from biological matrices, for example, 7-aminoflunitrazepam in urine [40]. MIPs used a target compound or a structurally similar molecule as a template during the polymerization reaction of extensively cross-linked polymers. During this process, specific cavities or recognition sites were formed, which were stereoselective against the compound of interest. Several drug studies were conducted with MIP sorbents in fiber and in-tube formats with excellent precision and limit of detection (LOD). Verapamil [41], propranolol [42,43], pindolol [43], and clenbuterol [44] were among the determined drugs. For more information about various chemical coatings, the reader is referred to other studies [6,45].

3.4 CALIBRATION METHODS

Principles of SPME based on the equilibrium established between the analyte in the sample and the extraction phase have been described. However, for many analytes the time required to reach equilibrium is long and extraction at the equilibrium condition is not feasible. This applies particularly to more hydrophobic compounds, whereas polar compounds reach the equilibrium between sample and fiber coating fast. For these analytes, *in vivo* sampling at equilibrium is still possible [19]. The length of time required to reach equilibrium depends not only on the properties of the analyte but also on the thickness of the extraction phase; the thinner the extraction phase, the faster the achievement of equilibrium. By using thin (<5 μm) PPY coatings, Wang et al. [46] were able to achieve equilibrium within 2 minutes when extracting verapamil from urine samples. Musteata et al. [21] used an artificial circulatory system and showed a significant difference in equilibration time of 2 and 30 minutes when 10 μm PPY and 60 μm PEG fibers were used, respectively, for the extraction of benzodiazepines from blood.

To decrease the time of sampling, preequilibrium extraction is preferable. To ensure precise and accurate quantitative measurements, several calibration methods have been developed [6,47]. Each method has both advantages and disadvantages; therefore, the selection of appropriate calibration should take into consideration all aspects of an experiment. Among the available methods, traditional calibration techniques, that is, "external standard," "internal standard," and "standard addition," can be used for both equilibrium and preequilibrium extractions. For the preequilibrium approach, it should be remembered that the amount of analyte extracted is related to the time when agitation/convection remains constant [6,47].

3.4.1 Traditional Calibration Methods

Biological fluids including whole blood, plasma, urine, or tissues are complex but reproducible; thus, external standard calibration can be successively used for the determination of analyte concentrations in living systems. External standard (calibration curve) is not a labor-intensive method and requires the preparation of a series of standard solutions in the sample matrix and the subsequent analysis of these standards at the same extraction conditions as unknown samples. Once the calibration curve is obtained, the concentration of an analyte in the unknown sample can be calculated with the equation of the calibration curve. External calibration used for *in vivo* SPME pharmacokinetic studies has provided accurate and precise data, which corroborate with conventional methods using blood sample withdrawal [21,23,26]. In cases where a matrix-matched blank sample is not available or the composition of the matrix is unknown, standard addition calibration can be used. In this method, a known amount of standard needs to be added to the sample containing an unknown concentration of the target analyte. In subsequent steps, this mixture is analyzed and the dependence concentration of standard versus response is plotted. The extrapolation of response to zero defines the original concentration of the analyte in the absence of standard. The concentration of the analyte can be calculated from the following equation:

$$C_s = \frac{A_s}{s} \tag{3.6}$$

where A_s is the peak area of analytes in the sample and s is the slope of the line plotted by the peak areas and the concentration of standard added. The main disadvantage of this method is the large number of samples needed to obtain precise results. On the other hand, this calibration enables compensation of the matrix effect. It must also be considered that for solid or heterogeneous matrices, the mass transfer mechanism of standards can be different from that of native analytes, which means that in such cases preequilibrium extraction is not suitable. Another option for calibration is internal standard addition. In this case, the selection of an appropriate compound is crucial to ensure the benefit of the method, that is, compensation of the matrix effect and losses of analyte during sample preparation as well as injection volume when GC or LC is employed for analysis. An internal standard should have very similar properties to the target analyte (partition coefficient, extraction time profile, etc.) to mimic its behavior during extraction, but at the same time the internal standard should be different enough to be well separated chromatographically. When an internal standard is used for the improvement of quantitative data, the optimum condition must apply to both the target analyte and the internal standard to ensure the satisfactory and robust extraction of both compounds [6,47].

3.4.2 Kinetic Calibration Methods

When an experiment is performed at preequilibrium, extraction is a kinetic process. The theoretical model describing the entire kinetic process of SPME based on the diffusion-controlled mass-transfer process was proposed by Ai [48,49]

$$n = \left[1 - \exp(-at)\right]\frac{K_{fs}V_fV_s}{K_{fs}V_f + V_s}C_0 \tag{3.7}$$

where a is a rate constant that is dependent on the extraction phase, HS and sample volumes, the mass-transfer coefficients, the distribution coefficient, and the surface area of the extraction phase.

According to this model, it can be assumed that the amount of analyte extracted at both preequilibrium and equilibrium conditions can be determined using the aforementioned equation, since there is a linear, proportional relationship between the amount extracted and the initial concentration of the analyte in the sample during the entire process of extraction. On the basis of this model, two kinetic calibration approaches were proposed: "in-fiber standardization" technique (on-fiber standardization [OFS]) [50,51] and "standard-free kinetic calibration" [52]. The first of these methods, in-fiber standardization, uses the isotropic relationship between absorption and desorption and requires the preloading of the standard on the fiber to monitor its desorption occurring when the extraction of analyte of interest from the sample matrix is performed. In this case, selection of the appropriate standard is crucial; the standard needs to be characterized by properties similar to the target compound (e.g., isotopically labeled analog) to meet the criteria of extraction/desorption symmetry. When Equation 3.7 is simplified, the amount of analyte extracted at preequilibrium can be expressed as follows:

$$n = \left[1 - \exp(-at)\right]n_e \tag{3.8}$$

where n_e is the amount of analyte extracted at equilibrium. The amount of standard desorbed from the fiber Q can be described as follows:

$$Q = \left[1 - \exp(-at)\right]q_0 \tag{3.9}$$

where q_0 is the amount of the standard preloaded on the fiber.

The isotropy of extraction and desorption can be verified by adding the rate of absorption and the rate of desorption; when the constant a is the same for both processes, the sum of n/n_e (or Q/q_e) must be close or equal to 1:

$$\frac{n}{n_e} + \frac{Q}{q_0} = 1 \tag{3.10}$$

Finally, the initial concentration of the analyte in the sample can be calculated from Equation 3.11:

$$C_0 = \frac{q_0 n}{K_{fs} V_f (q_0 - Q)} \tag{3.11}$$

This calibration method was successfully used for pharmacokinetic studies of benzodiazepines in Beagle dogs [21,27] and rats [23] and carbamazepine (CBZ) in mice [53]. Concentrations of the drugs obtained with in-fiber standardization method showed close agreement when validated against external calibration used for equilibrium SPME and conventional plasma analysis. To simplify the multi-time-point calibration method, the "single-time-point" calibration approach was proposed [54]. Similar to other kinetic calibration methods, this approach is based on the isotropy between adsorption and desorption of an analyte. Since the amount of the compound remaining on the fiber after a certain time of exposure to the sample matrix (Q) is constant, at each time point it can be assumed that Q/q_0 will be constant as long as a is constant (see Equation 3.9), which in practice means constant sampling conditions including temperature, flow rate/agitation, extraction phase, sample matrix, and analyte. It is important to mention that a is concentration independent; thus, even drastic changes observed during PK studies do not affect calibration results. Although the single-time-point calibration method simplifies the calibration process, at least three or four desorption steps must be conducted to achieve a precision that is comparable with other methods [54]. Also, the sampling time (length of extraction and desorption) should be long enough to ensure a reproducibly detectable amount of analyte. The recommended range of n/n_e and Q/q_0 ratios is 30%–70%. For example, for in vivo PK studies of benzodiazepines a 2 minute sampling time was used to meet the requirements.

For situations where isotopically labeled compounds are too expensive or unavailable, the "dominant desorption" approach can be used as an alternative. In this technique, fiber needs to be preloaded with the target analyte at a concentration fourfold higher than the potential amount extracted from the sample. This ensures a constant rate of desorption [55]. Since desorption and extraction must be performed simultaneously with the use of two separate fibers, it is essential to keep the fibers at a distance that prevents contamination of the blank fiber with the analyte desorbed from the preloaded fiber, which can lead to false-positive results. Also, the aforementioned single-time-point approach is applicable when a standard is unavailable. In such a case, the desorption procedure that is needed to determine the rate of desorption/adsorption should be performed with the use of the preloaded target analyte before administering the drug [54]. Another method that requires preloading of only one standard is the so-called "one-calibrant kinetic calibration" [56]. This approach offers the calibration of multiple analytes based on the relation between diffusion of the calibrant and target analyte in the sample matrix. However, the method seems to be very useful especially for multidrug analysis, monitoring of drug metabolism, or untargeted metabolome profiling. Its major limitation applies to diffusion coefficients of the analytes, in particular biofluids and tissues, whose values cannot be easily found in the literature. Kinetic preequilibrium extraction is more suitable for this calibration, as diffusion coefficients for similar molecular weights of analytes are close to each other. When preloading of standard is undesirable because of the risk of negative effects on the organism or because the availability of a suitable standard is limited, standard-free calibration method can be a suitable alternative. This can also be a good solution in fast *in vivo* analysis when loss of the preloaded standard is too small to be detected or when several compounds are measured simultaneously [6,52]. In this calibration approach named "double extraction method," two subsequent samplings need to be performed in short time intervals to ensure the same extraction conditions and a constant sampling rate. The amount of analyte extracted at equilibrium can be calculated from Equation 3.12:

$$\frac{t_2}{t_1} \ln\left(1 - \frac{n_1}{n_e}\right) = \ln\left(1 - \frac{n_2}{n_e}\right) \tag{3.12}$$

where n_1 and n_2 are the amounts of the analyte extracted at the sampling times t_1 and t_2, respectively. When n_e is known, the initial concentration of the analyte in the sample can be calculated from Equation 3.4 or 3.5. Musteata et al. [23] used this calibration for determining the pharmacokinetic profile of diazepam and its two major metabolites in the whole blood of Beagles. The results were compared with in-fiber standardization approach and conventional blood withdrawal followed by plasma protein precipitation. Results obtained with both kinetic calibration techniques showed a good correlation with standard method as well as with parameters published in the literature.

Another standard-free approach, diffusion-based calibration, was initially used for on-site time-weighted-average sampling [6,47,57]. The "diffusion-based interface" (DBI) can be additionally applied to *in vivo* bioapplications such as pharmacokinetic studies or tissue drug distribution and accumulation [58,59]. Equation 3.13 presents the dependence of analyte concentration on molecular diffusion coefficient, the amount of analyte extracted, sampling rate, time, temperature, and extraction phase geometry:

$$C = \frac{n \ln\left(b + \frac{\delta}{b}\right)}{2\pi L t D_L} \tag{3.13}$$

where b is outside radius of the fiber coating, δ is thickness of the boundary layer, D_L is diffusion coefficient of the analyte in the sample matrix, and L is length of the coating [60]. Thickness of the boundary layer can be calculated as follows:

$$\delta = 9.52\left(\frac{b}{Re^{0.62} Sc^{0.38}}\right) \tag{3.14}$$

where *Re* is Reynolds number and *Sc* is Schmidt number. The critical requirement for this model is to keep the thickness of boundary layer constant by strictly controlling sample velocity and convection. It must also be emphasized that the DBI model is suitable only for a short sampling time, where it can be assumed that the analyte concentration in the extraction phase is zero. This ensures a high concentration gradient of the analyte across the boundary layer.

In addition, external calibration for *in vivo* measurement can be performed, as the matrix associated with the leaving system is complex but reproducible. Thus, equilibrium extraction values are identical for the same tissue under homogenized and *in vivo* conditions. Also, it appears that in some cases kinetic preequilibrium could be used because the distribution constant values are similar for both homogenized and live conditions [59]. This similarity could be verified before the *in vivo* experiment.

3.5 DETERMINATION OF FREE DRUG CONCENTRATION, PLASMA PROTEIN BINDING, AND BLOOD-TO-PLASMA CONCENTRATION RATIO

3.5.1 DETERMINATION OF FREE DRUG CONCENTRATION BY SPME

One of the highly critical issues during investigations on living systems is information about the free concentration of a drug or some other biologically active compound, since only this fraction can diffuse through the biological membranes or interact with receptors and induce a physiological/pharmacological reaction. Thus, only free concentration of drugs truly correlates with the activity of pharmaceutics. The ratio of free and bound drugs depends on the equilibrium between the drugs and macromolecules, mainly proteins, which serve as transporters and reservoirs for the drugs and many other exogenous and endogenous compounds. The drug–protein interaction can be determined both *in vitro* and *in vivo*, and this is one of the basic steps in the drug development procedure. However, in clinical practice many factors can affect this interaction and result in the displacement of a drug in its binding site by the competitive binding of coadministered pharmaceutics. Moreover, a number of pathological conditions, including renal failure, inflammation, liver diseases, malignancy, malnutrition, genetic and metabolic diseases, trauma, and surgery, can alter the formation of drug–protein complexes. For this reason, measurement of free instead of total drug concentration, as well as plasma protein binding (PPB), is more valuable from a pharmacological point of view. The protein level, particularly albumin, which is the most abundant transporting protein in mammals, is different in neonates and infants [61], which concludes that therapeutic monitoring of free drug concentration is particularly important for pediatric patients. The changes of free concentration may not be significantly influenced for low protein binding drugs, but in the case of highly bound pharmaceuticals (>90%–95%) even small variations in plasma protein concentration or binding competition may result in a fewfold rise of free fraction.

Recently, a review on free drug concentration monitoring was published [62]. In this review, the author emphasizes the strengths and weaknesses of sampling and sample preparation methods suitable for this purpose. The standard methods used for the determination of free concentration or overall PPB in complex matrices are ultrafiltration (UF), equilibrium dialysis, and ultracentrifugation; however, the latter two techniques are mainly used for research investigation. Affinity chromatography, surface plasmon resonance, capillary electrophoresis, parallel artificial membrane (PAM), calorimetry, and spectroscopy were classified as methods that were rather convenient for isolated protein binding studies. The main disadvantages of these conventional techniques include problems with nonspecific interactions to the membrane, binding equilibrium shift due to the removal of drug molecules by separation devices, low accuracy and precision when low samples volume are used, formation of artifacts, long equilibration times, or low throughput [62].

Whereas PPB using both *in vivo* and *in vitro* techniques is mainly used in clinical practice as a part of PK studies or therapeutic drug monitoring, the determination of drug–protein binding using isolated proteins is one of the first steps in drug development practice. Such investigations bring qualitative as well as quantitative information about drug–protein interaction, show the number of available specific and nonspecific binding sites, give information about the affinity of a particular protein toward a studied drug by calculating binding (affinity) constants, and provide information about possible interactions between different drugs and physiological compounds. Thanks to the diversity of geometries and the variety of calibration methods, SPME is applicable for both purposes: overall PPB and isolated protein studies. The amount of an analyte extracted by SPME fiber is directly proportional to its free concentration in the sample:

$$C_{\text{free}} = \frac{m}{f_{\text{c}}} \tag{3.15}$$

where C_{free} is the free concentration of analyte remaining in the solution, m is the amount of analyte extracted by the fiber, and f_{c} is fiber constant (for liquid coatings it is the product of K_{fs} and V_{f} and for solid coatings the product of K_{es}^{s}, which is the partition coefficient when the sorbent is far from saturation, and S_{e}, which is the surface concentration at equilibrium). As explained in Section 3.2 f_{c} can be easily determined by extraction from standard solutions without a binding matrix (for biofluids, usually a phosphate buffer at pH 7.4 is preferred), because in such matrix-free solutions total and free concentrations are equal. The calculation of fiber constant can be accomplished by multiple or one-point calibration approach [63]. In the former, series of standards at increasing concentration need to be prepared in a binding-free matrix. To eliminate interfiber variability, individual calibration curve should be plotted for each fiber. Although this method is more labor extensive than one-point calibration, the throughput of analysis can be increased when using multifiber SPME devices, which will be discussed in Section 3.10. The slope of linear regression equals the value of fiber constant. The difference in fiber constants allows for determining the interfiber variation within the batch used for the study. If good interprobe reproducibility for the batch is known a priori, a calibration curve can be prepared for this set of fibers. A good regression coefficient obtained for the calibration curve should confirm small fiber-to-fiber variations. As an alternative to multipoint calibration, the one-point approach can be performed. For this purpose, one concentration of the standard should be prepared in a matrix-free solution and utilized for extraction by using all fibers dedicated for the experiment. Standard concentration should be selected from the middle of the calibration curve to ensure good precision and accuracy across the used concentration range. In their binding studies, Vuckovic and Pawliszyn [63] compared the results of multi- and one-point calibration. The values of fiber constant obtained with both approaches were plotted against each other. The analysis performed with Student's *t*-test showed no statistical difference between the data, suggesting that one-point calibration is able to compensate the interprobe variability from the analysis. The slope of regression line, 0.988, also confirms data integrity [63].

3.5.2 Determination of PPB by SPME

When free concentration of a drug is already known, PPB can be determined from the following equation:

$$PPB\% = \frac{C_{total} - C_{free\ plasma}}{C_{total}} \times 100\% \qquad (3.16)$$

where C_{total} is the difference between the initial concentration of the drug in plasma (before extraction) and the amount extracted by SPME [64]. In conditions of negligible depletion, C_{total} can be considered as the initial concentration. In case total (initial) concentration is unknown (when PPB is calculated in clinical samples), C_{total} can be calculated by using the matrix-matched calibration method. Please refer to Section 3.4 describing calibration methods for more details. Musteata et al. [64] performed the plasma binding studies of five well-studied drugs, ibuprofen, warfarin, verapamil, propranolol, and caffeine, that differ in physicochemical and binding properties. Two types of coating, PPY and PDMS, were used for the experiment. Extraction was performed at static conditions at equilibrium (10 and 180 minutes for PPY and PDMS fibers, respectively). All experiments were carried out at 37°C, because temperature significantly influences the binding equilibrium. The authors also investigated the influence of sample dilution and plasma pH on PPB values obtained by SPME measurements. When comparing the results obtained using diluted plasma with undiluted plasma, appropriate correction factors were introduced [64]. Since different methods for pH control reported in the literature resulted in irreproducible PPB, Musteata et al. proposed new solutions to produce stable pH, with no effect on PPB value. These included incubation of plasma with 10% CO_2 or dilution of plasma with isotonic phosphate buffered saline (PBS). As expected, the pH range obtained by using the proposed method was very close to physiological values and, in consequence, did not alter protein binding when compared with the untreated matrix. The effect of plasma samples' dilution on binding is known to be due to the decrease in protein concentration, which can be corrected by using appropriate equations. In the discussed article [64], it was also reported that dilution itself changes the pH and ionic strength of plasma, which likely results in the differences in PPB reported in studies where sample dilution was used. As proved by the authors, for the samples where dilution led to physiological pH (tenfold dilution with PBS) the values of PPB for studied drugs were in close agreement with the literature data regarding undiluted samples. Further comparison of the free concentration [expressed by $-\log_{10}(C_{free})$ and converted from PPB at certain total concentration] obtained from the studies with literature data was facilitated. Despite large variations in binding values being reported in the literature, SPME data corroborated with the previously published results and provided excellent accuracy and precision of the data.

PPB can also be a valuable parameter when muscle protein binding needs to be calculated with no employment of lethal sample preparation methods [29]. A comparison between PPB and percentage of drug binding to muscle proteins (MPB%) can be made when two variables are considered: (1) protein content of different tissues and (2) relative hydrophobicity of the drug [29]. On the basis of MPB%, the total concentration of the analyte in the tissue sample can be obtained according to the following equation:

$$\text{MPB}\% = \left(\frac{C_{total} - C_{free}}{C_{total}} \right) \times 100\% = \left(1 - \frac{C_{free}}{C_{total}} \right) \times 100\% \qquad (3.17)$$

This, in turn, can be used for the evaluation of bioaccumulation factor (BAF), which is commonly measured with the use of lethal sample preparation approaches [29].

3.5.3 Determination of Blood to Plasma Concentration Ratio by SPME

One of the key parameters that needs to be determined in PK studies is blood to plasma concentration ratio. In daily practice, drug concentration is calculated on the basis of plasma analysis, which does not always directly reflect concentration in whole blood. In situations where the drug tends to penetrate inside red blood cells or bind with the erythrocyte membrane, the blood to plasma concentration ratio is greater than unity. In practice, this means that plasma concentration overestimates the drug level in blood. The reverse situation occurs when the drug is excluded from erythrocytes and the blood to plasma concentration ratio is less than unity. Converting data obtained from using conventional sample preparation methods (protein precipitation, SPE, etc.) to blood concentration enables the understanding of the PK and pharmacodynamics of a drug. Moreover, for new drug candidates information about partitioning of the compounds inside red blood cells is necessary to predict potential hematotoxicity. The most current methods for the determination of blood to plasma concentration ratio *in vitro* and *ex vivo* are relatively time consuming and difficult to be automated. SPME offers an easy approach to determine this parameter: the spiked blood sample should be divided into two aliquots, and subsequently plasma should be separated from one of these aliquots. In the next step, extraction with SPME should be performed on both samples under the same conditions. Finally, when the amount of drug extracted is calculated on the basis of the obtained results, the blood to plasma concentration ratio can be determined as the ratio of the slopes of calibration curves plotted for whole blood and plasma, respectively. To increase the throughput of the analyses, parallel SPME experiments can be performed on both matrices with the use of automated systems (see Section 3.10). It is also worthwhile mentioning that SPME extraction provides a more efficient sample cleanup when compared with conventional methods (no matrix effect was reported for *in vivo* blood sampling by SPME [53]), which can be particularly important for whole-blood studies, where greater complexity of the matrix can provide ion suppression when a mass spectrometer is used for analysis.

3.6 *IN VIVO* PHARMACOKINETIC STUDIES

3.6.1 MONITORING OF RAPID CONCENTRATION CHANGES AND HETEROGENEOUS DISTRIBUTION OF DRUGS: TIME- AND SPACE-RESOLVED SAMPLING

Many compounds, which show their great therapeutic activity *in vitro*, have to be rejected in later stages of investigations due to their poor pharmacokinetic properties. A drug candidate must be characterized by good bioavailability and a desirable half-life ($t_{1/2}$), which are dependent on a number of factors during ADME processes; thus, *in vitro* data cannot be extrapolated to *in vivo* conditions with sufficient confidence and *in vitro* drug activity must be verified in living systems. Before a drug candidate is subjected for human trials, it needs to be tested in animal models [1]. To obtain reproducible and good-quality data, the method used for sample collection from a live organism must ensure minimal disturbance for the studied system [4]. These needs can be addressed by the use of microsampling and microextraction methods. Their feasibility in pharmacokinetic applications was considered by Musteata in his recent review [4]. Among the discussed microsampling methods, the two most common techniques of UF and MD are presented. UF used for *in vivo* sampling is a relatively easy approach that overcomes the problem of complex calibration and obtains analytes without sample withdrawal; however, it requires exerting negative pressure, which can alter the physiological balance between various tissue compartments and lead to the damage of cells and capillaries. This may result in the overestimation of tissue drug concentration. UF probes must be implanted in tissues with high fluid turnover so that fluid loss can be tolerated. For this reason, the method is not appropriate for small animal studies. Also, UF seems to not be a suitable method for highly protein bound drugs, due to the clogging of the device's membrane [4]. Although MD is a widely accepted method for both *in vivo* and *in vitro* analyses, it also has some drawbacks. The first disadvantage, which limits its applicability for quantitative studies, is related to complicated calibration and the necessity of using pumps to facilitate the flow of perfusates. The second disadvantage arises from the fact of dynamic sampling representing time-average concentrations, which may not be the most informative especially in cases of short-lived compounds [4]. The complexity of all contributing factors in drug ADME demands that researchers evaluate the concentration profile of drugs and metabolites versus time as well as determine their bioaccumulation and toxicity in different organs. Since drug and metabolite concentrations may change rapidly, the sampling method should have good temporal resolution, which allows the monitoring of these changes in real time. On the other hand, heterogeneity of different tissues corresponds to different distribution coefficients and partitioning ratios; thus, the sampling method should be able to distinguish gradient concentrations of the analyte between particular zones. In two subsequent subchapters, examples using time- and space-resolved SPME are described in detail. The following paragraph discusses fundamental considerations.

Temporal resolution of SPME (TR-SPME) is defined as the capacity to accurately determine sample concentrations at specific points of the time continuum and to clearly resolve two different concentrations in rapid succession [65]. In practice, this means that the sampling time must be shorter than the interval between the time points and minimum sampling time (t_{min}) and must be sufficient to extract the amount of analyte needed to provide a quantifiable and reproducible instrument signal. A number of factors affect temporal resolution ($1/t_{1/2}$): instrumental sensitivity, Q_{in}; sample conditions (initial concentration C_0, concentration change rate b, agitation conditions, etc.); and fiber properties (coating dimension and chemistry determining partition coefficient). The relationship between these factors and time is described by the following equation:

$$t_m = \frac{\sqrt{C_0^2 V_f^2 K_{fs}^2 a^2 + 2abV_f K_{fs} Q_{in}} - C_0 V_f K_{fs} a}{abV_f K_{fs}} \tag{3.18}$$

where a is a rate constant that is dependent on the extraction phase, HS and sample volumes, the mass-transfer coefficients, the distribution coefficient, and the surface area of the extraction phase (see Equation 3.7). The dependence between minimum sampling time and minimum fiber length that can extract a quantifiable amount of the analyte (l_m) can be expressed as follows:

$$t_m l_m = \frac{Q_{in}}{SK_{fs} C_s a} \tag{3.19}$$

where S is the surface area of the fiber. During the selection of appropriate sampling duration, it should also be considered that an overly lengthy sampling time can affect time resolution and consequently overestimate the results by providing time-average instead of real-time concentration. Maximum sampling time (t_{max}) can be calculated with the following two equations:

$$t \le \frac{0.2}{a} = 0.067 t_e \tag{3.20}$$

and

$$t \le \frac{C_0}{10b} \tag{3.21}$$

where t_e is equilibrium time. Whichever time is shorter based on the aforementioned equations should be selected for the experiment. In fact, the authors also draw attention to the situation in which the minimum sampling time calculated on the basis of Equation 3.18 or 3.19 is longer than the maximum sampling time chosen on the basis of the aforementioned equations. In such a case, a modification in method design, that is, increasing the extraction phase area and volume or using a more sensitive instrument, should be employed [65]. The experimental verification of theoretical consideration of TR-SPME was conducted by monitoring the dynamic changes occurring during the binding of CBZ and fluoxetine (FLX) to bovine serum albumin (BSA) [65]. With the use of a system consisting of an infusion pump and a

Hamilton syringe, which delivered a constant flow rate of drugs to a solution containing drugs and BSA (equilibration between drugs and protein was established previously during overnight incubation), Zhang et al. [65] compared changes in free concentrations of drugs obtained by SPME and MD. The sampling points for SPME were 10, 20, 30, 45, 65, 90, 120, 150, 180, 210, and 240 minutes, whereas simultaneous MD sampling was performed at the first six time points (higher concentration of free concentration) to minimize the disruption of binding equilibrium. To evaluate the appropriate sampling duration, 2, 5, and 10 minute durations were investigated. The precision of the developed mass-transfer model for SPME was confirmed by obtaining very close agreement between theoretical calculations and experimental data. The authors discussed in detail the influence of each sampling duration time on the results obtained at particular sampling points for the dynamic system. Validation against MD showed consistency between the results obtained for CBZ by both methods, whereas the determination of FLX concentration was not possible with MD because of the high hydrophobicity of the drug [65].

It has already been demonstrated that there is a negative correlation between minimum sampling time and minimum coating length; thus, none of these two parameters can be considered separately. Both t_m and l_m can be reduced when a highly sensitive instrument is used for detection and/or when the affinity between coating and analyte is high, which is expressed by a high K_{fs} value. The distribution of analyte in a heterogeneous sample resulting in a concentration gradient is dependent on the diffusion length of the analyte molecules within the sample matrix. This parameter can be described by the solution in one dimension of the diffusion equation (Fick's first law of diffusion):

$$x^2 = 4Dt \qquad (3.22)$$

where x is a distance of molecule migration; t is the time of duration of molecule migration via diffusion; and D is the diffusion coefficient, which is defined by the amount of substance diffusing across a unit area through a unit of concentration gradient per unit time [66]. To translate this theory into practice, the distance of molecule migration must be shorter than the minimal fiber length to allow the fiber to cover the concentration in the representative microregion. To verify the proof of this concept, space-resolved tissue sampling was studied using a multilayer plant (onion bulb), as well as fish muscle and adipose tissue simultaneously [67]. In their experiment, the authors used a specially designed, segmented coating, enabling simultaneous extraction from two closely located regions. Coatings immobilized on stainless steel wire (C18 and PDMS) were cut into 1 mm long segments separated by 5 and 4 mm spaces (for C18 and PDMS, respectively). Before extraction, in both cases fibers were preloaded with an appropriate deuterated compound to obtain accurate and precise quantitative analysis. For the onion bulb study diazepam was injected into the center of an onion bulb from the stem side 4 hours prior to sampling, whereas for the fish study rainbow trout were exposed to nine pharmaceuticals and other bioactive compounds at different lipophilicities for 8 days. SPME data was validated against two conventional methods: MD and liquid extraction (LE). Two holes made with a 22 gauge needle were subjected to segmented SPME fiber and MD probe. After 5 minutes, both SPME and MD probes were removed from the plant; the SPME fiber was quickly rinsed with water and it underwent desorption. To distinguish between analytes extracted by two separate segments of coating, desorption was conducted in two steps: in the first step the volume of desorption solvent was adjusted to cover only the lower part of the sorbent, and in the second step the level of fresh portion of the solvent was increased to desorb diazepam from the upper segment. Five minutes sampling time was selected on the basis of *in vitro* evaluation of diffusion-controlled mass transfer; the authors investigated migration velocity through different layers of the matrix at two temperatures to ensure quantitative analysis of the drug extracted during preequilibrium sampling. For fish sampling, an SPME fiber was placed in the tissue under anesthesia with the use of a needle to guide the way that the first segment of the coating was embedded inside the muscle, whereas the other was located in adipose fin tissue. Subsequently, the needle was removed and the fish was placed in freshwater for 8 minutes; thus the total sampling time was 10 minutes. Similar to onion sampling, fibers were quickly rinsed with water and subjected to two-step desorption. To compare the results of SPME study with MD and LE data for onion and fish sampling, respectively, the analyte concentrations in these matrices were calculated. Plant studies showed a higher level of diazepam concentration in the inner layer (closer to the site of injection) compared with the distant layer. The values of concentration calculated by MD were twofold higher than those in the distant layer and twofold lower than those in the inner layer. This suggested that the concentration obtained from the MD experiment was weighted average because of the long size of the probe. Therefore, to verify this theory the authors performed additional experiments with a longer SPME fiber, covering the same distance as the MD probe, in parallel to MD sampling. This experiment showed that concentrations of diazepam were in close agreement between the two methods. The studies revealed that SPME offers a quantitative capability that is comparable to MD but provides better spatial resolution. Whereas the results of plant studies were validated against MD, the results of *in vivo* preequilibrium fish sampling were compared with *ex vivo* equilibrium SPME and LE method. On the basis of free concentration in both tissues obtained from the SPME experiment, the authors calculated the distribution coefficient between fin adipose tissue and muscle (K_{am}^{true}), which were parameters that reflected the relative bioavailability of the pharmaceuticals. From LE data, the authors calculated the apparent distribution coefficient, which was expressed as the ratio between total concentration of the analyte in the adipose tissue and that in the muscle ($K_{am}^{apparent}$). The calculated free concentration from *in vivo* data corresponded to the SPME equilibrium extraction, validating the *in vivo* methodology that was used. It was also found that K_{am} values calculated on the basis of both SPME and LE correlated with log (K_{ow}) values, indicating greater partitioning of higher K_{ow} analytes into the adipose tissue, which is relatively lipid rich when compared with muscle tissue. With both of the used techniques, five out of nine compounds were detected in fish, whereas the remaining four were only found in water. As explained, this could be caused by their lack of bioconcentration potential [66].

In their following study, Zhang et al. used the same experimental design but with minor modification to increase the throughput of desorption. After extraction, the upper segment of the coating was moved to the top of the wire, which was subsequently cut into two equal pieces, thus enabling simultaneous desorption of the analyte from all segments [67]. To minimize the influence of interfiber variation on the overall results of *in vivo* experiments, lorazepam was used to evaluate probe-to-probe reproducibility. To monitor the bioconcentration of pharmaceuticals in semisolid tissues, distribution constant (K_{fs}) between the fiber and different tissues is required. To calculate this parameter, the total concentration of the analyte in the tissue needed to be known. In this experiment, the authors used total concentration of the analyte in the sample calculated from the LE *in vitro* experiment, because this conventional method was used for validation purposes. However, it must be emphasized that the nonlethal nature of SPME can be independently used in bioaccumulation or drug distribution studies, and in such cases gel medium, which mimics real tissue, can be used to calculate distribution constant and/or diffusion coefficient as demonstrated in the previous study on diazepam distribution in onion [66]. It was also proposed that the development of a database containing information about such parameters would significantly facilitate SPME experiments and increase throughput the analysis.

3.6.2 Blood Sampling with *In Vivo* SPME

SPME is a relatively new technique in the *in vivo* pharmacokinetic field, although the first *in vivo* PK study dates back to 2003. The main objective of this experiment by Lord et al. [24] was to design a new SPME probe for the *in vivo* analysis of drug concentration in the circulating blood of living animals. The introduction of PPY as the extraction phase addressed some limitations of the sorbents available at the time, which were suitable for LC applications, that is, Carbowax/Templated Resin (CAR/TPR). The PPY coating was prepared by electrodeposition on the stainless steel wire acting as a supporting fiber core from aqueous solution. Detailed information on probe preparation can be found in the original study [24] as well as in the recently published protocol on *in vivo* SPME sampling [68]. The main advantages of PPY fibers were full compatibility with solvents used in LC applications; good reproducibility; feasibility to capture semivolatile, nonvolatile, and more polar compounds; and the thin extraction phase enabling the achievement of equilibrium between the bulk of sample and sorbent in a short time. However, the reduced thickness of coatings showed some drawbacks including possible competition between analytes and displacement in adsorption sites and limited applicability for the drugs with a high degree of protein binding (low concentration of the free fraction). Prepared fibers underwent sterilization by autoclaving at 121°C and 13.5 psi for 20 minutes in PBS buffer. A subsequent preconditioning procedure was also performed under sterile conditions. The studies were conducted on three male Beagle dogs. SPME probes were introduced into the circulation via standard in-dwelling catheters (one cephalic and one saphenous) used in veterinary practice. At each time point, authors placed three PPY probes into the cephalic vein to obtain average data. This was dictated by the high interfiber variation reported on the basis of *in vitro* evaluation. The experiment was performed within a total time period of 3 weeks. Individual data points were taken in 1-week time intervals for all three animals. The same set of probes was reused in each dog, unless damage was noted and fiber replacement was required.

Simultaneously with SPME, conventional blood sampling was used (1 mL volume was withdrawn from saphenous and periodically from cephalic catheter before SPME probes were removed). These blood samples were used for validation purposes and were subjected to conventional analysis (protein precipitation) of benzodiazepine concentrations as model drugs. The extraction time of 30 minutes was chosen on the basis of an *in vitro* experiment where the worst-case scenario of sample agitation, that is, static mode, was introduced. This approach ensured the achievement of equilibrium conditions despite possible fluctuations of blood flow during the *in vivo* experiment. Twenty seconds was selected as the time when maximum desorption was observed. This allowed for obtaining desorption efficiencies of 94% for diazepam and nordiazepam and 92% for oxazepam and lorazepam. Interestingly, it was observed that there was no difference in the amount extracted over the equilibration time between static versus agitated mode. As explained, the influence of agitation on the thickness of boundary layer and consequently on the rate of extraction might occur under 5 minutes and could not be seen, since according to theoretical calculations equilibrium for this probe should be achieved in less than 1 minute [24]. Another important finding was the influence of fiber preconditioning on extraction time profile. Lord et al. [24] reported that for nonconditioned probes equilibrium was reached for the first time within 5 minutes and, subsequently, the amount of drugs started increasing after 30 minutes of extraction from the sample matrix. The described phenomenon was not observed when the coating was soaked in 50% methanol with 0.1% formic acid for 30 minutes or rinsed with 100% methanol for 15 seconds prior to extraction. This suggested the occurrence of the swelling effect and indicated the importance of careful evaluation of preconditioning conditions during the method's optimization step, which successfully prevented problems with swelling. Moreover, as reported, the selection of a proper pretreatment method significantly increased (ca. 2.5-fold) the sensitivity of the assay when 30 minutes extraction time was applied [24]. Selection of equilibration conditions allowed the authors to achieve high sensitivity (LOD < 0.4 ng/mL for all studied drugs) and showed the possibility of short preequilibrium extraction. As the author explained, diazepam was selected as the model for this study because the metabolism of this drug is well known and understood. This allowed not only the validation of data obtained with SPME probes against conventional assay but also the verification of the pharmacokinetic results with the literature. Authors reported slight differences between plasma and whole-blood concentration, although this was explained by the blood to plasma partitioning ratio being different for diazepam than one; the agreement between both assays improved when the appropriate conversion factor was used. The comparison of final SPME data with plasma analysis demonstrated good correlation within experimental error and consistency with previously published results. This first *in vivo* pharmacokinetic experiment with SPME presented the three most fundamental advantages of the method: (1) extraction of small molecules using a blood draw–free approach, (2) simultaneous monitoring of the drug and its metabolites in a living system after extraction on one sorbent, and (3) the possibility of calculating both total and free concentrations of analytes using the appropriate calibration method. Furthermore, one of the method's developmental steps was to optimize conditions of SPME probe storage, which would allow performing analytical procedures in a different facility.

It was found that after sampling and a quick rinse of probes in water (10–20 seconds) followed by air drying, fibers could be stored at room temperature for 1–24 hours without significant loss of analytes. This indicates that the use of SPME for *in vivo* sampling eliminates problems of transportation and storage of biohazardous materials and minimizes contact of analytical personnel with biological samples.

Questions and suggestions raised in this preliminary study were further investigated by many researchers using *in vivo* as well as *in vitro* approaches. As explained previously, long extraction times can affect time resolution of the analysis and for this reason may not be suitable for monitoring short-lived species or for applications where determination of rapid changes in drug concentrations is of interest. A follow-up *in vivo* experiment in Beagles addressed some of the drawbacks reported in the previous experiment, such as long sampling time and impractical use of interface for manual desorption [21]. Musteata and Pawliszyn [69] proposed a new sampling device based on hypodermic needles (Figure 3.4).

Two types of biocompatible coatings were used for the experiment: PPY with hydrophilicity improved by the addition of triethylene glycol and PEG bound to octadecyl silica (C18). SPME probes were assembled in 2-inch-long hypodermic needles so that the coating remained inside the needles between subsequent extractions and was exposed to the blood flow only for the short time of sampling. The hypodermic needle served as fiber protection and enabled easy access to the vein by piercing the PRN adapter, which sealed the catheter. This needle assembly became a prototype of the currently available commercial SPME device. After the extraction procedure, fibers were quickly rinsed with water and stored on dry ice until the following day, when the fibers were desorbed in 20 μL of acetonitrile: water mixture placed in plastic inserts. This approach was more practical than having a manual desorption interface mounted in the position of the sample loop on a six-port injection valve of an LC instrument, which was used by Lord et al. [24]. In their studies, Musteata et al. [21] were the first to use the on-fiber standardization calibration method, as discussed in Section 3.4.2, to enable precise and accurate measurement at preequilibrium (30 seconds and 1 minute for PPY and PEG fibers, respectively). To compare the results of preequilibrium kinetic calibration with the one obtained at equilibrium conditions with external calibration, authors performed *in vitro* experiments with the use of an

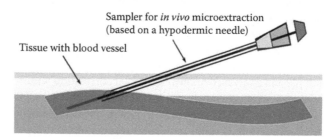

FIGURE 3.4 **(See color insert.)** Schematic representation of the SPME device for *in vivo* monitoring of drug concentrations in the veins of large laboratory animals (Beagles): total length of the device is 8 cm. (Reprinted from *Journal of Biochemical and Biophysical Methods*, 70, Musteata, F. M., and J. Pawliszyn, *In vivo* sampling with solid-phase microextraction, 181–193, Copyright (2006), with permission from Elsevier.)

artificial circulatory system. All blood samples used for *in vitro* experiments were incubated in 10% CO_2 to create pseudophysiological conditions. These *in vitro* data revealed that the equilibrium time for PPY fibers was significantly shorter than that for PEG (2 vs. 30 minutes). This was not surprising since PPY was a thin solid adsorbent coating with low capacity, whereas PEG was characterized as an absorbent coating with high capacity and a thicker extraction phase (in this case, six times). Similar to what occurred previously, the drug concentration was calculated for total as well as free fractions. Based on free concentration values, authors observed fast *in vivo* conversion of diazepam to oxazepam and their further triple- and double-exponential elimination. In contrast, for nordiazepam constant free concentration during the first hour followed by single-exponential elimination was reported. As in the previous studies by Lord et al. [24], SPME data was validated against the protein precipitation method performed on the samples collected with traditional blood withdrawal. However, in this case the authors did not use a correction factor to improve the correlation between these two data because of the great variability between the values of the blood to plasma distribution ratio reported in the literature. In spite of this, all presented results corroborated very well with those of conventional analysis [21].

To evaluate the influence of interfiber variation and blood flow velocity variation on the SPME process, Schubert et al. [22] used an artificial vein system consisting of commercially available heart and lung machine components. Analysis of an antibiotic drug, linezolid, was conducted with the use of a drug concentration corresponding to the therapeutic range in humans. Blood properties (lack of hemolysis, no increase in lactate dehydrogenase, and constant composition of the matrix) were monitored during the entire experiment to ensure physiological conditions and clinical relevancy. Authors reported no influence of flow rate on the amount of drug extracted at the used equilibrium time. In turn, the time required to reach equilibrium was shorter in circulating blood versus off-line (static) conditions. This outcome was expected, since blood circulation in living systems provides good agitation in decreasing the boundary layer and enables fast equilibration. The used blood flow rate in the studied case mimicked that of *in vivo* conditions, and the equilibration time for linezolid was less than 5 minutes [22]. Similar to previous studies with PPY probes, high interfiber variation was noted, particularly at the upper range of concentration. This issue was overcome by reusing the probes, although in clinical practice, where strict hygienic rules apply, such a solution would not be acceptable and single-use fibers with good fiber-to-fiber reproducibility would have to be used. Aside from the technical value of these studies, the authors drew attention to an important point regarding the clinical applicability of SPME: they explained the selection of the antibiotic for their studies by the fact that consistent control of antibiotic drug levels requires frequent blood withdrawals, which can cause hypovolemia and anemia in critically ill patients, babies, and infants [22]. It was also mentioned that in such cases, to avoid blood withdrawal MD could be used as a method similar in principle, although it seems to be more invasive and, accordingly, less suitable for the extraction of antibiotic or antimycotic agents from blood.

Dog studies continued with the aim of expending the kinetic calibration approach to the entire PK profile of benzodiazepines over the time of 8 hours and correlating whole-blood SPME data with whole-blood analysis and conventional sample preparation methods [27]. In fiber standardization calibration technique used as a proof-on-concept at two sampling points for fast *in vivo* sampling by Musteata *et al.* [21] was explained by Zhang et al. [27] in further detail and applied in all animals at all data points. It was previously mentioned that diazepam and its metabolites partition between plasma and red blood cells: Lord et al. [24] in their calculations used a conversion factor to correlate results obtained from whole blood by SPME and plasma protein precipitation data, and Musteata et al. [21] compared SPME data with conventional assay without taking into consideration the red blood cell/plasma partitioning ratio. In the currently described work [27], Zhang et al. calculated the binding constant to the whole-blood matrix, K_b. The constant reflects the binding of the drug to all matrix components including blood cells, proteins, and other macromolecules. The binding constant can be calculated according to Equation 3.15:

$$K_b = \frac{C_{free}}{C_0} \tag{3.23}$$

Based on the relation between initial concentration of the drug in the given matrix and the product of KfsVf (see Section 3.2) K_b can be expressed as follows:

$$K_b = \frac{slope_1}{slope_2} \tag{3.24}$$

where $slope_1$ and $slope_2$ are products of $K_{fs}V_f$ calculated for the extraction of standards from buffer and whole blood, respectively. They found the partitioning to plasma to be 59%, 63%, and 68% for diazepam, nordiazepam, and oxazepam, respectively. In parallel to preequilibrium extraction with the use of the OFS calibration method, the authors used equilibrium extraction with external calibration as a comparative SPME approach as well as acetonitrile addition as a traditional technique. However, considering the results obtained for blood to plasma partitioning, the authors correlated SPME results with the conventional method conducted on whole-blood samples. The correlation coefficient ranged between 0.97 and 0.99, which confirms the excellent agreement between all three methods.

Extensive evaluation of the performance of new PEG-C18 fibers used in the study and comparison of their properties with those of PPY probes was studied *in vitro* using a flow system and *in vivo* in Beagles by Es-haghi et al. [26]. Authors compared extraction efficiency, linearity, and LOD for diazepam and its metabolites for a wide range of concentrations. On the basis of an extraction time profile in all studied matrices (whole blood, plasma, and PBS), the authors selected equilibrium as an extraction mode. To decrease interfiber variation during analysis, they preselected fibers due to the results of extraction from one standard solution. Fibers showing low fiber-to-fiber relative standard deviation (RSD) were chosen for further experiments. Analysis of the extraction time profile revealed that all three studied drugs achieved equilibrium faster at higher flow rates (320 vs. 75 mL/min), although flow rate did not change the amount of analyte extracted at equilibrium, which agreed with the results presented by Schubert et al. [22]. The time required to reach equilibrium was longer in the case of both biological fluids when compared with the buffer due to the differences in viscosity; but because even for these matrices equilibration was achieved fast enough (5 minutes) to perform *in vivo* analysis, this mode was selected for the studies. The linear dynamic range for all three analytes in matrix-free solution (PBS) was 1–2000 ng/mL, in plasma 3–2000 ng/mL, and in whole blood 4–5000 ng/mL. The lower sensitivity observed for plasma and blood is a result of the high protein binding of diazepam, nordiazepam, and oxazepam, although for all matrices the dynamic range covered the expected concentration ranges for *in vivo* studies, including very high and very low concentrations of the drug and its metabolites during all parts of pharmacokinetic studies. This result shows the advantage of using high-extraction-capacity coatings and high sensitivity of the assay when equilibrium extraction is used. Although fast preequilibrium extraction offers more convenience due to a shorter time of sampling and allows to obtain more data points and, consequently, gives more detailed information about drug metabolism, for the first and last sampling time points equilibrium extraction may be more suitable in enabling quantitative analysis of both parent drug and the metabolites at even very low concentrations. Another key advantage of absorption coatings is the lack of competition effect, which could occur in complex matrices such as plasma or whole blood. Results reported by Es-haghi et al. [26] showed no competition among the drug, its metabolites, and/or matrix components even at high concentrations. As previously explained, most SPME fibers have higher affinity toward hydrophobic compounds, which have high free concentrations in living systems, and lower affinity toward hydrophilic species, which form complexes with macromolecules. Thus, their free concentrations remain low. Since SPME coatings interact with free molecules in solution, it is unlikely that the overall uptake of the compounds leads to fiber saturation or competition effect.

The dog studies showed excellent suitability of SPME to *in vivo* pharmacokinetic analysis; however, PK is mainly performed in small rodents, that is, rats and particularly mice, due to the accessibility of various strains and genetic mutants. Nevertheless, the use of small animals brings scientists to a very challenging issue in PK rodent studies, which is limited in sample volume. Regulatory agencies, such as the Animal Care and Use Committee, approve the withdrawal of blood samples less than 20% of their total volume, which is about 1.85–2 mL per 25 g mouse. To obtain a full PK profile with the use of traditional sample preparation techniques requiring blood withdrawal, a large number of animals need to be sacrificed. Aside from ethical considerations, this can be a problem when precious and rare species are of interest. Alternatively, several catheterized rodents dedicated to serial blood collection can be used in experiments; however, in such cases replacement of blood volume may be necessary, because of high risk of tissue anoxia or hemorrhagic shock [70]. This in turn can result in sample dilution and as a consequence lead to erroneous data interpretation. A lot of effort is being put into inventing a method that would allow reducing the number of animals used in the drug discovery process. Among the proposed approaches partially addressing this issue are "cassette dosing" (CD), "cassette-accelerated rapid rat screen" (CARRS), and "snapshot PK." CD utilizes the simultaneous administration of several compounds to each animal used in the study; thus evaluation of PK profiles can be obtained for a few compounds in one series of experiments, but the high probability of drug–drug interaction at all stages of ADME may lead to inappropriate results [71]. Also, the preparation of usable dosing formulations is complex and challenging. CARRS eliminates the problem of drug–drug competition, but it is applicable only to rats and gives partial PK parameters over a period of 6 hours; however, it is still sufficient to differentiate the drug candidates that exhibit poor bioavailability [71]. Another method used in both rat and mouse PK experiments is snapshot PK [2]. With this protocol, the number of animals used is two per compound and four 50 μL blood samples are collected from each animal over a time period of 5 hours. Although this approach allows for significant time saving and increases the throughput of the analysis, it does not provide full PK parameters. Despite the proposed dosing schedules, all proposed techniques require sample removal, which is followed by conventional sample preparation and extraction techniques.

To make SPME an appropriate tool for PK rodent studies, the sampling device and method of probe assembling required modification to become applicable to such experiments. First results of PK studies in rats were published by Musteata et al. [23] in 2008. Because the previously designed SPME sampling devices based on hypodermic needles were not suitable for direct insertion into blood vessels, the authors used a custom-designed sampler consisting of a flexible thin wire coated with a biocompatible extraction phase and housed inside two concentric hypodermic tubes. The role of the inner tube was to reinforce the wire, whereas the outer tube was used to pierce the sampling interface and to protect the coating. A special interface was built to access the blood vessel. Musteata et al. evaluated the feasibility of three different materials for interface construction: Teflon, stainless steel, and polyurethane (Figure 3.5). The latter was found to be the best as it did not induce blood clotting in the interface.

During *in vivo* experiments, it was found that free flow through the interface did not provide sufficient pressure to keep blood movement through the interface for an extended period of time; thus, a syringe pump was used to draw and push blood through the interface. The general design of the interfaces is based on a "Y" adapter,

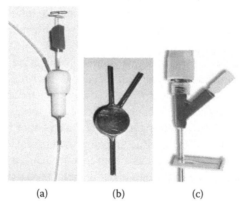

(a)   (b)   (c)

FIGURE 3.5 Interfaces for *in vivo* SPME: (a) Teflon Y adapter, custom made; (b) stainless steel Y adapter, custom made; (c) BD Saf-T-Intima™ closed intravenal catheter system. (Reprinted from *Journal of Pharmaceutical and Biomedical Analysis*, 47, Musteata, F. M., I. de Lannoy, B. Gien, and J. Pawliszyn, Blood sampling without blood draws for *in vivo* pharmacokinetic studies in rats, 907–912, Copyright (2008), with permission from Elsevier.)

in which the lower tube ("blood in") was attached to the carotid artery catheter prior to dosing with the small metal connector. Catheters were surgically inserted into the neck of three animals used in the experiment. The upper tube of the interface ("blood out") was either recirculated to the carotid artery catheter, which allowed automatic return of arterial blood with the use of heart action as a pumping force, or was connected to the syringe (Figure 3.6). The push/pull cycle with the use of the syringe allowed precise control of the flow rate through the interface, thus enabling accurate and repeatable results when kinetic calibration was employed. The blood flow induced by the syringe had a slower than normal flow rate in the rat carotid artery and provided minimal disturbance for the organism.

The authors investigated the PK profile of diazepam and its two metabolites nordiazepam and oxazepam as model compounds with well-known and documented metabolism [23]. SPME was performed at equilibrium (2 minutes) and preequilibrium (40 seconds for standard on-fiber calibration, and 20 and 40 seconds for double extraction method). After the extraction, fibers were quickly rinsed with water and stored at –20°C until the next day for analysis. Desorption of analytes from probes was completed within 1 minute, so the total time of sampling and sample preparation was ≤3 minutes. To validate the method, blood samples were removed at each time point for conventional analysis. All procedures were carried out in sterile conditions. SPME devices were autoclaved at 121°C and 14 psi for 30 minutes, and the probes were single use. The linear range, sensitivity, and

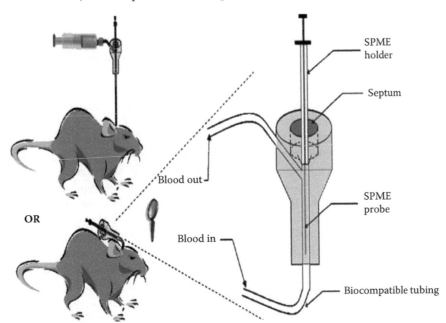

FIGURE 3.6 *In vivo* SPME study on rats: placement of SPME devices and interface connection to the carotid artery. (Reprinted from *Journal of Pharmaceutical and Biomedical Analysis*, 47, Musteata, F. M., I. de Lannoy, B. Gien, and J. Pawliszyn, Blood sampling without blood draws for *in vivo* pharmacokinetic studies in rats, 907–912, Copyright (2008), with permission from Elsevier.)

reproducibility of the assay were investigated in whole blood and PBS to calculate free concentrations of the compounds. All samples used for *in vitro* method development were incubated at 38°C in 10% CO_2 to mimic *in vivo* conditions. The obtained linear dynamic range (3–750 ng/mL for total concentration and corresponding 0.18–48 ng/mL for free concentration) allowed the monitoring of changes in drug and metabolite profiles over a time period of 6 hours and the estimation of full pharmacokinetic parameters using both equilibrium and kinetic calibration methods. For diazepam the authors used a two-compartment model to analyze mean concentration change versus time, whereas for nordiazepam and oxazepam a noncompartmental model was used. In the case of all three compounds, total blood concentration calculated by SPME showed excellent agreement with plasma precipitation data (correlation coefficient: 0.95–0.99). No correction factor was applied to convert plasma to blood concentration, because the blood to plasma partitioning ratio in rats was reported to be close to one. Also, all pharmacokinetic parameters determined for diazepam, that is, area under the plasma concentration versus time curve, distribution and elimination half-lives [$t_{1/2}$ (α) and $t_{1/2}$ (β), respectively], total body clearance (CL), area under the first moment curve, mean residence time, and steady-state volume of distribution (V_{ss}), were very close to the values obtained from conventional analysis and those reported in the literature. Additionally, with good agreement of total concentration, SPME demonstrated a great advantage over the conventional approach by offering a measurement of free concentration, which is a valuable parameter, especially in the case of highly bound drugs such as benzodiazepines.

Successful studies with rats motivated researchers to further miniaturization of the sampling device and pharmacokinetic studies in mice [53]. In one study, Vuckovic et al. [53] monitored the concentration profile of CBZ and its active metabolite, carbamazepine-10,11-epoxide (CBZEP). CBZ was selected for the study because it is known for interpatient variation in its free concentration, with the total concentration remaining stable at the same time. Thus, therapeutic monitoring of CBZ methods suitable for the measurement of free drug concentration is required. In their study, the authors used a sampling device similar to the one previously described by Musteata et al. [23] but smaller in size to reduce the total internal volume of the interface to 125 µL. In this study, a prototype of a new biocompatible octadecyl silica (C18) coating, which are currently commercially available, was used. As mentioned in Section 3.3 these probes have very good interfiber reproducibility, which allows for a single use of the probe but still requires a long equilibration time. In their parallel studies, Vuckovic et al. evaluated the performance of C18 probes and determined the time required to reach equilibrium at static condition (the worst-case scenario) as ≥1200 minutes [72]. Thus, the time of preequilibrium extraction selected for *in vivo* analysis was 2 minutes. Push/pull action with a syringe (approximately 10 cycles per 2 minutes of sampling) was used as the agitation method. Sampling was performed over a period of 4 hours. To enable precise quantification during preequilibrium extraction, the standard on-fiber calibration approach was used. To ensure a reproducible procedure of standard loading, CBZ-d10 was preloaded on the fiber simultaneously with preconditioning of the coating. After sampling, fibers were immediately rinsed and then stored on dry ice or in the freezer (–20°C) until analysis. In the laboratory, analytes underwent 60 minutes desorption with vortex agitation. The precision, accuracy, and linear dynamic range of the method were evaluated *in vitro* using the same sampling procedure (2 minutes extraction time, syringe agitation) that was used during the *in vivo* experiment, with the exception that the sampling device was placed inside the vial with mouse whole blood instead of being connected to the catheter. The obtained results of this validation met all the criteria of a bioanalytical assay. The linear ranges in whole blood were 1–2000 ng/mL and 1–150 ng/mL for CBZ and CBZEP, respectively, which covered the expected range of concentration of both targeted compounds, although as reported by the authors linear dynamic range was not limited by the extraction capacity of the fiber and could be extended if required [53]. Good time resolution is a critical factor that allows the monitoring of rapid changes of drug/metabolite concentration. During *in vitro* studies, the authors simulated a few different scenarios of such situations and concluded that the developed SPME method is able to track rapid fluctuations of drug concentration reasonably well. In this study, two validation methods were used: automated serial blood collection and terminal blood collection. To be able to compare the SPME results obtained for whole blood with those obtained for traditional methods, determination of the blood to plasma partitioning ratio was performed. In the results of PK studies, the authors reported rapid formation of CBZEP and apparent differences between individual mice in the concentration of this metabolite. Statistical evaluation of the obtained PK parameters (CL, V_{ss}, and $t_{1/2}$) showed no difference between SPME and two conventional techniques (Figure 3.7).

The overall comparison of the three methods showed the following advantages and disadvantages: during studies with serial blood sampling, the total volume of blood withdrawn over the entire experiment was 450 μL, which exceeded the recommended 20% of total blood volume and required replacement with heparin saline solution to avoid adverse effects. The volume of removed blood at each sampling point could be further reduced to 5 μL; but this would result in difficulties in sample preparation and lead to low precision and accuracy, whereas with the optimized condition (50 μL) serial blood sampling gave the most precise data in the study of CBZ PK. Terminal blood sampling required a large number of animals for the experiment (a total of 33 mice) and did not allow the observation of interindividual variabilities. SPME enabled obtaining a full PK profile for each mouse used in the study without sacrificing or providing significant disturbance to the animal; it allowed the monitoring of interanimal differences in metabolism rate and drug/metabolite concentration over the entire experiment. The precision of the SPME experiment was comparable to that of terminal blood sampling (50% vs. 60%); however, in the case of SPME, the precision was mainly affected by interindividual changes in the free concentration of the drug and did not influence the results of total blood concentration determined by conventional methods, but it had a direct impact on SPME data. However, for therapeutic purposes, information about free concentration is of high importance, especially because it was reported that for humans the unbound fraction might vary

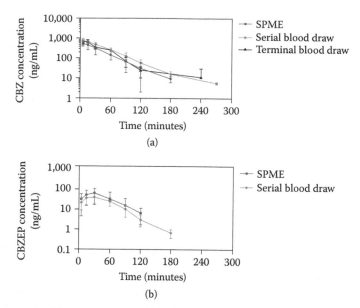

FIGURE 3.7 Mean (±standard deviation) concentration versus time profiles of carbamazepine (CBZ) (a) and the formed metabolite, carbamazepine-10,11-epoxide (CBZEP) (b), following 2 mg/kg intravenal administration of CBZ to mice: samples were taken by serial SPME sampling ($n = 7$ mice), serial automated blood draws ($n = 3$ mice), or terminal blood draws (3 mice/time point). SPME and serial blood sampling measured whole-blood concentrations, whereas terminal sampling measured plasma concentrations. (Reprinted from *Journal of Chromatography A*, 1218, Vuckovic, D., I. de Lannoy, B. Gien, et al., *In vivo* solid-phase microextraction for single rodent pharmacokinetics studies of carbamazepine and carbamazepine-10,11-epoxide in mice, 3367–3375, Copyright (2010), with permission from Elsevier.)

between 0.15 and 0.3 for CBZ and 0.33 and 0.67 for CBZEP. Thus, the possibility of monitoring the free concentrations of a drug and its metabolite by SPME is very beneficial. The option of evaluation of full PK profile using the same animal is particularly advantageous when valuable species of animals like transgenic mice are used in a study [53]. It was also proved in a parallel metabolomics study [73] that the *in vivo* SPME method allows the capturing of unstable, short-lived compounds; thus, it could be successfully used for the monitoring of active but not stable drugs and their metabolites, which is difficult when traditional techniques are used.

Recently, new designs of SPME interface for rodent sampling were proposed (Figure 3.8a and b), along with a sampling device based on luer lock adaptors compatible with standard peripheral venous catheters for large animals and human sampling (Figure 3.8c).

The most recent PK studies of fenoterol and methoxyfenoterol in rats [74] were performed by utilizing an automated sampling system connected to an implanted catheter. This allowed the precise control of blood flow (which was returned to the animal) through the interface, increasing the reproducibility of the method.

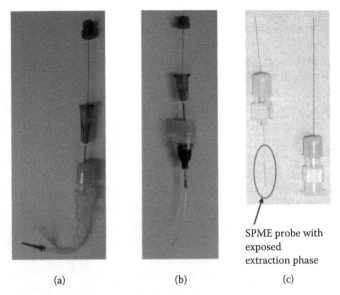

SPME probe with
exposed
extraction phase

(a) (b) (c)

FIGURE 3.8 SPME sampling interfaces for *in vivo* studies of rodent (a and b) and large animals or humans (c).

3.6.3 Tissue Sampling with *In Vivo* SPME

While blood sample withdrawal can be minimized to some extent using conventional sampling methods followed by analysis with solvent extraction, tissue sampling to monitor drug distribution is more challenging. To avoid sample withdrawal, one of the microextraction methods can be used. As mentioned previously, among currently established microextraction methods MD is the leading one. Presently, with suitable sampling devices, SPME has also been demonstrating its applicability in this field. SPME was used for the first time for *in vivo* tissue PK studies in 2006 by Nakajima et al. [75]. The authors monitored the PK profile of toluene in the brains of freely moving mice after 30 minutes of nasal inhalation. One week prior to sampling, a guided cannula (inner diameter of 0.51 mm, and 2.5 mm length) was surgically inserted under anesthesia into the area of the hippocampus. The position of the cannula/fiber was further confirmed by microscopic observation of slices of removed brains. For SPME sampling followed by GC/MS analysis, StableFlex fibers (PDMS/divinylbenzene [DVB]) were chosen because of their flexible core, which prevents fiber breakage. The SPME fiber was inserted into the brain tissue through the cannula for 2 minutes extraction before inhalation (at baseline) and immediately after 30 minute exposure to toluene. The animals were exposed to four concentrations of toluene: 0.9, 9, 50, and 90 ppm. In the case of 50 ppm, concentration sampling was repeated at 10 and 60 minutes after exposure. The results showed that there was no significant difference in toluene levels before and after exposure to 0.9 ppm concentration of toluene, whereas a significant increase of the detected compound was found for 9, 50, and 90 ppm concentrations. To confirm the respiratory origin of toluene in the brain, additional sampling was performed on the mice with artificially induced respiratory arrest, which were exposed to toluene at 9 ppm concentration. No difference in toluene level was found in the hippocampus before and after exposure. The investigation of toluene elimination from the brain showed that concentration significantly decreased 10 minutes after exposure, and the level of toluene came back to baseline values after the next 50 minutes. This time dependence was in accordance with results previously reported in the literature. The authors also referred to MD studies [76] in which toluene could not be determined because of problems with adsorption to the MD membrane. As the main disadvantage of the new method, Nakajima et al. pointed out that SPME results cannot be converted to the conventional calculation approach where concentration of toluene per unit brain weight is used because of the nonlethal nature of the sampling. In turn, the authors emphasized the suitability of SPME sampling for pharmacokinetic studies of chemicals in the brain as well as toxicological, especially neurotoxicological, studies of volatile organic compounds (VOCs) in dead people after accidental exposure to VOCs [75].

The same research group investigated the influence of toluene on the hippocampal level of amino acid neurotransmitters in mice [77]. Similar to previous studies, SPME was used for the *in vivo* sampling of toluene from mice brains, whereas the level of extracellular neurotransmitters was monitored by MD. For both methods, guide cannulas were implanted in the hippocampus of mice in the same locations. After the recovery period, mice were administered with toluene by intraperitoneal (i.p.) injection. MD perfusion of artificial cerebrospinal fluid started 3.5 hours before the collection of microdialysate, which was performed at 30 minute intervals for the next 4 hours. For SPME experiment, four 2 minute extractions were performed at 0, 30, 60, and 120 minutes after toluene exposure. In this study, the authors determined the sensitivity of the fiber *in vitro* using a water bath (37°C) and used the linear relationship between *in vitro* and *in vivo* extractions to convert the amount of toluene extracted by the fiber (nanograms per SPME) to estimate values in the brain (micrograms per milliliter). On the basis of SPME data, it was found that the level of toluene was significantly higher at all measured points when compared with the baseline value. The peak concentration was reached at 30 minutes after toluene injection, and it returned to the initial level after 2 hours. In parallel MD analysis, it was noticed that 30 minutes after the administration of toluene levels of glutamate and taurine in dialysate rapidly increased and dropped to the previous values after the next 30 minutes. No changes were observed in glycine and γ-aminobutyric acid (GABA). The faster return to basal level of amino acid neurotransmitters compared with toluene was explained by two possible factors: (1) lipophilic nature of toluene and its prolonged elimination from adipose tissue; and (2) ability of a body "buffer system" to maintain the physiological level of neurotransmitters by, for example, the uptake of glutamate by glial cells and its conversion to nontoxic amino acid. This was the first study in which the toluene time profile was indicated in the brains of living animals following toluene i.p. injection [77]. It was shown that precise and repeatable results can be obtained by using a short sampling time from brain tissue. Although SPME and MD were used to measure the levels of different compounds, it can be assumed that SPME can provide better time-resolved sampling, offering short extraction, whereas MD requires a longer time of sample collection and, in consequence, is less feasible to monitor rapid changes in concentration.

Even though both SPME and MD were already used for *in vivo* tissue sampling, the first *in vitro* and, most importantly, *in vivo* comparative study between the methods was performed in 2008 by determining pesticide levels in jade plants [78]. This investigation revealed that *in vivo* tissue sampling with SPME provides the same sensitivity, accuracy, and precision as MD but offers some additional advantages including convenience for in-field sampling because of its simplicity and there being no need for additional equipment (pump, syringe) and power supply. Also, quantification of the obtained results is less complex. The authors also reported the existence of carryovers in the case of MD, which was likely caused by the adsorption of the studied compounds to the hydrophobic surface of the membrane, whereas biocompatible SPME coatings did not create such difficulties [78].

Recently, SPME was validated against MD in pharmacokinetic studies of CBZ in the brains of freely moving rats [79]. In the experiment, simultaneous sampling from the left and right hemispheres of the brain was performed. SPME probes with 7 mm biocompatible mixed-mode coating were placed in the cortex and the striatum, whereas an MD probe was implanted between these locations. The concentration of CBZ was monitored over 4 hours in 30 minute intervals during steady and dynamic states (during infusion and on discontinuation). The sampling times were 10 and 30 minutes for SPME and MD, respectively. Cudjoe et al. obtained spatially resolved concentration differences between striatum and cortex with the use of SPME, whereas in the case of MD two separate experiments would be necessary. For qualitative purposes, OFS with a deuterated analog was used. Comparison of SPME and MD data obtained during both steady state and dynamic changes in drug concentration showed very close agreement between both methods and confirmed the suitability of SPME for PK studies in tissue. Additionally, the authors found variations in CBZ concentration at steady state for different dosing vehicles with nearly similar infusion flow rates, which suggest that this state time may be dependent on the type of dosing vehicle used.

Up to this point, SPME mainly utilized guided cannulas designed for MD or needle assembly devices for *in vivo* tissue sampling. Recently, a new tissue sampling device was presented by Togunde et al. [80]. This sampler has already been successfully used for fish studies; additionally, with some minor modifications, it seems to be a very promising tool for clinical application. The principle of the device is based on syringe, with the fiber hidden inside the device for storage (Figure 3.9). For sampling, the tissue is pierced with the needle, which acts as a guide for the fiber. Once the needle is fixed at the appropriate depth, the fiber is released and the needle along with the entire device is removed. After the preoptimized time, the fiber can be manually retracted from the tissue, rinsed, and stored for further desorption and analysis. More details on various aspects of tissue bioanalysis by *in vivo* SPME, including environmental applications can be found in recent review [81].

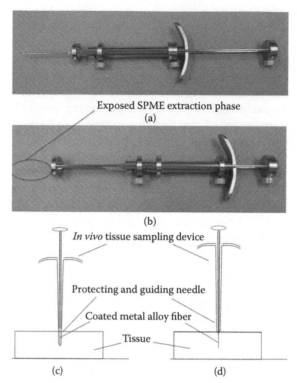

FIGURE 3.9 Device for direct *in vivo* tissue sampling with SPME: (a) and (c) show the device with SPME probe hidden inside the hypodermic needle ("ready for use" and storage); (b) and (d) show the device with exposed SPME coating (during sampling). (From Togunde O.P., H. Lord, K.D. Oakes, M.R. Servos, and J. Pawliszyn, Development and evaluation of a new *in vivo* solid-phase microextraction sampler, Journal of Separation Science, 2013, 36, 219-23. Copyright, with permission from Wiley.)

3.7 APPLICATION OF SPME IN OTHER STAGES OF DRUG DEVELOPMENT

3.7.1 CELL CULTURE STUDIES BY SPME

Cell culture models are often used in drug discovery to investigate *in vitro* drug absorption, identify the metabolites of a drug candidate, assess the therapeutic effect of the drug on certain cell lines, or evaluate the potential toxicity of the drug. Verification of such properties at an early stage of drug development allows substantial time and cost savings. SPME, predominantly in the HS mode, can be successfully used in different applications employing cell cultures.

Recently, Sofrata et al. [82] used SPME for studies of antimicrobial properties of the roots of *Salvadora persica* (miswak), a plant known to be used in traditional medicine for the oral hygiene tool as chewing sticks. The effect of extracts of *S. persica* roots on plaque bacteria was previously reported; however, different preparation protocols resulted in varying contents of the extracts and precluded identification of the compounds, which exhibited antimicrobial activity. In the discussed studies, volatile compounds from the cut root of miswak placed in a covered beaker were extracted by a PDMS/DVB fiber for 15 minutes. Analytes were subsequently analyzed by GC/MS. With the use of transmission electron microscopy, the authors observed a strong bactericidal effect of essential oils from the roots against oral pathogens involved in periodontal disease and other gram-negative bacteria. They reported no changes or growth inhibition in colonies of gram-positive bacteria. The identification of compounds in the extracts pointed at benzyl isothiocyanate as a major antibacterial component. Further studies with synthetic benzyl isothiocyanate confirmed its bactericidal activity against gram-negative bacteria, which is the dominant flora in periodontitis.

The antimicrobial activity of the vapor generated by a combination of cinnamon and clove oil against four gram-negative and four gram-positive bacteria was also studied using SPME [83]. In this study, several strains of microorganisms were exposed to a mixture of cloves and cinnamon essential oils (EOs); in each case, the quantitative antibacterial effect on the cultured microorganisms was evaluated by the determination of minimum inhibitory concentration, reduction concentration, and the fractional inhibition concentration index. SPME polyacrylate (PA) fibers were used for the extraction of clove EOs, whereas PDMS fibers were used for the extraction of cinnamon as well as a mixture of cinnamon and clove EOs. The vapor exhibited a more significant effect against seven out of eight bacteria than direct contact of EOs with the cultures. This difference was explained by the physicochemical properties of the antimicrobial agents; the microbial effect in direct contact was likely caused by more hydrophilic and less soluble species, whereas the active volatile compounds present in vapor phase had a nonpolar nature and did not diffuse through the agar medium. Quantitative analysis enabled the assessment of the synergistic and concentration-dependent effect of cloves and cinnamon vapor on tested bacteria. On the basis of GC/MS analysis, the authors proposed a list of 35 compounds found in the atmosphere and generated by the investigated EOs. Eugenol and phenolic compounds found in the vapor were previously reported in the literature for their highly antimicrobial and antifungal (eugenol) activity.

Recently, the time-resolved analysis of the extracellular metabolomic responses of E. coli to cinnamaldehyde was investigated by Hossain et al. [84]. Additionally to monitoring of content of the head space of the studied system at various phases of bacteria growth and determination of the minimum inhibitory concentration (MIC) of cinnamaldehyde the authors proposed using SPME as a delivery tool of active compound. This showed the potential of SPME to be used for extraction of biologically active compounds *on site*, followed by transport and storage of the analyte on fibre and subsequent transfer of the compounds to studied cell culture. The proposed approach, although present for bacterial culture and cinnamaldehyde, can be extended to different cell cultures used in the process of discovery of new drugs. Moreover, the protocol of the study was automated, thus increasing significantly the throughput of the analysis and making the SPME approach very convenient tool for investigation of the biological effect of active compound on metabolic pathways of cell culture.

Various analytes, including drugs and their metabolites, can be directly extracted from biological matrices by using a different format of SPME devices. Among them, in-tube SPME was proved as a high-throughput method for the determination of verapamil and its metabolite, which also indicate pharmacological properties in cell culture media, plasma, and urine [85]. The authors monitored the pharmacokinetic profile of this calcium channel blocker in rats using a PPY-coated fused silica capillary for extraction followed by LC/MS analysis. This 60 cm capillary column was connected between the injection needle and injection loop of the autosampler, which controlled cycles of sample aspiration and ejection through the capillary to extract targeted compounds. The extracted analytes were then desorbed by redirecting the mobile phase through the capillary column by switching the six-port injection valve from the load to the inject position. Validation of the method (RSD < 5%, and good LOD and linear dynamic range for all studied analytes) confirmed the applicability of the proposed in-tube SPME assay for the pharmacokinetic study of verapamil. The results showed that the metabolic profile of plasma was similar to the data of incubations with hepatocytes isolated from the animals. The main metabolite found was identified as norverapamil, and the other metabolites were reported as the O-demethylverapamil isomers D-702 and D-703 and the dealkylverapamil metabolites D-715 and D-620. In urine samples, dealkylverapamil D-617 was the major metabolite observed; norverapamil and D-620 were also found in substantial amounts, whereas D-715 and D-702 were detected as glucuronide conjugates. The presented protocol showed that in-tube SPME can be a suitable technique offering precise data and a high-throughput analysis of various complex biological samples in routine drug development processes.

The in-tube SPME approach was also used for investigating the carcinogenic properties of 4-(methylnitrosamino)-1-(3-pyridyl)-1-butanone and several metabolites in hepatocyte cultures [86]. Although the authors did not discuss the obtained results from a biological point of view, the developed method showed accurate validation, confirming the usefulness of the technique for studies of organ-specific metabolism using cell cultures.

As a preliminary step to *in vivo* experiments, Kemper et al. [87] studied the metabolic clearance of geranyl nitrile (GN) and citronellyl nitrile (CN), two fragrant components used in consumer and personal care products, by using mouse, rat, and human hepatocytes [87]. Automated HS-SPME was used for the determination of analyte concentration changes in time. For calibration purposes, heat-inactivated liver cells spiked with various concentrations of standards were used. A comparison of data obtained for all three species revealed that the rate of metabolism significantly differed in mice, rats, and humans. The clearance of GN and CN in mice was, respectively, five and two times more rapid than in rat hepatocytes. Two out of three human donors exhibited slower metabolism of both terpenes, whereas in the third case the kinetics was similar to the rate found in rats. With the use of species-specific relative liver weight and hepatocellularity, the obtained data was scaled up to organ level and subsequently used for estimating whole-body intrinsic clearance. The predicted *in vivo* clearance was much higher than normal hepatic blood flows in all species, which suggested the blood flow–limited metabolic clearance of both the studied compounds. In addition, the authors also studied the biotransformation of the compounds with the use of conventional techniques. This data did not enable the finding of a metabolic basis for differences in genotoxicity.

Monitoring an endogenous metabolite profile induced by external stimuli or changes of physiological conditions may bring forth valuable information about disease mechanisms and lead to the invention of new methods of pharmacotherapy. Bianchi et al. [88]

studied the influence of various hypoxia conditions on levels of 17β-estradiol (17-BE) and 2-methoxyestradiol (2-MEOE) in granulosa cells. Both 17-BE and 2-MEOE are known as potential modulators of angiogenesis, which plays a critical role in the development of cancer. For the studies, ovarian swine granulosa cells were used as the model, since they are primarily involved in angiogenesis regulation. Harvested cells were incubated at 37°C under a humidified atmosphere for 24 hours and then subjected to normoxic, hypoxic, or anoxic conditions. The authors used a homemade octadecyl silica fiber to extract target analytes from the HS, and then one-step on-fiber derivatization was performed. After thermal desorption in a GC injector, 17-BE and 2-MEOE were analyzed with a mass spectrometer. The validation indicated a very good lower limit of quantification and range of linearity, as well as precision of the method; this is especially important for the determination of endogenous compounds, which are usually present in biological matrices at low concentrations. The increased production of both substances was reported under hypoxic and anoxic conditions. This data indicates the potential determination of 17-BE and 2-MEOE in controlling angiogenesis and an improved knowledge about the genesis of hormone-dependent cancers.

Another interesting study with the use of a cell culture model was conducted by Poli et al. [89]. The authors developed an SPME-GC/MS method for monitoring the concentration of styrene oxide (SO) in exposed neuroblastoma cells. SO is a highly reactive, neurotoxic volatile compound with proven carcinogenic properties in animals and potential carcinogenic properties in humans. It exhibits a particular effect on the nervous system; thus, neuroblastoma was selected for the study as a model cell line. For the SPME experiment, two types of sorbents, carboxen/PDMS and polyacrylate, were used. Method development included the optimization of extraction and desorption times, effects of salt on extraction efficiency, and temperature selection. The results showed that the determined SO level induced apoptotic cell death, and the critical levels obtained were in agreement with the hypothesis of styrene-induced neurotoxic effects after long-term low-level exposure *in vivo*.

The employment of SPME in parallel with flow cytometric analysis for pharmaco-metabolomics studies on the effect of cisplatin on human non-small cell lung carcinoma A549 cell lines enabled the finding of new potential biomarkers of cisplatin-induced apoptosis and necrosis [90]. The cells were incubated with various concentrations of cisplatin over 24 hours and then the viability of the cells were measured and correlated with the results of VOC concentration obtained on the basis of SPME-GC/MS analysis. The cytometric data indicated that cisplatin might cause dose-dependent induction of apoptosis in the studied cell lines. Moreover, it was observed that a shift in the necrotic phase occurs above a certain drug concentration. Very good correlation (R^2 of 0.9884–0.9991) was found between these results and the concentration of nonanal, 1,3-bis(1,1-dimethyl)-benzene, and 2,6-bis(1,1-dimethyl)-2,5-cyclohexadiene-1,4-dione, which were calculated on the basis of SPME measurements. The fourth compound, decane, was found to be possibly related to the necrotic pathway of cell death. The presented work evidenced that SPME is a sensitive, nondestructive, time-saving, and reliable tool for differential analysis of necrosis and apoptosis by determining volatile biomarkers of cell death.

The presented HS-SPME studies are an excellent example of the suitability of the method for identification of volatile compounds, which show biological activity toward different types of pathogens, whereas in-tube SPME applications support the feasibility of this technique also for drug discovery studies employing cell culture. Currently in our research group, there are ongoing studies on the metabolome profile changes of various cell lines induced by external drug stimuli by using DI-SPME with biocompatible fibers.

3.7.2 IN VITRO DRUG–PROTEIN BINDING STUDIES BY SPME

To study the binding parameters in a system, both free and total concentrations must be known. Detailed theoretical explanations of ligand–receptor binding using SPME can be found elsewhere in the literature [91], whereas in this chapter general concepts and procedures are discussed. To obtain pairs of corresponding values C_{free} and C_{total}, three methods were proposed: (1) method of multiple standard solutions, (2) method of successive extractions, and (3) method of successive additions. In all cases, the calculation of free and total concentrations is performed on the basis of extraction from the solution containing ligand and receptor. In the first approach, a series of solutions containing known concentrations of ligand and protein needs to be prepared. For each data point, a separate standard solution is required. After equilibration between ligand and receptor is reached, an SPME fiber is used to extract a small portion of the analyte. Subsequently, free concentration is calculated using Equation 3.15. In this case, total concentration may be easily calculated as the difference between the initial concentration of the drug and the amount extracted by the fiber. When negligible depletion occurs, total concentration can be considered to be equal to the initial one. The preparation of the solution for this method can be modified by introducing ligand on the fiber to the solution containing a known amount of receptor (a known amount of drug must be previously preloaded from the binding-free solution). Consequently, calculations can be made by analyzing the amount of analyte remaining on the fiber after drug delivery to the receptor-containing solution [91]. The method of successive extractions is based on the analysis of a single solution containing a known concentration of the receptor and a known concentration of the drug, which decreases after each extraction. Briefly, the highest concentration of the ligand is added to the solution, and SPME extraction is conducted after equilibration. As with the previous study, the drug can be delivered to the receptor solution on the SPME fiber; once the equilibrium between the receptor and the drug desorbed from the fiber is established, the amount of drug remaining on the fiber (and the indirect amount remaining in the solution) can be calculated. In subsequent steps, the fiber is used to extract the ligand from the solution, which now contains a lower concentration of the drug. The formulas allowing the estimation of the number of extraction steps necessary to cover a certain concentration range as well as fraction of the ligand extracted in subsequent steps can be found elsewhere in the literature [91]. The third proposed method of successive additions is similar in principle to the method of successive extractions; however, here the lowest concentration of the analyte is present in the solution with a receptor and in subsequent steps the drug is delivered to the solution either by introduction on the preloaded fiber or by standard pipetting. The main advantage of the method of successive additions over the method of successive extractions is the possibility of covering a wider range of concentration, since the researcher is not limited to the extraction efficiency of the fiber determining the magnitude of concentration change in the solution at each step [91].

When free and total concentrations of the ligand are known, the number of drug molecules bound per one molecule of receptor (B) can be expressed as follows:

$$B = \frac{C_{\text{total}} - C_{\text{free}}}{C_{\text{m}}} \tag{3.25}$$

where C_{m} is receptor concentration. If we consider the complex where the ligand occupies several binding sites at different affinities, then each binding site B_i characterized by the site binding constant K_i can be described by the following equation:

$$B_i = \frac{K_i C_{\text{free}}}{1 + K_i C_{\text{free}}} \tag{3.26}$$

Therefore, by combining the aforementioned divagations, the receptor as a whole can be expressed as follows:

$$B = \frac{C_{\text{total}} - C_{\text{free}}}{C_{\text{m}}} = \sum_{i=1}^{b} B_i = \sum_{i=1}^{b} \frac{K_i C_{\text{free}}}{1 + K_i C_{\text{free}}} \tag{3.27}$$

When the number of binding sites and their affinities are known, SPME can be used not only for the determination of initial free and total concentrations of the drug but also for the calculation of receptor concentration [91]. In the first situation, when receptor concentration is known, drug concentrations can be determined by single measurement, whereas in the case where receptor concentration is unknown successive extractions need to be performed. For detailed protocols and equations, please refer to the work by Musteata and J. Pawliszyn [91].

Binding studies with SPME were performed even before direct extraction from complex matrices like plasma or blood became a common approach. In 1999, Yuan et al. [92] investigated the binding of alkylbenzenes to BSA by HS-SPME followed by GC analysis. In their calculations, the authors considered partitioning the compounds between four phases: fiber coating, HS, aqueous phase, and dissolved macromolecule. It was demonstrated that analyte partitioning in HS can be ignored when the Henry's law constant and the ratio between HS volume and sample volume are sufficiently small. Since most of the compounds that bind with protein in biological matrices are semi- or nonvolatile, and biocompatible sorbents were introduced to SPME, direct immersion became an effective alternative to the HS mode.

One of the first binding studies with DI-SPME was conducted with diazepam and human serum albumin (HSA) as the model system [93]. The authors used two methods for the preparation of drug–protein solutions: in the first approach the drug was spiked with protein solution and after equilibration SPME with PDMS fibers was performed, whereas in the second method the drug was introduced to the protein solution on the fiber and the amount remaining on the fiber was further analyzed. On the basis of the obtained results, a Scatchard plot was employed to obtain the number of binding sites and binding constants. This widely accepted method of determining binding parameters was used for validation of the method proposed by the authors; binding parameters were calculated from the ratio of fiber extraction in the absence and presence of protein. The value of equilibrium constant obtained by the proposed calculation method was 9.77×10^5 and 1.23×10^6 M^{-1} for the first and second methods of calibration curve preparation, respectively. By analogy, the values determined for the Scatchard plot were 1.159×10^6 and 1.25×10^6 M^{-1}, respectively. This study demonstrated the feasibility of SPME for drug–protein binding studies and successfully introduced a method for determining binding parameters from a single small-volume solution of receptor by using a fiber for drug delivery [93].

DI-SPME was also used for the determination of binding estradiol, an endogenous hormone, which in the organism can be found bound with serum albumin [94]. Because this compound is present in biological matrices at very low concentrations, it is often used in its radiolabeled form for the analysis. Also, in the discussed study liquid scintillation counting (LSC) was used for detection. Moreover, prior to incubation with the drug, the protein was stripped with dextran-coated charcoal to remove potential remains of bound endogenous estradiol. The obtained value of binding constant was 8.9×10^4 M^{-1} with a standard deviation of 1.6×10^4 M^{-1}, which correlated well with the literature data. Reproducibility of the assay evaluated from a total of nine measurements was 18%. Although binding studies are more often conducted using LC or GC systems coupled with MS, radiolabeling still offers some advantages when a studied protein is not available in pure form and an investigation on a raw matrix containing endogenous ligands must be performed. Also, studies of binding competition between endogenous compounds and drugs can be successfully conducted with LSC. As demonstrated by Heringa et al. [94], radioassays can be effectively coupled with SPME for the determination of binding parameters.

Displacement of endogenous estradiol by xenoestrogens in its binding places of serum albumin was also studied by SPME radioassay [95]. As discussed by the authors, inefficient stripping of endogenous steroids from biological matrices can lead to misinterpretation of the results when a tested xenobiotic displaces these endogenous compounds in their protein binding sites. The released hormones may activate biochemical pathways, which will be interpreted as an estrogenic response induced by the tested xenobiotic. In their studies, Heringa et al. investigated the stripping efficiency of charcoal and measured if estradiol is displaced from its protein binding sites by octylphenol. The first-step experiments with BSA and fetal calf serum were carried out to verify the applicability of charcoal assay for binding studies. In the later step, *in vitro* assay with the use of 5% fetal calf serum and human embryonal kidney cells culture to investigate the competition between studied compound and estradiol was performed. On the basis of the obtained results, the authors concluded that charcoal stripping efficiently removes endogenous ligands; thus the initial hypothesis on "artificial estrogenity" induced by released hormones was rejected. The authors also drew attention to a possible shift of binding equilibrium caused by charcoal removal of not only the free fraction but also some part of the bound fraction, which was not an issue in the case of SPME [95].

Musteata and Pawliszyn [91] demonstrated the applicability of RAM (liquid C18 coating with a cutoff of ca. 15 kDa) fibers for binding studies of diazepam and isosorbide dinitrate (ISDN) to HSA [91]. After equilibrium extraction (8 minutes for both compounds), analytes were desorbed in a homemade interface serving as a desorption chamber for the SPME fiber. The programmed autosampler switched the mobile phase flowing through the chamber after 2 minutes of static desorption, allowing elution of the desorbed compounds. For determining diazepam binding, the authors used drug:HSA molar ratio sufficient for analysis of interaction in both high- and low-affinity binding sites. For the evaluation of binding parameters, the authors tested three different models: stoichiometric, one class of binding sites, and two classes of binding sites. By comparing values of the residuals, the dependencies between independent parameters, and the correlation coefficient the best fit were found for three binding sites model and the corresponding stoichiometric binding constant equation (correlation coefficient 0.09992). The values of binding parameters were in good agreement with the literature data. However, the results showing a three-site binding model did not support the previous diazepam-HSA studies with SPME [93] in which a one-site binding model was proposed. This discrepancy was explained by the fact that Yuan et al. used a narrow drug:HSA molar ratio range, which led to the observation of diazepam interaction only in the high-affinity site. For studies of ISDN binding to HSA, the authors used three methods to obtain C_{total} and C_{free} pairs: method of multiple solutions, successive extractions, and successive additions. Although by using all three methods different regions of the binding curve were observed, because of different ligand:protein molar ratios obtained, in all cases the binding parameters were determined for the model assuming three identical binding sites, which fit with the experimental data with a correlation coefficient of 0.9928. The method of successive additions resulted in a higher RSD being calculated for the binding constant (17.2%), because of experimental errors associated with both standard loading to the fiber and measurement of the remaining amount of the drug on the fiber after equilibration with HSA. For the two other methods, RSD was below 6%. It is important to mention that this study showed the suitability of both negligible (in case of diazepam) and nonnegligible depletion (ISDN) SPME for binding studies. Also, the advantage of SPME was shown, where high-affinity coatings such as RAM or immunosorbents can be used to increase the sensitivity of the assay when high binding drugs are of interest, over dialyses, which is limited in terms of buffer affinity.

Recently, Vuckovic and Pawliszyn [63] published the results of binding studies of diazepam to HSA using an automated SPME system. In the studies, C16-amide-coated particles were used for the preparation of fibers. The robotic PAS Concept 96 autosampler that was used enabled performing the SPME experiment simultaneously with 96 fibers. The detailed description of the device is given in Section 3.10. In their studies, the authors used 12 fibers for evaluation of extraction time profile (six for extraction from a receptor-free solution, and six from solutions containing albumin) to increase the throughput and precision of the analysis. An equilibrium time of 30 minutes was used for the experiment. To obtain the binding curve, a relatively small number of fibers (seven) was employed, which resulted in higher experimental error when compared with manual SPME studies. However, it must be emphasized that increasing the number of experimental data points would enhance the precision of the method without sacrificing the overall time of the analysis. The obtained results fit the one-site binding model, with a regression coefficient of 0.991 and good correlation of the binding constant to the literature data [63].

Another interesting drug–protein binding study was presented by Theodoridis [96]. By using commercially available PDMS and PA fibers, Theodoridis investigated the binding of seven pharmaceuticals to HSA and validated the results against UF as the established binding assay. To meet the condition of negligible depletion, the extraction time selected was shorter than the time of SPME equilibrium for most of the studied analytes. The concentration of free fraction of three drugs (naproxen, paclitaxel, and haloperidol) was below the limit of detection, and in these cases binding to protein was assumed as exhaustive. The percentage of other drugs bound with HSA was compared with literature data, which showed good agreement, with the exception of nortriptyline, which exhibited a lower value than expected. However, as explained, the differences were not attributed to the extraction process but rather to errors caused by detection carried out at a low, not selective, UV region. Cross-validation performed with the three drugs of naproxen, quinine, and ciprofloxacin showed good agreement in free concentration/percentage of binding in two later cases, whereas naproxen was not detected in the ultrafiltrate and, similar to SPME studies, was considered as exhaustively bound.

The comparison of carbamazepine binding study performed by SPME coupled to UV-VIS and MS/MS detector with two spectroscopic methods, fluorescence and nuclear magnetic resonance (NMR) was published recently [97]. The results showed complementary nature of the used techniques, because of their different principles; SPME allowed obtaining overall information about interaction between the drug and the studied macromolecule, while fluorescence and NMR enabled observing high and low affinity binding sites, respectively. One of the main values of SPME, which differentiates this techniques from other approaches is the possibility of studying drug-protein binding in complex matrices, including *in vivo* blood analysis. On the other hand, the precise indication of the binding domains in protein structure as well as assigning drug's atoms or groups participating in the interaction is withheld for fluorescence and NMR, respectively. The advantages and disadvantages of all three methods discussed in the article explained the benefits of combining various techniques in ligand-protein studies for various applications.

3.8   MONITORING OF METABOLOME CHANGES: TARGETED AND UNTARGETED METABOLOMICS

Metabolomics study is complementary to genomics, transcriptomics, and proteomics. It provides information about metabolite changes in living systems induced by various stimuli including pathogens, drugs, and the environment. Metabolomics can be considered either as "global profiling" or as a targeted analysis of endogenous compounds. This "omic" approach has drawn significant interest in medicine and related sciences over the past few years, because it has been found to be a very promising tool for biomarker discovery, especially for cancer diagnosis and treatment. Moreover, monitoring organism response to applied treatment as well as recognizing biochemical pathways that change during a pathological condition gives valuable insight into the knowledge about pathogenesis of certain diseases. Metabolomics appears to be particularly helpful in personalized medicine, since it is affected not only by genetics but also by environmental factors and thereby gives a unique pattern of global metabolome for each individual. Therefore, therapy tailored as a result of pharmacometabolomics would significantly improve the efficiency of treatment and, indeed, this attitude has been gaining recognition in clinical practice. Although there is a significant increase in published works on this subject, sample preparation remains a crucial step in metabolomics. There are several established protocols in this area, but they employ mainly solvent extraction and for this reason cannot be used for *in vivo* studies. In consequence, after sample withdrawal metabolism must be stopped immediately to obtain a true snapshot of the metabolome. Depending on the method used and the time elapsed from the moment of sampling to metabolism quenching, the content of the sample may change and no longer be representative. In this Section 3.8 we introduce breath and skin metabolomics studies done by HS-SPME and discuss a few examples of the most recent findings of DI-SPME employed for metabolome profiling.

3.8.1   METABOLOMIC STUDIES WITH *IN VIVO* SPME

By analogy to pharmacokinetic studies, the first metabolomics SPME experiments followed by GC instruments analysis and coupled with various types of detectors were performed on volatile compounds. Among all SPME analyses addressed to medical purposes, two have been successfully introduced in clinical practice to date: breath and, more recently, skin analyses. Breath analysis has been used for diagnostic purposes for years, mostly for determination of acetone levels in patients with diabetes or for monitoring dietary fat loss and assessing the harmlessness of weight loss, as well as in epileptic patients on ketogenic diets used to treat intractable seizures [98]. When SPME was already known for its excellent feasibility to extract volatile compounds for HS, with minor modification, Grote and Pawliszyn [99] used commercial SPME holders for sampling several compounds from human breath (Figure 3.10). The authors performed targeted analysis to determine levels of acetone, ethanol, and isoprene in healthy volunteers.

Successful validation of the method encouraged other researchers to improve the sampling device and test different sorbents for targeted as well as untargeted analyses of animal [100,101] and human [7–16] exhaled breath. The greatest number of breath studies with SPME relates to volatile biomarker discovery in patients with lung cancer. To mention one, Gaspar et al. [10] investigated differences between lung cancer patients and healthy controls in a total group of 28 individuals. The studied subgroups included healthy smokers and nonsmokers, and cancer nonsmokers and cancer smokers (among this group, patients were distinguished between

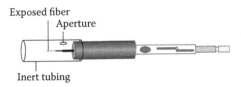

Exposed fiber

Aperture

Inert tubing

FIGURE 3.10   SPME device modified for breath analysis. (Reprinted with permission from [99, 587–596]. Copyright [1997] American Chemical Society.)

those under chemotherapy and those not subjected to chemotherapy). SPME followed by GC/MS measurement and multivariate analysis of the obtained results showed different metabolomics profiles between healthy volunteers and cancer patients. Although some of the studied individuals appeared as misclassified cases on the principal component analysis (PCA) plot, posterior investigation explained these results. Obvious visual clustering was noticed for nonsmokers and smokers separately, whereas in cases of cancer patients smoking history had no effect on the metabolic pattern. Buszewski et al. [102] studied differences in the breath of nonsmokers and active and passive smokers. Their results revealed that the main contributing compounds in differentiation between smokers and nonsmokers were saturated hydrocarbons. Whereas in healthy, nonsmoking volunteers the content of these compounds was 32%, it turned out to be predominant in the rest of the studied group. Apart from these compounds, several additional compounds including acetonitrile, furan, and 2-butanone were identified in the breath of both passive and active smokers. A significant number of hydrocarbons was also found in the studies performed on a group of 65 patients at different stages of lung cancer versus 31 healthy volunteers [7]. Among 103 detected and 84 identified compounds, the authors proposed 8 as potential biomarkers of lung cancer. Another study employing the on-fiber derivatization approach aimed to propose four aldehydes (pentanal, hexanal, octanal, and nonanal) as biomarkers of lung cancer [12]. Sensitivity and specificity of the developed method were comparable with those of conventional soluble serum markers and computed tomography imaging.

The potential of SPME breath analysis as a diagnostic tool was also demonstrated in the recognition of microbial pathogens [9,16]. In both of the discussed studies, the experimental focus was on the development of a simple and robust method for determination of targeted analytes known a priori as diagnostically valuable. Syhre et al. [9] used SPME in the determination of 2-pentylfuran in few different pathogens: *Aspergillus fumigatus, Fusarium* spp., *A. terreus, A. flavus*, and *A. niger*. After confirming the production of the compound *in vitro* with cell cultures, analyses were performed on human subjects with cystic fibrosis, with and without colonization by *A. fumigatus* and other pathogens, as well as healthy volunteers. The results obtained confirmed the suitability of 2-pentylfuran as a good biomarker of lung colonization by fungi and the applicability of SPME method as a diagnostic tool in breath analysis. The other study on microbial pathogens [16] concerned the monitoring of methyl nicotinate in smear-positive patients with tuberculosis. Method development, which used *in situ* derivatization of methylate nicotinic acid to methyl nicotinate, resulted in excellent sensitivity, aiming at detection of the compound in patients' breath at low femtomole/mole levels. Although the investigations were performed on a small cohort, statistical analysis showed a significant difference between patients with tuberculosis and healthy controls.

Most of the breath studies used Tedlar bags for sample collection, followed by extraction with SPME fibers. Fairly recently, breath and urine analysis in type-2 diabetes patients as well as healthy volunteers under fasting conditions was conducted with the use of particle-packed needles [103]. The newly proposed HS method was used for the determination of acetone concentration in both sampled matrices. The results indicated good correlation between exhaled and urine acetone concentrations in fasting individuals. The level of acetone in the breath of type-2 diabetes patients correlated with the level of glycosylated hemoglobin in blood (acetone was not detected in urine samples, which was confirmed by conventional test strips). The obtained results evidenced that the proposed method can be successfully used as an alternative method for the diagnosis and monitoring of diabetes and/or ketoacidosis.

SPME was also applied for the metabolomic analysis of skin volatiles [17,18, 104–107]. Rizanskaia et al. [17] developed the "skin-patch" method (Figure 3.11), which was further validated *in vitro* and *in vivo* before being utilized for a proof-of-concept study in patients with carcinoma lesions. This sampling patch was cut from a commercial silicone elastomer sheet, washed and conditioned according to developed protocol. During sampling, the patch was placed on the skin in a previously selected skin area, then covered with a cotton wool pad, and finally secured with microporous tape to the skin surface. After the desired sampling time, the patch was removed and placed in a thermal desorption tube for analysis by GC/MS. The proposed method was subjected in their later study to monitor the profiling of volatiles in skin cancers, fibrotic skin disorders, wound healing and infection, as well as human odor profiles. Two years later, the same group presented results of their work in the identification of biomarkers of chronic wounds [18]. Since the studied lesions were of mixed etiology, the obtained data showed VOC profiles for individual participants rather than biomarkers for specific pathogens, but the authors successfully identified numbers of compounds, what enabled distinguishing wound lesions, boundary areas, and control areas. The studies showed good reproducibility of the proposed approach and its suitability for noninvasive studies of chronic wounds and likely other skin disorders as well.

An alternative skin-sampling device was proposed by Soini et al. [105]. The device is based on a stir bar coated with PDMS sorbent and fixed on a roller device (Figure 3.12). To improve reproducibility of the data and evaluate the influence of

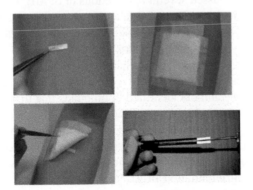

FIGURE 3.11 Demonstrating the application of a single skin patch *in vivo*: top left, skin patch is laid on the area of interest; bottom left and top right, the skin patch is covered with a cotton wool pad and secured with microporous tape to the skin surface, allowing the participant to move naturally throughout the sampling procedure; bottom right, the skin patch is placed and immediately sealed inside an empty Silcosteel thermal desorption tube ready for analysis. ({17}Reproduced by permission of The Royal Society of Chemistry.)

FIGURE 3.12   Schematic of the roller device designed for stir bar surface sampling: arrows show the attachment points of the stir bar between the jaws of this device. (Reprinted with permission from [105, 7161–7168]. Copyright [2006] American Chemical Society.)

storage time and condition, the authors impregnated the extraction phase with the internal standard. The method was utilized for the extraction of human skin, plants, bird feather, and fingerprint from a solid surface. The authors indicated the feasibility of the method for not only human skin sampling and forensic analysis but also agriculture and biological field studies.

In vitro HS-SPME was recently used for the differential analysis of volatile compounds of melanoma and nevi [106]. Abaffy et al. [106] performed fuzzy logic analysis on the results obtained from fresh and frozen tissue samples and found 12 potential signatures of human melanoma with 90% specificity and 89% sensitivity.

3.8.2   Monitoring of Cell Metabolome Changes by SPME

We have previously discussed several applications of SPME for cell culture studies, where the pharmacological effect of a drug or some other active compound was investigated. In the current section, we summarize the applicability of SPME for biomarkers' discovery by using cell culture.

Acevedo et al. [107] investigated the changes of VOC profiles in cultured human fibroblasts exposed to different culture conditions. On the basis of HS-SPME coupled to GC/MS, the authors identified compounds that discriminated skin cells cultured as monolayers on plastic (two dimensional [2D]) and cells encapsulated in the alginate matrix (three-dimensional [3D]) system. This finding was correlated with visual observations of fibroblasts and likely with their physiological properties. The authors pointed at ethylhexanol and benzaldehyde as major compounds contributing to the differentiation of 2D and 3D cell culture systems. Although no reports were found on further correlation analysis between released VOCs and physiological properties of the fibroblast, it could be a fast and easy method for first-stage determination of applicability of the harvested tissue for clinical applications such as grafting.

Another successful application of SPME for profiling volatile metabolites of cells to find potential biomarkers was performed by Zimmermann et al. [108].

Two cell lines, SW480 and SW1116, representing two different stages of colon cancer (grade IV, Duke B and grade II, Duke A, respectively) underwent comparative analysis against each other as well as against the normal colon epithelial cell line—NCM460. Commercial Carbowax/DVB SPME fiber was exposed to the atmosphere of a cell culture flask for 40 minutes and then thermally desorbed in the injector of the GC instrument. Sampling with SPME allows avoiding disturbance to the atmosphere and at the same time changes the representation of the sampled metabolites, which could happen if an air stream was removed continuously or when discontinuous sampling of a certain gas was applied while using different HS sampling methods. Identification of the obtained mass spectra revealed that for the first time methyl dodecanoate, undecan-2-ol, and nonan-2-one correlated with human metabolism. An analogy to bacteria cell metabolism suggested that an enzyme such as secondary-alcohol oxygenase could be involved in the anabolism of the latter compound. Comparison between metabolome profiles of SW480, SW1116, and normal colon cells showed that SW480 line exclusively produced pentadecan-2-one, whereas SW1116 produced 3-methylbutan-1-ol and heptan-1-ol. These potential biomarkers were not previously recognized in human metabolism, thus indicating that certain metabolic pathways were not possible. However, based the literature review the authors found biochemical reactions in other species where these compounds participate in. Consequently, potential pathways differentiating metabolism of studied stages of colon cancer and healthy epithelial cells were identified.

Extensive studies on celiac disease in children (T-CD) who remained on a gluten-free diet for at least 2 years involved investigation of microbiota and metabolic recognition [109]. The authors used culture-dependent and culture-independent methods to fulfill the first objective of the work, whereas the latter part was performed with SPME-GC/MS and proton NMR (^{1}H NMR). For SPME method, after 10 minutes of sample equilibration at 45°C, VOCs were extracted from the HS of the vial containing fecal sample for 40 minutes. Subsequently, 5 minutes of desorption was followed by GC/MS analysis. NMR analysis was carried out on fecal and urine samples. In total, 107 VOCs were detected and identified in both matrices, with lower levels in urine. The profile of metabolites differed among children. The significant statistical difference was found between celiac disease and the healthy control group. Major differences were found in the level of short-chain fatty acids, particularly isocaproic, butyric, and propanoic acids. The change of butyric acid is of special interest since it plays an important role in regulation of cell proliferation and differentiation of colon epithelial cells. Also, the levels of some alcohols were found higher in T-CD children versus healthy controls. ^{1}H NMR studies were in agreement with SPME-GC/MS data. The obtained results showed that the profile of VOCs was affected by changes in microbiota in celiac disease. Several compounds (e.g., ethyl acetate, octyl acetate, glutamine, and short-chain fatty acids) were suggested as signatures of the disease. The parallel microbiology and metabolomic studies also proved that a gluten-free diet lasting for 2 years is not sufficient to restore natural microbial flora in children.

A similar experimental design was used later by Maccaferri et al. [110] to monitor the influence of rifampicin on colonic microbiota in active Crohn's disease. By utilizing the cell culture model, the authors observed that Rifaximin did not affect the overall composition of gut microbiota but increased bifidobacteria, giving potential benefits to the host. Moreover, parallel metabolomics studies showed the increase of short-chain fatty acids, propanol, decanol, nonanone, and aromatic organic compounds and the decrease of ethanol, methanol, and glutamate.

3.8.3 DI-SPME-LC/MS FOR METABOLOMICS

The first *in vivo* metabolomics experiment using DI-SPME combined with LC/MS for animal blood analysis was done by Vuckovic et al. [73]. During *in vitro* experiments, the authors evaluated the properties of 42 coatings to select the one most appropriate for global screening [19]. The primary aim was to find a sorbent covering a broad range of analytes differing in physicochemical properties, especially compounds at various polarities. A total number of 36 metabolites from different chemical classes and log *P* ranging from −7.9 to 7.4 were selected for the experiments. Among all tested coatings, three were found to be the most suitable: mix mode (C18+benzenesulfonic acid), polar-enhanced polystyrene–DVB, and phenylboronic acid. The first of these three coatings was selected for the further development of rapid *in vivo* sampling method. The authors performed extraction from human plasma and compared the results with two standard methods commonly used for metabolomics analysis: UF and protein precipitation. In the latter case, two precipitating solvents, methanol and acetonitrile, were used. For SPME, it was found that for polar compounds equilibrium is reached in ≤ 5 min. Utilizing short sampling time has two advantages: (1) sample composition is unlikely to change during this time and (2) external equilibrium calibration can be used for quantification of these compounds, which equilibrate in such a short time. For the later eluting compounds where equilibrium is not reached, one of the kinetic calibration approaches can be used. However, for 24 hour extraction, authors did not observe the displacement of any analytes and any changes in the amount of analyte extracted were attributed to metabolite degradation. For analysis, Vuckovic et al. used two chromatographic columns, reversed phase and hydrophilic interaction liquid chromatography (HILIC), as well as two polarity modes of Orbitrap mass analyzer to cover the maximum number of metabolites. A comparative analysis for all methods revealed that with the use of SPME total coverage was 1592–3320 features, for solvent precipitation 2082–3245, and for UF 2093–2686. The corresponding precisions of the methods (RSD) were 11%–18%, 8%–19%, and 20%–22%, respectively. It was also discussed that SPME provides the most balanced coverage of the compounds and gives coverage of polar species complementary to solvent precipitation, whereas UF suffers from the loss of hydrophobic features. The evaluation of absolute matrix effect for SPME showed that ionization was affected in <5% and <20% of all tested compounds in the case of reversed-phase and HILIC methods, respectively. The suppression was mainly attributed to citric acid and compounds coeluting with the same retention time window, which was a result of the presence of sodium citrate used as anticoagulant. Two additional aspects were compared: analytical sensitivity and the possibility of calculating active (free) concentration of the metabolites. With respect to the first issue, for both conventional techniques the instrument response was much higher when compared with SPME; however, with the use of Orbitrap as the mass analyzer this can be considered as advantageous feature of SPME, because in this instrument the detector is filled with a set number of ions during each scan; thus, peaks at high intensities may affect the presence of low abundance signals. Moreover, very intense signals may cause ion suppression of ions with species with close retention times and consequently affect quantitative data. By analogy to targeted analyses, SPME enables calculation of both free and total concentrations of the analytes, whereas each of the conventional methods gives only one of these possibilities [19]. After *in vitro* development, fibers were used for *in vivo* proof-of-concept study [73]. Eight mice were used for analyzing the influence of CBZ dosing on blood metabolome profile. Animals were divided into two groups: one was subjected to CBZ administration and the second served as a control. Results obtained for 2 minute SPME sampling were compared with data obtained for *ex vivo* SPME, protein precipitation, and UF. The most interesting finding was the detection of over 100 molecular features only by *in vivo* method. This led to the conclusion that *in vivo* extraction allows the capturing of unstable and/or short-lived species, which are undetectable with the use of blood-drawing techniques. As an example, the authors demonstrated the presence of β-nicotinamide adenine dinucleotide (β-NAD) and adenosine monophosphate (AMP), whose identity was confirmed by the use of an authentic standard. Moreover, *in vivo* data did not contain some of the compounds detected *in vitro*, which evidences the degradation of some substances under *in vitro* conditions. Determination of glutathione to oxidized glutathione ratio showed significant differences between *in vivo* and *in vitro* data. For the *in vivo* method the ratio was 2.5, which is in excellent agreement with the expected value, whereas for solvent precipitation, UF, and *ex vivo* SPME the ratios were 0.001, 0.005, and 0.2, respectively. Although PCA showed good separation of dosed and nondosed animals, interanimal variability was clearly observed. As previously mentioned for PK, this emphasizes the benefit of using a single animal for drug studies [73].

The investigations of metabolome changes in cardiac surgical patients induced by applied pharmacotherapy and surgery with the use of cardiopulmonary bypass revealed alteration in the lysophospholipids, triacylglycerols, mediators of platelet aggregation, and linoleic acid metabolites profile [111]. Additionally, two of the patients showed different response to the treatment. Interindividual variability may have a different background, starting from genetics through enzymatic polymorphism to environmental factors, therefore monitoring patient's reaction and applying personalized therapy may increase the chance for successful treatment and minimize risk of its adverse effects.

The potential of DI-SPME-LC-MS analysis for saliva metabolic profiling was presented recently [112]. Non-invasive sampling with two types of biocompatible coatings was performed to identify the influence of diet on the metabolome of this biofluid. Although the proof-of-concept study involved small number of samples, unsupervised statistical analysis showed good separation between the days as well as between the individuals. The obtained results indicate that saliva can be used as an alternative matrix for prognostic/diagnostic purpose to determine disease biomarkers or to monitor the progress of the treatment in non-invasive manner.

In vivo SPME followed by LC/MS analysis was recently used to monitor changes in neurotransmitter levels in rats induced by deep brain stimulation, a technique used for the treatment of neurodegenerative diseases [113]. The authors observed the time profile of two amino acids, glutamic acid and GABA, and two monoamines, serotonin and dopamine, in both left and right hemispheres. Sampling with the use of 4 mm mix-mode sorbent was performed at equilibrium (30 minutes), which was evaluated prior to the experiment during *in vitro* studies. Biological matrices are structurally heterogeneous, and compound distribution is uneven along the tissue; thus, for estimation of extraction time profile, brain tortuosity and fluid volume within the extracellular space were taken into account. The overall time of *in vivo* analysis was 4 hours, 2 hours before to obtain a stable basal level followed by 2 hours after deep brain. The results of *in vivo* studies revealed only a fourfold increase in serotonin level, whereas no changes were observed for other studied neurotransmitters. The coefficient of variation for neurotransmitters ranged from 3% to 25% of basal concentrations [113].

3.9 *IN VITRO* DRUG ANALYSIS AND THERAPEUTIC DRUG MONITORING

The number of SPME-GC and SPME-LC applications subjected for drug analysis is continuously increasing. In this chapter, we present only a few examples demonstrating the feasibility of *in vitro* SPME assays for monitoring the concentration of various drugs in biological matrices.

When pharmacotherapy is applied, special attention needs to be paid to elderly patients, children, and neonates, because of changes in drug metabolism [114], distribution, and elimination [115], which affect the pharmacokinetic profile determined during the drug discovery process. In case of neonates, newborns, and children, methods utilizing a small sample volume are preferable [116]. One of the extensively studied options is dried blood spot (DBS) analysis. This technique is not a new approach, since it was used in newborn screening for inherited metabolic disorders in the early 1960s [117]. Although DBS is very cost-effective and addresses problems of sample preparation, transportation, and storage, the reproducibility and accuracy of the method can be problematic due to the influence of various blood parameters such as hematocrit on the obtained results. Also, a small amount of analyte captured on the paper used as transportation medium requires highly sensitive instruments such as mass spectrometers for analysis. This in turn necessitates sufficient sample cleanup to prevent problems with ionization, which is challenging to achieve with cellulose-based cards. Another issue related to DSB is the extraction of drugs from sampling paper and the stability of analytes on DBS [116, 117]. Recently, a thin-film format of SPME was proposed as a transportation and extraction medium for blood spot analysis [118]. Contrary to other proposed DBS and DBS-like assays, the approach described here (extracted blood spot [EBS]) employs all the steps of SPME, which mimics *in vivo* SPME sampling. These preliminary data showed interesting and promising steps toward new modifications of DBS analysis, which can address some of the issues of the conventional approach including sample cleanup and instability of the compounds. Moreover, in the same article Mirnaghi and Pawliszyn proposed direct coupling of EBS with mass spectrometer using Direct Analysis in Real Time (DART) as an excellent tool for rapid determination of target compounds.

In-tube SPME method was developed for the determination of interferon alpha$_{2a}$ from human plasma [38]. The drug is used for the treatment of patients with human immunodeficiency virus (HIV) or hepatitis C virus (HCV). The relationship between the drug concentration in blood and viral responses can be postulated; however, commonly used analytical methods based on bioassays, immunoassays, or electrophoretic separation are time consuming and expensive. The validation of the SPME method proposed by Chaves et al. [119] resulted in excellent precision (coefficient of variation [CV%] ranged between 3.1 and 7.9%), accuracy (106%–114%), limit of quantitation (LOQ) (0.06 MIU·mL^{-1}), and selectivity. The utilization of automation allows the completion of the measurement (sample preparation and LC analysis) within 25 minutes. Another in-tube and fiber SPME method suitable for the determination of benzodiazepines from serum [37] and whole blood [32] and urine [33], respectively, also provided suitable robustness and reproducibility of the assays.

Therapeutic drug monitoring is often applied for patients administered with antidepressant medications; therefore, new analytical approaches, including SPME, are still under development. Cantu et al. [120] proposed SPME-LC-UV with an additional diode-array detector method for the measurement of tricyclic antidepressants and anticonvulsants on plasma. Selection of appropriate coating, extraction and desorption time, temperature, ionic strength, and pH as well as investigation of the influence of plasma protein on method efficiency allowed the achievement of good interassay precision (RSD < 15%). LOD, quantification, and linear dynamic range were satisfactory and enabled the determination of the concentration of anticonvulsants and antidepressants from therapeutic to toxic levels.

Polypropylene fibers were also used for the determination of several antidepressants in plasma samples [121]. The authors studied the influence of various factors including temperature or ionic strength on the extraction process, as well as the possible interference of 24 different drugs to verify the selectivity of the assay. Successful validation was performed on plasma samples spiked with mirtazapine, citalopram, paroxetine, duloxetine, FLX, and sertraline. Subsequently, this method was used for monitoring drug concentration in geriatric patients 12 hours after drug administration. Concentrations of all studied drugs were within the range of therapeutic values.

Another SPME approach for the determination of tricyclic antidepressants was introduced by Alves et al. [122]. These authors used a commercial PDMS/DVB coating to analyze desipramine, imipramine, nortriptyline, amitriptyline, and clomipramine in plasma. A 30 minute extraction from plasma adjusted to alkaline pH was followed by a 20 minute desorption in the mobile phase at 50°C in the laboratory-made SPME-LC interface. For optimization of the experiment, the authors used factorial planning. The constructed empirical models were adjusted with 96%–98% of explained deviation, which allowed for adequate data set comprehension. With the use of electrospray ionization, the authors obtained an excellent detection level (0.1 ng/mL), good precision, accuracy, and linearity of the method.

Buszewski et al. [35] utilized SPME-LC/MS for the measurement of metoprolol, propranolol, oxprenolol, mexiletine, and propaphenone. The adrenolytic drugs were extracted from plasma samples by lab-made PPY fibers. The authors achieved very good reproducibility in the biological matrix ranging between 2.5% and 4.5%. For all studied pharmaceuticals, LODs and LOQs at low ng/mL were obtained. With the use of SPME coupled with chiral HPLC, de Oliveira et al. analyzed enantiomers of ibuprofen in human urine [123]. This popular nonsteroidal anti-inflammatory drug undergoes chiral metabolic inversion, resulting in stereoselective PK parameters. For this reason, enantioselective methods need to be used to assess the contribution of certain processes in drug metabolism. In the described study, the authors proposed the SPME method for direct extraction of the drug from acidified urine samples to increase extraction efficiency. Three commercial sorbents, CAR/TPR, polyacrylate, and PDMS/DVB, were tested with the final selection of PDMS/DVB. Extensive method development was performed, including investigation of drug stability and possible interfering drugs. Finally, the validated assay was used for determination of (–)-(R)-ibuprofen and (+)-(S)-ibuprofen at different time intervals in healthy volunteers after oral drug administration. The obtained profile was in agreement with previously reported data and confirmed the chiral metabolic inversion that occurs in ibuprofen metabolism.

Clenbuterol is a β-adrenergic drug used to treat pulmonary diseases; however, because of its anabolic steroid-like properties, it had begun to be used illegally as a growth promoter in feed for farm animals. Due the fact that clenbuterol can lead to various side effects when used in long-term or high-dose administration and because it is used as a doping agent in sports, monitoring of urinary and/or serum levels of this drug is recommended. Aresta et al. [124] presented a simple sample preparation and extraction SPME method followed by LC-UV analysis for determination of clenbuterol in both matrices. The authors used 60 minute extraction at 20°C with the use of commercial PDMS/DVB fibers from samples adjusted to pH 12 to enhance the recovery of this basic compound. The linearity was obtained as 5–500 ng/mL and 10–500 ng/mL for serum and urine samples, respectively, with very good inter- and intraday precisions (RSD < 6% and RSD < 10%, respectively).

The monitoring of a drug level is of special importance mostly in the case of pharmaceuticals with a narrow therapeutic concentration range, highly protein-bound drugs, or antibiotics, whose antimicrobial activity strictly correlates to pharmacokinetic/pharmacodynamic dosing parameters. Szultka et al. [36] proposed a new method of preparing PPY fibers with an improved interprobe variation and subsequently showed their applicability for the measurement of three drugs belonging to different classes of antibiotics from PBS, plasma, and whole blood [36]. SPME extraction was performed from the matrices at static and flow conditions, with the use of a heart–lung machine in the latter case. The artificial vein system was operated in the flow rate range of 50–500 mL/min, and the concentration of the spiked drugs was adjusted according to therapeutic values. The obtained results of method validation showed the feasibility of the developed method for pharmacokinetic studies of linezolid, daptomycin, and moxifloxacin *ex vivo* and its potential for *in vivo* application. In addition to PPY sorbent, the same group also presented performances of polythiophene (PTh) coatings for the determination of multiresistant *Staphylococcus aureus* (MRSA) antibiotics and validated the results against the conventional solid-phase extraction (SPE) method [125]. Other than good validation results, the new PTh coatings offered a 50% higher recovery of the tested drugs.

Several selective serotonin-reuptake inhibitors (venlafaxine, fluvoxamine, mirtazapine, FLX, citalopram, and sertraline) were analyzed by using GC/MS after SPME extraction from urine. The comparison of two calibration methods—by the use of standards spiked with the used matrix or water—did not show statistically significant differences, thus suggesting their interchangeable application in the proposed assays [126]. In addition to the targeted compounds, the authors were able to extract and putatively identify some metabolites of the studied drugs in samples of the patients treated with these medications.

Determination of verapamil was conducted to demonstrate the novel approach of SPME/surface-enhanced laser desorption/ionization-ion mobility spectrometry (SPME/SELDI-IMS) for the analysis of drugs or other molecules [20]. Wang et al. used specially modified PPY fibers (SPME/SELDI fibers) for the extraction of analytes from urine samples. After a 2 minute immersion of the tip of the probe (the "laser tip") in the sample matrix, the fiber was inserted into the IMS and laser pulses were fired to introduce the analytes to IMS. With the optimized laser parameters, the extraction recovery was 80% and the limit of detection was 2 µg/mL. The new SPME/SELDI-IMS device offers very fast analysis, which can be particularly beneficial in cases of unstable metabolites, because the total time of analysis including the sample preparation step can be within minutes.

SPME was recently used in clinical studies to experimentally verify if the concentration of tranexamic acid (TA) after applying the theoretically predicted dosing regime met the expected level [127,128]. The studies were performed on a group of patients undergoing cardiac surgery with the use of cardiopulmonary bypass. Two SPME approaches were proposed: the first utilized manual extraction with commercial C18 fibers [127], whereas the second implemented an automated high-throughput system with a thin-film-geometry coating, increasing the recovery of the drug [128]. The method was successfully validated against two conventional sample preparation methods: solvent precipitation and UF [127]. The results showed significant interpatient variability in the concentration of TA during surgery. It was also reported that the therapeutic concentration of TA was maintained in all patients in the studied group, although not all patients met the target value, but at the same time the drug concentration significantly exceeded the levels proposed by the theoretical model in individual patients [128].

SPME assay based on HS extraction followed by GC/MS analysis was developed to monitor the therapeutic concentrations of selegiline, selective inhibitor of monoamine oxidase type B, and its metabolite, desmethylselegiline, in whole blood and urine [129]. The total time of the SPME procedure was 27 minutes (25 minutes extraction and 2 minutes desorption). The results showed that the proposed methods met the validation criteria and enabled a quantitative analysis of the drug and metabolite in samples obtained from the patients with Parkinson's disease.

HS-SPME was also used for determination of tramadol from plasma [130], several club drugs [131], as well as four anesthetics and three analgesics [132] from urine. In the latter studies, the clinical applicability of the developed method was successfully verified in clinical practice for monitoring drug concentration in patients who had undergone coronary bypass surgery.

3.10   AUTOMATION AND HIGH THROUGHPUT
IN VITRO AND *IN VIVO* SPME

Nowadays, in various areas of life automation is an essential requirement. For many years, sample preparation was a bottleneck for analyses; thus, high throughput became a key target in analytical method development. The same applies to SPME, where automated systems and online coupling with various analytical instruments revolutionized the time and precision of the method. The first autosampler, Varian Model 8100, was introduced in 1992 [133]. With time, further developments of automated and semiautomated solutions were proposed, although most of them were focused on analyses of volatiles and semivolatiles by using GC instruments. When SPME-LC analysis became more popular due to the increased availability of sorbents suitable for such applications, the low throughput of analysis drove the need for method automation. The modification of an injection valve to enable desorption of the compounds extracted by fiber as well as in-tube devices are discussed in this chapter. However, this approach still did not ensure sufficient throughput. The introduction of the first 96 SPME devices compatible with common 96-well plates significantly improved the robustness of the method. Subsequent steps of development of high-throughput systems for SPME analysis can be found elsewhere in the literature [45,134,135]. In the current chapter, we describe only the main principle of the robotic SPME system for LC analysis, followed by recent examples of its application for drug analysis.

3.10.1   MULTI-WELL SYSTEM FOR SPME-LC ANALYSIS

The automated system consists of the arm moving in the *XYZ* mode to transport the 96-fiber (thin-film, disk, in-tip) brush to subsequent stations with solvents subjected to particular SPME steps. The first station accommodates a 96-well plate filled with the preconditioning solvent, the second station accommodates samples to be analyzed, the third stage accommodates water used for washing droplets and salts remaining after extraction from the biological matrix, and the fourth stage accommodates 96-well plate with the optimized desorption solvent. All four stations are agitated, what enhances extraction rate and allows for shorter analysis without compromising the recovery. Additionally, the system is equipped in evaporation unit, which enables preconcentration of the analytes, what is particularly beneficial for analysis of trace compounds, or reconstitution of the solvent in case of its incompatibility with the optimized LC method. The entire system is operated by the software; thus no intervention of the researcher is required. The high-throughput 96 CONCEPT robotic system is suitable for utilizing different coating geometries, that is, fiber, thin film, or disk [136–139]. The validation of the device performances presented on benzodiazepines as model drugs showed very good precision and reproducibility of the data in spite of the geometry of the extraction phase used. Increased recovery obtained with thin-film coatings due to the larger surface area and convenient shape has gained much interest, and new coatings suitable for thin-film applications are being proposed [140]. The utilization of an automated system with the new coatings offers several benefits including time savings due to the possibility of analyzing multiple parallel samples, cost savings due to the reusability of the extraction phase (at least 140 extractions from plasma), and higher recoveries compared with fiber coatings, which is a fine solution for the extraction of trace analytes with satisfactory recovery. Moreover, the newly developed coatings for automated 96-blade system offer extraction of much wider range of analytes and cover also polar species, which were beyond the extraction possibilities of sorbents used to date for SPME analysis [140]. Typically, the preparation of samples for LC analysis by using the 96 CONCEPT system takes about 1 minute per sample [141]. Recently, the applicability of the robotic system was verified in clinical application for determining the concentration of TA performed on 94 samples, and the total time for sample preparation was <3 minutes [128]. Current work on the improvement of the presented automated system is focused on further increasing the throughput and possible online coupling to LC platforms. For detail information and discussion on high-throughput SPME in multi-well-plate format the reader is referred to recent review by Vuckovic [135].

3.10.2 MULTI-FIBER HANDLING DEVICE

Although preequilibrium extraction offers fast *in vivo* analysis, desorption is still a time-limiting step of the entire process. To improve the throughput of *in vivo* pharmacokinetic analysis and provide greater convenience during transportation and storage of the fibers used for the experiment, Musteata and Pawliszyn [69] introduced the first prototype of the multifiber desorption device built from three plates. The first plate served as fiber support, the second was used for sealing the SPME samplers before and after extraction to protect them from contamination, and the third plate was used for the desorption process after removing plate 2. Zhang et al. [27] then proposed a parallel desorption device based on a 96-well plate. This version of the device consisted of a commercial 96-well plate covered with a silicon compression mat and an outer poly(tetrafluoroethylene) (PTFE) cover with 96 holes aligned to fit 96 wells on the plate. The cover was tightened with four stainless steel clips to prevent leakage of desorption solvent during desorption. For transportation and storage, SPME fibers were retracted into the needle assembly and placed in the desorption device. In the analyzing laboratory, clips were released and all fibers were simultaneously moved onto a 96-well plate containing the desorption solvent [27]. Although this device was more practical in use, it still showed some disadvantages. The procedure of coating the exposure to the desorption solvent required special attention to prevent coating harm due to hitting the well bottom and the top part of the device being open, thus making it prone to accidental fiber damage. The design of the device was later modified to address these issues. A later version contained a special base housing the 96-well plate and a middle part, which supported and aligned SPME probes placed inside the wells. The middle part also ensured the uniform and safe adjustment of height of all fibers during the desorption step. Finally, the top part acted as a cover protecting the fibers from contamination and damage. In addition, to avoid any undesired movement of the probe and to simplify the exposure of the fiber coating into the wells, a flat plastic material was placed on the top of the device. All parts were tightened with four stainless steel clips as proposed earlier. The design aimed for the compatibility of the device with any laboratory agitator handling multiwell plates and for the efficient stabilization of the device during agitation, resulting in good reproducibility of the desorption process. For this purpose, the middle part of the device firmly gripped the agitator so as to prevent the drop of the device during movement. Very recently, the newest model of the device was proposed [142]. Cudjoe and Pawliszyn built the device based on the previous concept by using lighter plastic materials and simplifying the overall design. They also evaluated the properties of the device paying special attention to inter- and intrawell reproducibility. Nonuniform agitation in all wells would lead to variations of the amount of analyte desorbed from the fiber and consequently the lack of accuracy and reproducibility of the results. The uniformity of agitation was investigated at two agitation speeds, 500 and 1200 rpm. Variability of the results (RSD) reported for the higher speed was ≤7% and for the lower speed was ≥15%. Although a higher speed of agitation provided better results, one must remember that the agitation speed should be optimized while taking into consideration the selected volume of desorption solvent to prevent fluid splash and cross-contamination. Inter- and intrawell variations for all studied drugs were ≤ 12%. These results were compared with the data obtained for the interfiber variability test. Fiber-to-fiber variation ranged between 10% and 13%; thus, it was concluded that observed inter- and intrawell differences were determined by the variability of fibers and were not related to agitation differences between all wells. The results of comparative studies with conventional off-line desorption revealed that the introduction of the multifiber handling device allowed for higher throughput and lower RSD, lesser researcher intervention, and reduction of possible contamination during the preparation.

3.11   CONCLUDING REMARKS

To summarize the benefits and drawbacks of pharmacokinetic studies by *in vivo* SPME, SPME offers the possibility of blood-drawing-free sampling, which enables the repetitive sampling of the same animal. This in turn eliminates interindividual variability, provides more precise data, and reduces the cost of the studies. Also, at the early stage of drug discovery where availability of the drug is limited, a single-animal experiment can be beneficial. In addition, drugs could be delivered to the organ directly, thus reducing drug need and providing better targeted delivery. Good temporal resolution can be obtained with preequilibrium extraction and various calibration methods, particularly the kinetic calibration approach, thereby ensuring fully quantitative analysis when a short time of sampling is used. Presented examples of *in vivo* sampling of dogs, rats, and mice showed that all pharmacokinetic parameters can be obtained from a single-animal study. New sampling devices give the opportunity of sampling not only blood but also solid tissues. The small dimension of the fiber allows for very good spatial resolution, which permits monitoring of drug distribution and accumulation in various parts of the organ. Although there is still a limited availability of commercial biocompatible fibers for LC analysis, increasing number of lab-made coatings and their better performances opens new possibilities for *in vivo* studies. Highly selective immunosorbents allows for the detection of trace analytes or endogenous compounds presented in very low concentrations, whereas innovative biocompatible modifications offers nonselective extraction of small molecules. A new prototype of mix-mode probes presents the opportunity to monitor global changes in living systems induced by external stimuli such as drug administration or other forms of medical intervention, by covering a wide and balanced range of analytes in terms of their physicochemical properties. *In vivo* studies give a unique opportunity to capture unstable, fast-turnover metabolites or short-lived species, which are undetectable by using conventional sample preparation methods. This feature can have an impact on understanding the physiological and pathological pathways involving such compounds, which have not yet been fully defined. Moreover, this offers the chance to find specific biomarkers that cannot be detected in *ex vivo* studies due to their instability. *In vivo* analysis of volatile compounds in breath and skin has already confirmed the clinical utility of SPME for biomarkers discovery. The simplicity of the method can also be used during other stages of the drug development process, for example, cell culture studies or protein binding evaluations. Finally, in clinical practice therapeutic drug monitoring can also be performed with the use of SPME. In this case, the use of automated systems dedicated for on-line and off-line analyses increases the throughput of sample preparation compatible with direct blood and homogenized tissue analysis in addition to facilitating time savings similar to other automated systems.

ACKNOWLEDGMENTS

The authors would like to acknowledge Natural Sciences and Engineering Research Council of Canada (NSERC), Canada Research Chairs (CRC) programs and Supelco for financial support of the research projects.

REFERENCES

1. Lin, J. H., and A. Y. Lu. 1997. Role of pharmacokinetics and metabolism in drug discovery and development. *Pharmacological Reviews* 49 (4): 403–49.
2. Liu, B., J. Chang, W. P. Gordon, J. Isbell, Y. Zhou, and T. Tuntland. 2008. Snapshot PK: A rapid rodent in vivo preclinical screening approach. *Drug Discovery Today* 13 (7–8): 360–7.
3. Brunner, M., and O. Langer. 2006. Microdialysis versus other techniques for the clinical assessment of in vivo tissue drug distribution. *The AAPS Journal* 8 (2): E263–71.
4. Musteata, F. M. 2009. Pharmacokinetic applications of microdevices and microsampling techniques. *Bioanalysis* 1 (1): 171–85.
5. Oehr, P. 2006. 'Omics'-based imaging in cancer detection and therapy. *Personalized medicine* 3 (1): 19–32.
6. Pawliszyn, J., ed. 2009. *Handbook of solid phase microextraction*. 2–56,144–172. Chemical Industry Press, Beijing, China.
7. Ligor, M., T. Ligor, A. Bajtarevic, et al. 2009. Determination of volatile organic compounds in exhaled breath of patients with lung cancer using solid phase microextraction and gas chromatography mass spectrometry. *Clinical Chemistry and Laboratory Medicine* 47 (5): 550–60.
8. Mochalski, P., B. Wzorek, I. Sliwka, and A. Amann. 2009. Suitability of different polymer bags for storage of volatile sulphur compounds relevant to breath analysis. *Journal of Chromatography B-Analytical Technologies in the Biomedical and Life Sciences* 877 (3): 189–96.
9. Syhre, M., J. M. Scotter, and S. T. Chambers. 2008. Investigation into the production of 2-pentylfuran by *Aspergillus fumigatus* and other respiratory pathogens in vitro and human breath samples. *Medical Mycology* 46 (3): 209–15.

10. Gaspar, E. M., A. F. Lucena, J. Duro da Costa, and H. Chaves das Neves. 2009. Organic metabolites in exhaled human breath: A multivariate approach for identification of biomarkers in lung disorders. *Journal of Chromatography A* 1216 (14): 2749–56.

11. Song, G., T. Qin, H. Liu, et al. 2010. Quantitative breath analysis of volatile organic compounds of lung cancer patients. *Lung Cancer* 67 (2): 227–31.

12. Fuchs, P., C. Loeseken, J. K. Schubert, and W. Miekisch. 2010. Breath gas aldehydes as biomarkers of lung cancer. *International Journal of Cancer* 126 (11): 2663–70.

13. Gong, Y., E. Li, G. Xu, et al. 2009. Investigation of propofol concentrations in human breath by solid-phase microextraction gas chromatography–mass spectrometry. *Journal of International Medical Research* 37 (5): 1465–71.

14. Peng, G., U. Tisch, O. Adams, et al. 2009. Diagnosing lung cancer in exhaled breath using gold nanoparticles. *Nature Nanotechnology* 4 (10): 669–73.

15. Bajtarevic, A., C. Ager, M. Pienz, et al. 2009. Noninvasive detection of lung cancer by analysis of exhaled breath. *BMC Cancer* 9: 348.

16. Syhre, M., L. Manning, S. Phuanukoonnon, P. Harino, and S. T. Chambers. 2009. The scent of mycobacterium tuberculosis: Part II breath. *Tuberculosis* 89 (4): 263–6.

17. Riazanskaia S., G. Blackburn, M. Harker, D. Taylor, and C. L. P. Thomas. 2008. The analytical utility of thermally desorbed polydimethylsilicone membranes for in-vivo sampling of volatile organic compounds in and on human skin. *The Analyst* 133 (8): 1020–27.

18. Thomas, A. N., S. Riazanskaia, W. Cheung, et al. 2010. Novel noninvasive identification of biomarkers by analytical profiling of chronic wounds using volatile organic compounds. *Wound Repair and Regeneration* 18 (4): 391–400.

19. Vuckovic, D., and J. Pawliszyn. 2011. Systematic evaluation of solid-phase microextraction coatings for untargeted metabolomic profiling of biological fluids by liquid chromatography–mass spectrometry. *Analytical Chemistry* 83 (6): 1944–54.

20. Wang, Y., S. Nacson, and J. Pawliszyn. 2007. The coupling of solid-phase microextraction/surface enhanced laser desorption/ionization to ion mobility spectrometry for drug analysis. *Analytica Chimica Acta* 582 (1): 50–4.

21. Musteata, F. M., M. L. Musteata, and J. Pawliszyn. 2006. Fast in vivo microextraction: A new tool for clinical analysis. *Clinical Chemistry* 52 (4): 708–15.

22. Schubert, J. K., W. Miekisch, P. Fuchs, et al. 2007. Determination of antibiotic drug concentrations in circulating human blood by means of solid phase micro-extraction. *Clinica Chimica Acta; International Journal of Clinical Chemistry* 386 (1–2): 57–62.

23. Musteata, F. M., I. de Lannoy, B. Gien, and J. Pawliszyn. 2008. Blood sampling without blood draws for in vivo pharmacokinetic studies in rats. *Journal of Pharmaceutical and Biomedical Analysis* 47 (4–5): 907–12.

24. Lord, H. L., R. P. Grant, M. Walles, B. Incledon, B. Fahie, and J. B. Pawliszyn. 2003. Development and evaluation of a solid-phase microextraction probe for in vivo pharmacokinetic studies. *Analytical Chemistry* 75 (19): 5103–15.

25. Wu, J., and J. Pawliszyn. 2004. Solid-phase microextraction based on polypyrrole films with different counter ions. *Analytica Chimica Acta* 520 (1–2): 257–64.

26. Es-haghi, A., X. Zhang, F. M. Musteata, H. Bagheri, and J. Pawliszyn. 2007. Evaluation of bio-compatible poly(ethylene glycol)-based solid-phase microextraction fiber for in vivo pharmacokinetic studies of diazepam in dogs. *The Analyst* 132 (7): 672–8.

27. Zhang, X., A. Es-haghi, F. M. Musteata, G. Ouyang, and J. Pawliszyn. 2007. Quantitative in vivo microsampling for pharmacokinetic studies based on an integrated solid-phase microextraction system. *Analytical Chemistry* 79 (12): 4507–13.

28. Musteata, M. L., F. M. Musteata, and J. Pawliszyn. 2007. Biocompatible solid-phase microextraction coatings based on polyacrylonitrile and solid-phase extraction phases. *Analytical Chemistry* 79 (18): 6903–11.

29. Zhou, S. N., K. D. Oakes, M. R. Servos, and J. Pawliszyn. 2008. Application of solid-phase microextraction for in vivo laboratory and field sampling of pharmaceuticals in fish. *Environmental Science & Technology* 42 (16): 6073–9.

30. Musteata, F. M., M. Walles, and J. Pawliszyn. 2005. Fast assay of angiotensin 1 from whole blood by cation-exchange restricted-access solid-phase microextraction. *Analytica Chimica Acta* 537 (1–2): 231–37.

31. Mullett, W. M., and J. Pawliszyn. 2001. Direct LC analysis of five benzodiazepines in human urine and plasma using an ADS restricted access extraction column. *Journal of Pharmaceutical and Biomedical Analysis* 26 (2–6): 899–908.

32. Walles, M., W. M. Mullett, and J. Pawliszyn. 2004. Monitoring of drugs and metabolites in whole blood by restricted-access solid-phase microextraction coupled to liquid chromatography–mass spectrometry. *Journal of Chromatography A* 1025 (1): 85–92.

33. Mullett, W. M., and J. Pawliszyn. 2002. Direct determination of benzodiazepines in biological fluids by restricted-access solid-phase microextraction. *Analytical Chemistry* 74 (5): 1081–7.

34. Lambert, J. P., W. M. Mullett, E. Kwong, and D. Lubda. 2005. Stir bar sorptive extraction based on restricted access material for the direct extraction of caffeine and metabolites in biological fluids. *Journal Chromatography A* 1075 (1–2): 43–9.

35. Buszewski, B., P. Olszowy, T. Ligor, et al. 2010. Determination of adrenolytic drugs by SPME-LC-MS. *Analytical and Bioanalytical Chemistry* 397 (1): 173–9.

36. Szultka, M., R. Kegler, P. Fuchs, et al. 2010. Polypyrrole solid-phase microextraction: A new approach to rapid sample preparation for the monitoring of antibiotic drugs. *Analytica Chimica Acta* 667 (1–2): 77–82.

37. Mullett, W. M., K. Levsen, D. Lubda, and J. Pawliszyn. 2002. Bio-compatible in-tube solid-phase microextraction capillary for the direct extraction and high-performance liquid chromatographic determination of drugs in human serum. *Journal of Chromatography A* 963 (1–2): 325–34.

38. Chaves, A. R., B. J. Silva, F. M. Lancas, and M. E. Queiroz. 2011. Biocompatible in-tube solid phase microextraction coupled with liquid chromatography–fluorescence detection for determination of interferon alpha in plasma samples. *Journal of Chromatography A* 1218 (21): 3376–81.

39. He, J., Z. Liu, L. Ren, et al. 2010. On-line coupling of in-tube boronate affinity solid phase microextraction with high performance liquid chromatography-electrospray ionization tandem mass spectrometry for the determination of *cis*-diol biomolecules. *Talanta* 82 (1): 270–6.

40. Lord, H. L., M. Rajabi, S. Safari, and J. Pawliszyn. 2006. Development of immunoaffinity solid phase microextraction probes for analysis of sub ng/mL concentrations of 7-aminoflunitrazepam in urine. *Journal of Pharmaceutical and Biomedical Analysis* 40 (3): 769–80.

41. Mullett, W. M., M. Walles, K. Levsen, J. Borlak, and J. Pawliszyn. 2004. Multidimensional on-line sample preparation of verapamil and its metabolites by a molecularly imprinted polymer coupled to liquid chromatography–mass spectrometry. *Journal of Chromatography B, Analytical Technologies in the Biomedical and Life Sciences* 801 (2): 297–306.

42. Mullett, W. M., P. Martin, and J. Pawliszyn. 2001. In-tube molecularly imprinted polymer solid-phase microextraction for the selective determination of propranolol. *Analytical Chemistry* 73 (11): 2383–9.

43. Hu, X., J. Pan, Y. Hu, and G. Li. 2009. Preparation and evaluation of propranolol molecularly imprinted solid-phase microextraction fibre for trace analysis of β-blockers in urine and plasma samples. *Journal of Chromatography A* 1216 (2): 190–7.

44. Koster, E. H., C. Crescenzi, W. den Hoedt, K. Ensing, and G. J. de Jong. 2001. Fibers coated with molecularly imprinted polymers for solid-phase microextraction. *Analytical Chemistry* 73 (13): 3140–5.

45. Vuckovic, D., X. Zhang, E. Cudjoe, and J. Pawliszyn. 2010. Solid-phase microextraction in bioanalysis: New devices and directions. *Journal of Chromatography A* 1217 (25): 4041–60.
46. Wang, Y., S. Nacson, and J. Pawliszyn. 2007. The coupling of solid-phase microextraction/ surface enhanced laser desorption/ionization to ion mobility spectrometry for drug analysis. *Analytica Chimica Acta* 582 (1): 50–4.
47. Ouyang, G., and J. Pawliszyn. 2008. A critical review in calibration methods for solid-phase microextraction. *Analytica Chimica Acta* 627 (2): 184–97.
48. Ai, J. 1997. Solid phase microextraction for quantitative analysis in nonequilibrium situations. *Analytical Chemistry* 69 (6): 1230–6.
49. Ai, J. 1997. Headspace solid phase microextraction. Dynamics and quantitative analysis before reaching a partition equilibrium. *Analytical Chemistry* 69 (16): 3260–6.
50. Chen, Y., J. O'Reilly, Y. Wang, and J. Pawliszyn. 2004. Standards in the extraction phase, a new approach to calibration of microextraction processes. *The Analyst* 129 (8): 702–3.
51. Chen, Y., and J. Pawliszyn. 2004. Kinetics and the on-site application of standards in a solid-phase microextraction fiber. *Analytical Chemistry* 76 (19): 5807–15.
52. Ouyang, G., J. Cai, X. Zhang, H. Li, and J. Pawliszyn. 2008. Standard-free kinetic calibration for rapid on-site analysis by solid-phase microextraction. *Journal of Separation Science* 31 (6–7): 1167–72.
53. Vuckovic, D., I. de Lannoy, B. Gien, et al. 2011. In vivo solid-phase microextraction for single rodent pharmacokinetics studies of carbamazepine and carbamazepine-10,11-epoxide in mice. *Journal of Chromatography A* 1218 (21): 3367–75.
54. Zhang, X., A. Es-Haghi, J. Cai, and J. Pawliszyn. 2009. Simplified kinetic calibration of solid-phase microextraction for in vivo pharmacokinetics. *Journal of Chromatography A* 1216 (45): 7664–9.
55. Zhou, S. N., W. Zhao, and J. Pawliszyn. 2008. Kinetic calibration using dominant pre-equilibrium desorption for on-site and in vivo sampling by solid-phase microextraction. *Analytical Chemistry* 80 (2): 481–90.
56. Ouyang, G., S. Cui, Z. Qin, and J. Pawliszyn. 2009. One-calibrant kinetic calibration for on-site water sampling with solid-phase microextraction. *Analytical Chemistry* 81(14):5629–36.
57. Koziel, J., M. Jia, and J. Pawliszyn. 2000. Air sampling with porous solid-phase microextraction fibers. *Analytical Chemistry* 72 (21): 5178–86.
58. Yeung, J. C., D. Vuckovic, and J. Pawliszyn. 2010. Comparison and validation of calibration methods for in vivo SPME determinations using an artificial vein system. *Analytica Chimica Acta* 665 (2): 160–6.
59. Ouyang, G., K. D. Oakes, L. Bragg, et al. 2011. Sampling-rate calibration for rapid and nonlethal monitoring of organic contaminants in fish muscle by solid-phase microextraction. *Environmental Science & Technology* 45 (18): 7792–8.
60. Paschke, A., and P. Popp. 2004. Diffusion-based calibration for solid-phase microextraction of benzene, toluene, ethylbenzene, *p*-xylene and chlorobenzenes from aqueous samples. *Journal of Chromatography A* 1025 (1): 11–6.
61. Grandison, M. K., and F. D. Boudinot. 2000. Age-related changes in protein binding of drugs: Implications for therapy. *Clinical Pharmacokinetics* 38 (3): 271–90.
62. Musteata, F. M. 2011. Monitoring free drug concentrations: Challenges. *Bioanalysis* 3 (15): 1753–68.
63. Vuckovic, D., and J. Pawliszyn. 2009. Automated study of ligand–receptor binding using solid-phase microextraction. *Journal of Pharmaceutical and Biomedical Analysis* 50 (4): 550–5.
64. Musteata, F. M., J. Pawliszyn, M. G. Qian, J. T. Wu, and G. T. Miwa. 2006. Determination of drug plasma protein binding by solid phase microextraction. *Journal of Pharmaceutical Sciences* 95 (8): 1712–22.

65. Zhang, X., K. D. Oakes, D. Luong. et al. 2010. Temporal resolution of solid-phase microextraction: Measurement of real-time concentrations within a dynamic system. *Analytical Chemistry* 82 (22): 9492–9.
66. Zhang, X., J. Cai, K. D. Oakes, F. Breton, M. R. Servos, and J. Pawliszyn. 2009. Development of the space-resolved solid-phase microextraction technique and its application to biological matrices. *Analytical Chemistry* 81 (17): 7349–56.
67. Zhang, X., K. D. Oakes, S. Cui, L. Bragg, M. R. Servos, and J. Pawliszyn. 2010. Tissue-specific in vivo bioconcentration of pharmaceuticals in rainbow trout (*Oncorhynchus mykiss*) using space-resolved solid-phase microextraction. *Environmental Science & Technology* 44 (9): 3417–22.
68. Lord, H. L., X. Zhang, F. M. Musteata, D. Vuckovic, and J. Pawliszyn. 2011. In vivo solid-phase microextraction for monitoring intravenous concentrations of drugs and metabolites. *Nature Protocols* 6 (6): 896–924.
69. Musteata, F. M., and J. Pawliszyn. 2007. In vivo sampling with solid phase microextraction. *Journal of Biochemical and Biophysical Methods* 70 (2): 181–93.
70. Valenzano, K. J., L. Tafesse, G. Lee, et al. 2005. Pharmacological and pharmacokinetic characterization of the cannabinoid receptor 2 agonist, GW405833, utilizing rodent models of acute and chronic pain, anxiety, ataxia and catalepsy. *Neuropharmacology* 48 (5): 658–72.
71. Manitpisitkul, P., and R. E. White. 2004. Whatever happened to cassette-dosing pharmacokinetics? *Drug Discovery Today* 9 (15): 652–8.
72. Vuckovic, D., R. Shirey, Y. Chen, et al. 2009. In vitro evaluation of new biocompatible coatings for solid-phase microextraction: Implications for drug analysis and in vivo sampling applications. *Analytica Chimica Acta* 638 (2): 175–85.
73. Vuckovic D., I. de Lannoy, B. Gien, et al. 2011. In vivo solid-phase microextraction: Capturing the elusive portion of metabolome. *Angewandte Chemie International Edition England* 50 (23): 5344–8.
74. Yeung J. C., I. de Lannoy, B. Gien, et al. 2012. Semi-automated in vivo solid-phase microextraction sampling and the diffusion-based interface calibration model to determine the pharmacokinetics of methoxyfenoterol and fenoterol in rats. *Analytica Chemica Acta* 742: 37–44.
75. Nakajima, D., T. T. Win-Shwe, M. Kakeyama, H. Fujimaki, and S. Goto. 2006. Determination of toluene in brain of freely moving mice using solid-phase microextraction technique. *Neurotoxicology* 27 (4): 615–8.
76. Honma, T., and M. Suda. 2004. Brain microdialysis study of the effects of hazardous chemicals on the central nervous system 2. Toluene exposure and cerebral acetylcholine. *Industrial Health* 42 (3): 336–47.
77. Win-Shwe, T. T., D. Mitsushima, D. Nakajima, et al. 2007. Toluene induces rapid and reversible rise of hippocampal glutamate and taurine neurotransmitter levels in mice. *Toxicology Letters* 168 (1): 75–82.
78. Zhou, S. N., G. Ouyang, and J. Pawliszyn. 2008. Comparison of microdialysis with solid-phase microextraction for in vitro and in vivo studies. *Journal of Chromatography A* 1196–1197: 46–56.
79. Cudjoe, E. 2011. Pre-equilibrium in vivo solid phase microextraction for monitoring drug concentration changes in rat brain. Paper presented at Toronto Post-ASMS Discussion Group Meeting, Toronto, Canada.
80. Togunde O. P., H. Lord, K. D. Oakes, M. R. Servos, and J. Pawliszyn. 2013. Development and evaluation of a new in vivo solid-phase microextraction sampler. *Journal of Separation Science* 36: 219–23.
81. Cudjoe E., B. Bojko, P. Togunde, and J. Pawliszyn. 2013. In vivo solid-phase microextraction for tissue bioanalysis. *Bioanalysis* 4(21): 2605–2619.

82. Sofrata, A., E. M. Santangelo, M. Azeem, A. K. Borg-Karlson, A. Gustafsson, and K. Putsep. 2011. Benzyl isothiocyanate, a major component from the roots of *Salvadora persica*, is highly active against gram-negative bacteria. *PloS One* 6 (8): e23045.

83. Goni, P., P. Lopez, C. Sanchez, R. Gomez-Lus, R. Becerril, and C. Nerin. 2009. Antimicrobial activity in the vapour phase of a combination of cinnamon and clove essential oils. *Food Chemistry* 116 (4): 982–89.

84. Hossain, Z. S. M., B. Bojko, and J. Pawliszyn. 2013. Automated SPME–GC–MS monitoring of headspace metabolomic responses of E. coli to biologically active components extracted by the coating. *Analytica Chimica Acta*, In press.

85. Walles, M., W. M. Mullett, K. Levsen, J. Borlak, G. Wünsch, and J. Pawliszyn. 2002. Verapamil drug metabolism studies by automated in-tube solid phase microextraction. *Journal of Pharmaceutical and Biomedical Analysis* 30 (2): 307–19.

86. Mullett, W. M., K. Levsen, J. Borlak, J. Wu, and J. Pawliszyn. 2002. Automated in-tube solid-phase microextraction coupled with HPLC for the determination of N-nitrosamines in cell cultures. *Analytical Chemistry* 74 (7): 1695–701.

87. Kemper, R. A., D. L. Nabb, S. A. Gannon, T. A. Snow, and A. M. Api. 2006. Comparative metabolism of geranyl nitrile and citronellyl nitrile in mouse, rat, and human hepatocytes. *Drug Metabolism and Disposition: The Biological Fate of Chemicals* 34 (6): 1019–29.

88. Bianchi, F., M. Mattarozzi, M. Careri, et al. 2010. An SPME-GC-MS method using an octadecyl silica fibre for the determination of the potential angiogenesis modulators 17-beta-estradiol and 2-methoxyestradiol in culture media. *Analytical and Bioanalytical Chemistry* 396 (7): 2639–45.

89. Poli, D., M. V. Vettori, P. Manini, et al. 2004. A novel approach based on solid phase microextraction gas chromatography and mass spectrometry to the determination of highly reactive organic compounds in cells cultures: Styrene oxide. *Chemical Research in Toxicology* 17 (1): 104–9.

90. Pyo, J. S., H. K. Ju, J. H. Park, and S. W. Kwon. 2008. Determination of volatile biomarkers for apoptosis and necrosis by solid-phase microextraction–gas chromatography/mass spectrometry: A pharmacometabolomic approach to cisplatin's cytotoxicity to human lung cancer cell lines. *Journal of Chromatography B-Analytical Technologies in the Biomedical and Life Sciences* 876 (2): 170–4.

91. Musteata, F. M., and J. Pawliszyn. 2005. Study of ligand–receptor binding using SPME: Investigation of receptor, free, and total ligand concentrations. *Journal of Proteome Research* 4 (3): 789–800.

92. Yuan, H., R. Ranatunga, P. W. Carr, and J. Pawliszyn. 1999. Determination of equilibrium constant of alkylbenzenes binding to bovine serum albumin by solid phase microextraction. *The Analyst* 124 (10): 1443–8.

93. Yuan, H., and J. Pawliszyn. 2001. Application of solid-phase microextraction in the determination of diazepam binding to human serum albumin. *Analytical Chemistry* 73 (18): 4410–6.

94. Heringa, M. B., D. Pastor, J. Algra, W. H. Vaes, and J. L. Hermens. 2002. Negligible depletion solid-phase microextraction with radiolabeled analytes to study free concentrations and protein binding: An example with [3H]estradiol. *Analytical Chemistry* 74 (23): 5993–7.

95. Heringa, M. B., B. van der Burg, J. C. van Eijkeren, and J. L. Hermens. 2004. Xenoestrogenicity in in vitro assays is not caused by displacement of endogenous estradiol from serum proteins. *Toxicological Sciences : An Official Journal of the Society of Toxicology* 82 (1): 154–63.

96. Theodoridis, G. 2006. Application of solid-phase microextraction in the investigation of protein binding of pharmaceuticals. *Journal of Chromatography B-Analytical Technologies in the Biomedical and Life Sciences* 830 (2): 238–44.

97. Bojko B., D. Vuckovic, and J. Pawliszyn. 2012. Comparison of solid phase microextraction versus spectroscopic techniques for binding studies of carbamazepine. *Journal of Pharmaceutical and Biomedical Analysis* 66: 91–9.
98. Musa-Veloso, K., S. S. Likhodii, and S. C. Cunnane. 2002. Breath acetone is a reliable indicator of ketosis in adults consuming ketogenic meals. *The American Journal of Clinical Nutrition* 76 (1): 65–70.
99. Grote, C., and J. Pawliszyn. 1997. Solid-phase microextraction for the analysis of human breath. *Analytical Chemistry* 69 (4): 587–96.
100. Spinhirne, J. P., J. A. Koziel, and N. K. Chirase. 2003. A device for non-invasive on-site sampling of cattle breath with solid-phase microextraction. *Biosystems Engineering* 84 (2): 239–46.
101. Spinhirne, J. P., J. A. Koziel, and N. K. Chirase. 2004. Sampling and analysis of volatile organic compounds in bovine breath by solid-phase microextraction and gas chromatography–mass spectrometry. *Journal of Chromatography A* 1025 (1): 63–9.
102. Buszewski, B., A. Ulanowska, T. Ligor, N. Denderz, and A. Amann. 2009. Analysis of exhaled breath from smokers, passive smokers and non-smokers by solid-phase microextraction gas chromatography/mass spectrometry. *Biomedical Chromatography* 23 (5): 551–6.
103. Ueta, I., Y. Saito, M. Hosoe, et al. 2009. Breath acetone analysis with miniaturized sample preparation device: In-needle preconcentration and subsequent determination by gas chromatography–mass spectroscopy. *Journal of Chromatography B-Analytical Technologies in the Biomedical and Life Sciences* 877 (24): 2551–6.
104. Zhang, Z. M., J. J. Cai, G. H. Ruan, and G. K. Li. 2005. The study of fingerprint characteristics of the emanations from human arm skin using the original sampling system by SPME-GC/MS. *Journal of Chromatography B-Analytical Technologies in the Biomedical and Life Sciences* 822 (1–2): 244–52.
105. Soini, H. A., K. E. Bruce, I. Klouckova, R. G. Brereton, D. J. Penn, and M. V. Novotny. 2006. In situ surface sampling of biological objects and preconcentration of their volatiles for chromatographic analysis. *Analytical Chemistry* 78 (20): 7161–8.
106. Abaffy, T., R. Duncan, D. D. Riemer, et al. 2010. Differential volatile signatures from skin, nevi and melanoma: A novel approach to detect a pathological process. *Plos One* 5 (11): e13813.
107. Acevedo, C. A., E. Y. Sanchez, J. G. Reyes, and M. E. Young. 2007. Volatile organic compounds produced by human skin cells. *Biological Research* 40 (3) (2007): 347–55.
108. Zimmermann, D., M. Hartmann, M. P. Moyer, J. Nolte, and J. I. Baumbach. 2007. Determination of volatile products of human colon cell line metabolism by GC/MS analysis. *Metabolomics* 3 (1): 13–7.
109. Di Cagno, R., M. De Angelis, I. De Pasquale, et al. 2011. Duodenal and faecal microbiota of celiac children: Molecular, phenotype and metabolome characterization. *BMC Microbiology* 11: 219.
110. Maccaferri, S., B. Vitali, A. Klinder, P. Brigidi, and A. Costabile. 2011. Rifaximin modulates the colonic microbiota of patients with Crohn's disease: An in vitro approach using a continuous culture colonic model system—authors' response. *Journal of Antimicrobial Chemotherapy* 66 (5): 1194–5.
111. Bojko, B., M. Wasowicz, and J. Pawliszyn. 2013. Metabolic profiling of plasma from cardiac surgical patients concurrently administered with tranexamic acid. DI-SPME-LC-MS analysis. *Journal of Pharmaceutical Analysis*, In press.
112 Bessonneau V., B. Bojko, and J. Pawliszyn. 2013. Analysis of human saliva metabolome by direct immersion solid-phase microextraction LC and benchtop orbitrap MS. *Bioanalysis* 5 (7): 783–92.
113. Cudjoe, E., C. Hamani, and J. Pawliszyn. 2011. Solid-phase microextraction method for monitoring endogenous compounds in the pre-frontal cortex of freely moving rats during deep brain stimulation. Paper presented at Pittcon Conference.

114. de Wildt, S. N. 2011. Profound changes in drug metabolism enzymes and possible effects on drug therapy in neonates and children. *Expert Opinion on Drug Metabolism & Toxicology* 7 (8): 935–48.
115. Shi, S., and U. Klotz. 2011. Age-related changes in pharmacokinetics. *Current Drug Metabolism* 12 (7): 601–10.
116. Millership, J. S. 2011. Microassay of drugs and modern measurement techniques. *Pediatric Anesthesia* 21 (3): 197–205.
117. Li, W., and F. L. Tse. 2010. Dried blood spot sampling in combination with LC-MS/MS for quantitative analysis of small molecules. *Biomedical chromatography BMC* 24 (1): 49–65.
118. Mirnaghi F. M. and J. Pawliszyn. 2012. Reusable solid-phase microextraction coating for direct immersion whole-blood analysis and extracted blood spot sampling coupled with liquid chromatography–tandem mass spectrometry and direct analysis in real time–tandem mass spectrometry. *Analytical Chemistry* 84: 8301–9.
119. Chaves, A. R., B. J. Silva, F. M. Lancas, and M. E. Queiroz. 2011. Biocompatible in-tube solid-phase microextraction coupled with liquid chromatography–fluorescence detection for determination of interferon alpha in plasma samples. *Journal of Chromatography A* 1218 (21): 3376–81.
120. Cantu, M. D., D. R. Toso, C. A. Lacerda, F. M. Lancas, E. Carrilho, and M. E. Queiroz. 2006. Optimization of solid-phase microextraction procedures for the determination of tricyclic antidepressants and anticonvulsants in plasma samples by liquid chromatography. *Analytical and Bioanalytical Chemistry* 386 (2): 256–63.
121. Chaves, A. R., G. Chiericato Jr., and M. E. Queiroz. 2009. Solid-phase microextraction using poly(pyrrole) film and liquid chromatography with UV detection for analysis of antidepressants in plasma samples. *Journal of Chromatography B-Analytical Technologies in the Biomedical and Life Sciences* 877 (7): 587–93.
122. Alves, C., A. J. Santos-Neto, C. Fernandes, J. C. Rodrigues, and F. M. Lancas. 2007. Analysis of tricyclic antidepressant drugs in plasma by means of solid-phase microextraction-liquid chromatography–mass spectrometry. *Journal of Mass Spectrometry* 42 (10): 1342–7.
123. de Oliveira, A. R. M., E. J. Cesarino, and P. S. Bonato. 2005. Solid-phase microextraction and chiral HPLC analysis of ibuprofen in urine. *Journal of Chromatography B-Analytical Technologies in the Biomedical and Life Sciences* 818 (2): 285–91.
124. Aresta, A., C. D. Calvano, F. Palmisano, and C. G. Zambonin. 2008. Determination of clenbuterol in human urine and serum by solid-phase microextraction coupled to liquid chromatography. *Journal of Pharmaceutical and Biomedical Analysis* 47 (3): 641–5.
125. Olszowy, P., M. Szultka, P. Fuchs, et al. 2010. New coated SPME fibers for extraction and fast HPLC determination of selected drugs in human blood. *Journal of Pharmaceutical and Biomedical Analysis* 53 (4): 1022–7.
126. Salgado-Petinal, C., J. P Lamas, C. Garcia-Jares, M. Llompart, and R. Cela. 2005. Rapid screening of selective serotonin re-uptake inhibitors in urine samples using solid-phase microextraction gas chromatography–mass spectrometry. *Analytical and Bioanalytical Chemistry* 382 (6): 1351–9.
127. Bojko, B., D. Vuckovic, E. Cudjoe, et al. 2011. Determination of tranexamic acid concentration by solid phase microextraction and liquid chromatography–tandem mass spectrometry: First step to in vivo analysis. *Journal of Chromatography B-Analytical Technologies in the Biomedical and Life Sciences* 879: 3781–7.
128. Bojko, B., D. Vuckovic, F. M. Mirnaghi, et al. 2012. Therapeutic monitoring of tranexamic acid concentration: High throughput analysis with solid phase microextraction. *Therapeutic Drug Monitoring* 34: 31–7.

129. Kuriki, A., T. Kumazawa, X. P. Lee, et al. 2006. Simultaneous determination of selegiline and desmethylselegiline in human body fluids by headspace solid-phase microextraction and gas chromatography–mass spectrometry. *Journal of Chromatography B; Analytical Technologies in the Biomedical and Life Sciences* 844 (2): 283–91.
130. Sha, Y. F., S. Shen, and G. L. Duan. 2005. Rapid determination of tramadol in human plasma by headspace solid-phase microextraction and capillary gas chromatography–mass spectrometry. *Journal of Pharmaceutical and Biomedical Analysis* 37 (1): 143–7.
131. Brown, S. D., D. J. Rhodes, and B. J. Pritchard. 2007. A validated SPME-GC-MS method for simultaneous quantification of club drugs in human urine. *Forensic Science International* 171 (2–3): 142–50.
132. Raikos, N., G. Theodoridis, E. Alexiadou, et al. 2009. Analysis of anaesthetics and analgesics in human urine by headspace SPME and GC. *Journal of Separation Sciences* 32 (7): 1018–26.
133. Arthur, C. L., L. M. Killam, K. D. Buchholz, J. Pawliszyn, and J. R. Berg. 1992. Automation and optimization of solid-phase microextraction. *Analytical Chemistry* 64 (17): 1960–6.
134. O'Reilly, J., O. Wang, L. Setkova, et al. 2005. Automation of solid-phase microextraction. *Journal of Separation Sciences* 28 (15): 2010–22.
135 Vuckovic D. 2013. High-throughput solid-phase microextraction in multi-well-plate format. *Trends in Analytical Chemistry*, 45: 136–53.
136. Vuckovic, D., E. Cudjoe, D. Hein, and J. Pawliszyn. 2008. Automation of solid-phase microextraction in high-throughput format and applications to drug analysis. *Analytical Chemistry* 80 (18): 6870–80.
137. Cudjoe, E., and J. Pawliszyn. 2009. A new approach to the application of solid phase extraction disks with LC-MS/MS for the analysis of drugs on a 96-well plate format. *Journal of Pharmaceutical and Biomedical Analysis* 50 (4): 556–62.
138. Cudjoe, E., D. Vuckovic, D. Hein, and J. Pawliszyn. 2009. Investigation of the effect of the extraction phase geometry on the performance of automated solid-phase microextraction. *Analytical Chemistry* 81 (11): 4226–32.
139. Mirnaghi, F. S., Y. Chen, L. M. Sidisky, and J. Pawliszyn. 2011. Optimization of the coating procedure for a high-throughput 96-blade solid phase microextraction system coupled with LC-MS/MS for analysis of complex samples. *Analytical Chemistry* 83 (15): 6018–25.
140 Mirnaghi F. S. and J. Pawliszyn. 2012. Development of coatings for automated 96-blade solid phase microextraction-liquid chromatography–tandem mass spectrometry system, capable of extracting a wide polarity range of analytes from biological fluids. *Journal of Chromatography A* 1261: 91–8.
141. Vuckovic, D., E. Cudjoe, F. M. Musteata, and J. Pawliszyn. 2010. Automated solid-phase microextraction and thin-film microextraction for high-throughput analysis of biological fluids and ligand-receptor binding studies. *Nature Protocols* 5 (1): 140–61.
142. Cudjoe, E., and J. Pawliszyn. 2012. A multi-fiber handling device for in vivo solid phase microextraction-liquid chromatography-mass spectrometry applications. *Journal of Chromatography A* 1232: 77–83.

Identification and Detection of Antibiotic Drugs and Their Degradation Products in Aquatic Samples

Dror Avisar and Igal Gozlan

CONTENTS

4.1 INTRODUCTION

Freshwater is a natural resource that is scarce all over the world. Potable water in semiarid regions in particular is even scarcer, due to the rapidly increasing population. This population produces enormous amounts of wastewater (two-thirds of the annual water consumption per capita), which enters wastewater treatment plants (WWTPs) as raw sewage. WWTPs are not the only solution available for water reuse, although they are the most conventional engineering technology in use throughout the industrialized world, functioning as independent units for water reclamation. In the process, emphasis is placed on water quality, which is characterized by the chemical, physical, and microbiological constituents present in water. These common contaminants are generally removed during the wastewater treatment process, discharging an effluent of satisfactory quality for environmental targets such as irrigation of agricultural fields and municipal parks, artificial recharge to groundwater, and rehabilitation of rivers and streams (Figure 4.1).

To improve the quality of the effluent, quantitative and qualitative analysis of traditional constituents is commonly carried out, although in the past 20 years a new generation of organic micropollutants has arisen. These nano/micropollutants are residues of pharmaceutical compounds and their metabolites, which originate in human, agricultural, and industrial sources (Figure 4.1). The issue of pharmaceuticals, especially antibiotics, in the aquatic environment has raised increasing concern in recent years. Human and veterinary pharmaceuticals are a group of "emerging" contaminants (Richardson, 2004), some of which are produced and increasingly used in large volumes each year. The amounts produced are approaching quantities similar to those of pesticides and other organic pollutants. Residues of these biologically active compounds can enter the aquatic environment via transport pathway emissions during manufacture, disposal of unused or expired medicines, human and animal excretion in urine and feces, direct discharge of aquaculture products, and spreading of animal manure (Gentili, 2006) (Figure 4.1).

The majority (25%–75%) (Kulshrestha et al., 2004) of pharmaceutical compounds enter aquatic systems after ingestion and subsequent excretion as nonmetabolized parent compounds or as degradation products (DPs) via the wastewater

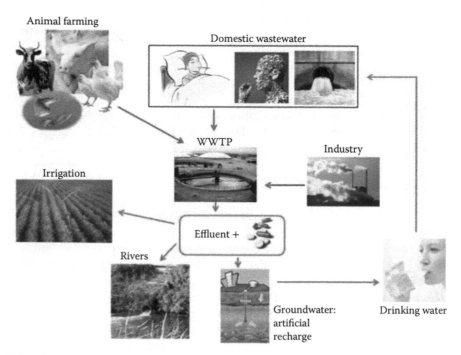

FIGURE 4.1 (See color insert.) Sources of pharmaceutical contaminants and their pathways into the aqueous environment.

treatment network (Derksen et al., 2004). Once they enter the WWTP, the pharmaceutical pollutants are transported via the multiple barrier stages of the plant. When eventually discharged into the environment, they have been only partially (20%–80%) eliminated (or not eliminated at all) in the water treatment process (Halling-Sørensen et al., 1998; Heberer, 2002). Several studies have shown that WWTPs do not remove these drugs sufficiently and their DPs are discharged to various aquatic environments (such as irrigation water, streams, rivers, groundwater, and drinking water) at low concentrations ranging from nanograms per liter to micrograms per liter (Zuehlke, et al., 2004). Being a new field of research, scientists throughout the world face a major problem with regard to finding available data. Having said that, several studies that were conducted throughout Europe have confirmed the occurrence of pharmaceutical residues (e.g., antiphologistic, blood lipid regulators, anti-inflammatory drugs, and antibiotics) in surface water and groundwater. In Germany, Heberer et al. (1998) detected trace contaminants of several pharmaceutical residues in surface water. They found clofibric acid, diclofenac, and ibuprofen at levels of up to 1.9 µg/L in water that had passed through the WWTPs in Berlin. In another part of Germany (Frankfurt/Main), elimination of different drugs was calculated during passage through a municipal WWTP (Ternes, 1998). It was found that the level of clofibric acid had decreased by 51%, that of diclofenac by 69%, that of ibuprofen by 90%, and that of carbamazepine by only 7%. In Switzerland, Soulet et al. (2002) proved that only a small amount of clofibric acid was removed from water by the WWTP. Moreover, in two cases out of the six examined, the concentration in the effluent was higher than the concentration in the influent (i.e., raw sewage). Similar results in domestic wastewater in the United States were described by Garrison et al. (1976). They found concentrations of clofibric acid in raw wastewater of 0 and 0.8 µg/L and in activated sludge effluent of 1 and 2 µg/L. The authors suggest that the clofibric acid could have been adsorbed on activated sludge particulate matter from earlier raw sewage effluents and partially desorbed during sampling. The higher concentration in the effluent could also be explained by the delay in water movement through the plant. Thus influent and effluent samples may not represent the same parcel of water. In a recent study, Tauxe-Wuersch et al. (2005) pointed out that clofibric acid was not degraded during the treatment process in Lausanne and Morges, Switzerland, regardless of the sampling period. Although clofibric acid is not usually eliminated from wastewater, its removal can be facilitated in some WWTPs by using the treatment processes outlined in their study.

The increasing use of these drugs and the formation of their DPs are increasing their presence in the environment. As mentioned earlier, elimination of these compounds by WWTPs was found to be rather low, and consequently effluent-containing antibiotic residues and their DPs may lead to the genetic selection of more harmful and resistant bacteria, which is a matter of great concern.

Additionally, due to their physicochemical properties (i.e., high water solubility), these compounds are able to survive physically by passing through natural barriers and may also reach groundwater and drinking water (Petrovic et al., 2005; Avisar et al., 2009a,b).

4.2   METABOLITES: PHASES I AND II

From the first moment they are introduced to the body, medical compounds are submitted to the metabolic system, which affects their pharmacokinetic behavior, concentration in the blood, and clinical effect on the target organ. Biotransformation is the main removal path of medical compounds from the body, in which their solubility increases and they are eliminated through bodily secretions. These metabolic processes considerably determine the appearance of pharmaceutical compounds including their DPs in raw wastewater that arrives at WWTPs.

It is agreed to classify the metabolites that are produced within the body into two stages: Phase I and Phase II (Halling-Sørensen et al., 1998; Al-Hadithi, 2007), which include oxidation–reduction and hydrolysis processes that convert the molecule to a more hydrophilic form. The drug hydrophilic process may occur directly by the substitution and addition of functional groups (e.g., hydroxylation) or by changing existing functional groups as follows (Al-Hadithi, 2007):

1. Reduction of ketones and aldehydes to alcohol
2. Oxidation of alcohol to acid
3. Hydrolysis of esters and amides to create amines, carboxylic acids, and alcohol groups
4. Reduction of azo and nitro compounds to create amine
5. Dealkalization of oxidized –S, –O, and –N to obtain –SH, –OH, and NH_2

Phase I products might be reactive and even more ecotoxic than the parent molecules (Isidori et al., 2005; Zhang et al., 2008). The most important enzyme in the metabolic system of Phase I is cytochrome P450, which is responsible for the oxidation process catalysis of drugs, especially in liver endoplasmic tissue. "Nicotinamide adenine dinucleotide phosphate (NADPH)–cytochrome P450 reductase" transfers electrons from NADPH to cytochrome P450. As a result, the cytochrome, which is reduced, serves as an electron carrier and participates in the oxidation process of the drugs.

In Phase II, the polar metabolites interact with organic molecules originating in the body, such as amino acid, glycin, sulfates, and glucoronic acid. The existing conjugation paths in Phase II are glucuronidation, acetylation, methylation, and glutathione conjugation. Although these paths signify the end of biological activity of the DPs, only conjugation with glutathione protects the body from toxication. For example, diclofenac passes first conjugation with glucuronic acid to form diclofenac 1-O-acetyl glucuronide, which is potentially toxic to the body. Then it transacylation of glutathione, during it exchange the glucuronic acid with glutathione. This process converts the compound to a nontoxic form.

The formation of a combination between glucuronides and glucoronic acid is the most common conjugation process due to the uridine diphosphoglucuronic acid (UDP)–glucuronic acid factor required for this process. This cofactor exists in all body tissues and, besides its participation in the synthesis of glycogen, it acts as a high-energy medium that responds to the mobility of glucuronic acid in the body. N-glucuronides are obtained from conjugations of amines (mostly aromatic), amides, and sulfonamides with UPD–glucuronic acids. Conjugations derived from acetylation characterize aromatics, amines, and sulfonamides. The cofactor acetyl CoA participates in this process. Conjugated metabolites are not clinically active and are mostly nontoxic; additionally, their solubility is raised and they are thus excreted via urine.

The biotransformation of medical compounds in the body, as described earlier, has an implication on the reactivity of drugs, their clinical activity, and their toxicity to consumers. In some cases, metabolites can have a similar activity as the original drugs. For example, the metabolite of benzodiazepine (a drug for the treatment of insomnia and anxiety) has effects long after the original drug is decomposed. The effect of antidepressant drugs like imipramine and amitriptyline is due to their DPs dsemethylimipramine and nortriptyline, respectively. In other cases, drugs that are called "prodrugs" become active only after there are metabolites in the body. The metabolic process might even change the medical property of a drug, like acetylsalicylic acid (aspirin), which is anti-inflammatory but affects antiplatelet activity. The metabolite salicylic acid has an anti-inflammatory property, but it does not affect blood platelets. Occasionally, a drug metabolite can activate another metabolite that might affect the functionality of another system in the body. This phenomenon is known for analgesic drugs like paracetamol, which can cause damage to the liver by the metabolism of N-acetylpbenzoquinone imine (Al-Hadithi, 2007).

In a WWTP, the aggregates that were obtained in the body during the conjugation process (Phase II) and the medical compounds including their DPs might be released in their free form. Thus, the concentration of medical compounds in the treated water might be higher than their concentration in the influent (Miao et al., 2005).

4.3 ANTIBIOTICS AND THEIR DEGRADATION PRODUCTS

Antibiotics are one class of antimicrobials (Table 4.1), a larger group that also includes antiviral, antifungal, and antiparasitic drugs. They are relatively harmless to the host and therefore can be used to treat infections. The term, coined by Selman Waksman, originally described only formulations derived from living organisms, in contrast to "chemotherapeutic agents," which were purely synthetic. Nowadays, the term antibiotic is also applied to synthetic antimicrobials, such as sulfa drugs. Antibiotics are small molecules with molecular weights less than 2000 g/mol. Some antibiotics are made from mold. Antibiotics are used daily. The ultimate goal of their use is to kill bacteria and prevent the growth of microorganisms. The numerous types of bacteria correlate to different types of antibiotics that have been developed to attack them (Minneman et al., 2005).

Antibiotics can be released to the environment in their original form or as their DPs after a pharmacokinetic process in the body. Both forms can pass through a degradation process in the environment.

Furthermore, the antibiotic DPs that are produced in the environmental conditions and can exist in some aquatic sources can be even more toxic than the parent antibiotics themselves.

Many studies describe them as unknown or even dangerous factors in the environment, but only a few describe their natural and chemical properties, especially

TABLE 4.1
Group of Known Antibiotics and Related Information on Each Type

Group	Antibiotics	pK$_a$ (25°C)	log K$_{ow}$	MW (g/mol)
Aminoglycosides	Neomycin	12.9	−3.70	614.6
	Streptomycin	NA	NA	581.6
	Kanamycin	7.2	NA	484.5
β-Lactams	Penicillin G	2.62	1.67	334.4
	Ampicillin	2.61	1.35	349.4
	Amoxicillin	2.68; 7.49; 9.63	−3.78 (log D at pI)	365.4
	Ceftiofur	2.62	0.54	523.6
Macrolides	Tylosin	13	3.41	917.1
	Tilmicosin	13.16	5.09	869.1
	Erythromycin	8.4	2.6	733.9
	Clarithromycin	8.4	3.24	747.5
	Azithromycin	8.91; 9.57	3.00	748.5
	Roxithromycin	9.06	2.44	836.5
	Oleandomycin	7.7	NA	785.9
Sulfonamides	Sulfamethoxine	6.69	0.42	310.3
	Sulfamethoxine	7.45	0.8	278.3
	Sulfanilamide	10.6	−0.62	172.2
	Sulfadimidine	7.6	0.89	278.3
	Sulfadiazine	6.4	−0.09	250.3
	Sulfapyridine	8.4	0.35	249.3
	Sulfamethoxazole	5.81; 2.0	1.04	253.3
Tetracyclines	Chlortetracycline	4.5	−0.17	478.1
	Oxytetracycline	4.5	−1.86	460.4
	Tetracycline	3.3–9.6	−0.78	444.1
Lincosamides	Lincomycin	12.9	0.86	406.5
Fluoroquinolones	Enrofloxacin	2.74	2.53	359.4
	Norfloxacin	8.68; 5.77	1.44	319
	Danofloxacin	2.73	1.85	357.4
	Sarafloxacin	6	NA	385.4
	Oxolinic acid	6.9	NA	261.2

Note: pK$_a$ = acidity constant; log K$_{ow}$ = octanol–water partition coefficient; MW = molecular weight; NA = not applicable.

how the degradation of antibiotics occurs in the environment (such as tetracyclines [Halling-Sorensen et al., 2002], sulfamides [Nabil and Al-Hadithi, 2007], amoxicillins [AMXs] [Deshpande et al., 2004; Lamm et al., 2009; Gozlan et al., 2010], erythromycin [Hirsch et al., 1999], and others).

Researchers in the United States (Lindsey et al., 2001; Zhu et al., 2001; Vogel et al., 2003; Batt et al., 2006), Germany (Hirsch et al., 1999; Sacher et al., 2001; Hamscher et al., 2002), and Denmark (Holm et al., 1995) have investigated the occurrence of antibiotics sourced from WWTP effluents, confined animal feeding operations (CAFOs), and landfill runoff. These studies did not find any detectable levels of tetracyclines in groundwater wells. In a study of CAFO lagoons and the groundwater downgradient from the lagoons, Zhu et al. (2001) found tetracyclines such as oxytetracycline (OTC), tetracycline, and chlortetracycline in 23 out of 26 lagoon samples but no detectable levels of tetracyclines in 11 groundwater samples. An investigation by Hamscher et al. (2002) found tetracycline and chlortetracycline residues in fields fertilized with liquid manure to be extremely immobile in soil. They found concentrations of 86.2 µg/kg (0–10 cm), 198.7 µg/kg (10–20 cm), and 171.7 µg/kg (20–30 cm) tetracycline and 4.6–7.3 µg/kg chlortetracycline (in all three sublayers) in soil treated with liquid manure containing 4.0 mg/kg tetracycline and 0.1 mg/kg chlortetracycline. OTC (Table 4.1) is an antibiotic commonly used for veterinary therapy in aquaculture, mainly as a growth promoter and to prevent infections (Coyne et al., 2004). Lalumera et al. (2004) found that 75% of the given drug is released from the body of a fish through secretions as a nonmetabolitic molecule and then directly discharged into the aquatic environment. Avisar et al. (2009) found OTC in groundwater wells located beneath fish ponds that were habitually treated with OTC as a growth promoter. Avisar indicated that due to specific hydrological conditions that developed in the vadose zone beneath the fish ponds (i.e., a preferential saturated flow path), a leaching to groundwater of the sorption-prone molecule (OTC) was shown.

The sulfonamides antibacterial group is considered one of the first antimicrobial drugs used and it paved the way for the antibiotic revolution in medicine. It has been intensively manufactured since the mid-1940s. Presently, only a few sulfonamide derivatives are still in clinical use because of developed bacterial resistance (Tilles, 2001). Sulfamethoxazole (SMX; $C_{10}H_{11}N_3O_3S$), a representative antibacterial compound of the sulfonamides group, is considered an emerging microcontaminant of concern due to its potential adverse effects on ecosystems and human health (Hirsch et. al., 1999; Lindsey et. al., 2001). The presence of antibiotic residues in groundwater was intensively investigated, mainly below effluent recharge ponds (Hirsch et al, 1999; Sacher et al., 2001), feedlots (Lindsey et al., 2001; Zhu et al., 2001; Hamscher et al., 2002; Batt et al., 2006), landfills (Holm et al., 1995), and riverbanks (Sacher et al., 2001; Vogel et al., 2005). In many cases, where geohydrological information is available, it can be seen that antibiotic residues were mainly detected in shallow phreatic aquifers 3–6 m below the ground surface (Krapac et al., 2002; Batt et. al., 2006). Avisar et al. (2009) reported that SMX had infiltrated through the unsaturated zone beneath an intensively effluent-irrigated field and was detected in a nearby deep drinking well (90 m).

4.3.1 DEGRADATION PRODUCTS OF SELECTED ANTIBIOTICS

Most of the studies worked on parent detection and not on their DPs. Particularly, there is an urgent need for the additional detection of DPs, as has been shown, for example, in the case of a penicillin group represented by AMX (Lamm et al., 2009; Gozlan et al., 2010; Perez-Parada et al., 2011) and the macrolides group represented by erythromycin (Huber et al., 2005). In most cases, the active parent compounds could not be found in various matrixes of water samples, although the main DPs that were chemically identified were detected in the environment. Thus, it is essential to investigate the behavior of antibiotic DPs in various aquatic environments.

4.3.1.1　Amoxicillin

AMX is a moderate-spectrum β-lactam antibiotic used to treat bacterial infections caused by susceptible microorganisms. It is usually the drug of choice within the class because it is better absorbed, following oral administration, than other β-lactam antibiotics. AMX is susceptible to degradation by β-lactamase-producing bacteria, and so it may be given with clavulanic acid to increase its susceptibility. AMX acts by inhibiting the synthesis of bacterial cell walls. It inhibits cross-linkage between linear peptidoglycan polymer chains that make up a major component of the cell walls of gram-positive bacteria. It is a moderate-spectrum antibiotic that is active against a wide range of gram-positive and a limited range of gram-negative organisms. AMX is similar to ampicillin in its bactericidal action against susceptible organisms during the stage of active multiplication. It acts through the inhibition of biosynthesis of cell wall mucopeptides (Minneman et al., 2005).

Amoxicillin belongs to a group of drugs that are excreted unmetabolized and was expected to present a heavy predicted environmental load, as it is one of the most consumed antibiotics in the world for both human and veterinary purposes.

Penicillin is a unique molecule with a fused β-lactam–thiazolidine ring system, wherein the strained β-lactam ring is susceptible to cleavage by a variety of reagents as well as some enzymes. The labile β-lactam ring of penicillins and other β-lactam antibiotics is characterized by its pronounced susceptibility to various nucleophiles (Figure 4.2), metal ions (Figure 4.3), oxidizing agents, or even solvents like water and alcohol. Some of these are reported in the literature (Blaha et al., 1976; Bentley and Southgate, 1988), but there are few reviews on the chemical and biological degradation of β-lactam. The instability of β-lactam antibiotics in solution was observed to be a major hurdle in the development of penicillin and other useful β-lactam antibiotics. Therefore, degradation or stability studies of β-lactam antibiotics have been of paramount importance not only for their market availability but also to evaluate their pharmacokinetic properties and adverse reactions. It is also interesting to note that the stabilities or rates of degradation of different members of β-lactam antibiotics *in vivo* as well as *in vitro* are quite different. However, the major pathways of their degradation have remained similar, leading to various breakdown products in a majority of the β-lactams. Furthermore, various antigenic determinants of penicillin responsible for its allergenicity have also been shown to occur from many of their DPs or through the formation of penicilloyl proteins.

The degradation of penicillins and cephalosporins has been studied in the presence of various metal ions, for example, mercury (Tutt and Schwartz, 1971), zinc (Ayim and Rapson, 1972; Navarrom et al., 2003) cadmium (Martinez et al., 1999), cobalt (Gutierez et al., 1998), and copper (Chatterjee et al., 1988). It was found that these ions catalyzed the rate of inactivation or hydrolytic opening of β-lactams.

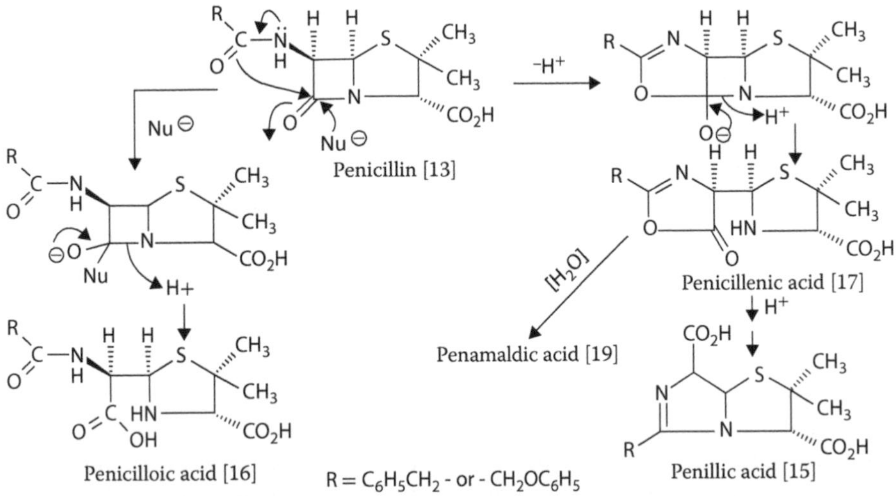

FIGURE 4.2 Nucleophilic ring opening and acid sensitivity of penicillin due to the influence of acyl side chain. (From Deshpande et al., *Current Science*, 87, 12–25, 2004.)

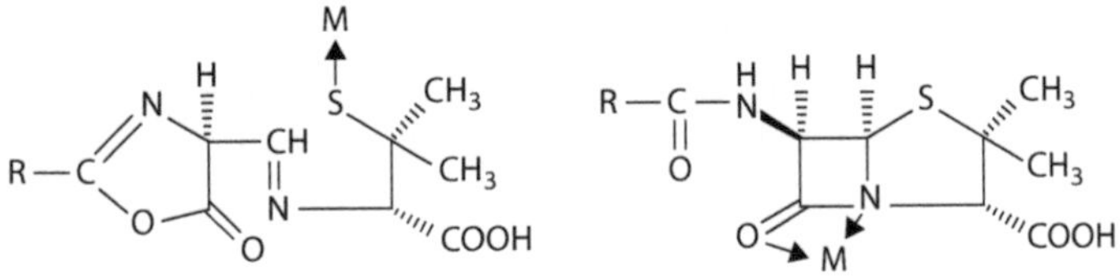

FIGURE 4.3 Degradation of penicillin via metal-ion catalysis to form metal mercaptide and chelate. (Deshpande et al., *Current Science*, 87, 12–25, 2004.)

Such metal ion–triggered assisted degradation was supposed to happen through the formation of a single intermediate substrate–metal complex or two intermediate substrate–metal complexes (Navarrom et al., 2003). It has been established that metal ions catalyze penicillin degradation via two routes, that is, first by interacting with the sulfur atom of the penicillenic acid intermediate to form the metal mercaptide (Figure 4.3) and second by complexing the penicillin with a metal ion to form the chelate (Figure 4.3).

It has been observed that the former pathway plays a major role during penicillin degradation by metal catalysis.

4.3.1.2 Degradation Products of Amoxicillin

Worldwide, there seems to be a misconception among environmental hydrochemists concerning the occurrence of aminopenicillins in the aquatic environment. There have been few attempts to detect traces of penicillins in general and aminopenicillins in particular, especially AMX, in environmental samples (Hirsch et al., 1998; Hirsch et al., 1999; Sacher et al., 2001; Ternes, 2001; Calamari et al., 2003; Christian et al., 2003; Cahill et al., 2004; Zuccato et al., 2005; Xu et al., 2007) and hospital wastewater (Lindberg et al., 2004; Lindberg et al., 2005). On the one hand, they are the most consumed antibiotics worldwide and are therefore expected to be found at relatively high concentrations in wastewater and surface water. Yet, on the other hand, these compounds are tremendously difficult to trace. Given that the β-lactam ring is susceptible to opening, it is likely that the parent compound undergoes a degradation process. If that is the case, they are almost impossible to trace and efforts should be placed on determining their DPs in the environment, which is a challenge that was never confronted before. Therefore, Lamm et al. (2009) undertook this task by attempting to detect the traces of one of the major DPs of AMX, its diketopiperazine-2′,5′ form (ADP), by analyzing wastewater samples (i.e., raw sewage and effluent). The results indicate that the diketopiperazine AMX may be present as two main isomers: (1) amoxicillin-2(S)-piperazine-2,5-dione and (2) amoxicillin-2(R)-piperazine-2,5-dione (Valvo et al., 1997; Nagele and Moritz, 2005). Furthermore, additional DPs, such as AMX penicilloic acid, AMX penilloic acid, and phenol hydroxipyrazin, can be obtained together with AMX-S-oxide (Figure 4.4) (Gozlan et al., 2010).

ADPs were detected in secondary effluents from several WWTPs and are therefore constantly discharged to the environment in WWTP effluents used for agricultural irrigation. Lamm et al. (2009) was the first to evidentially prove the occurrence of a relatively stable form of AMX, its diketopiperazine-2′,5′ derivative, in wastewater and secondary effluent samples. In addition, Perez-Parada et al. (2011) also found diketopiperazine-2′,5′ and AMX penicilloic acid in wastewater and rivers.

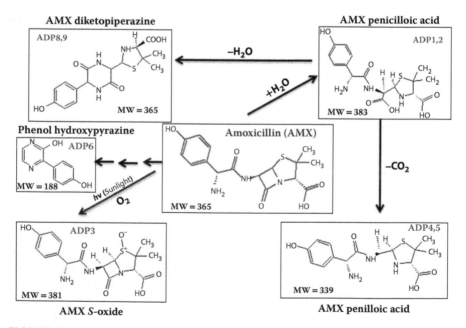

FIGURE 4.4   Suggested degradation pathway of amoxicillin in aqueous media.

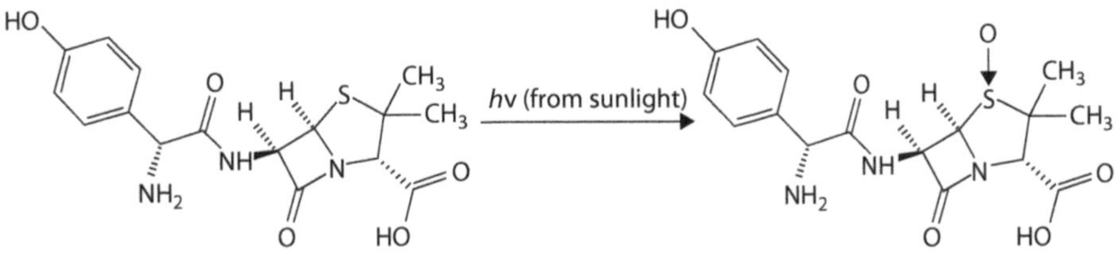

FIGURE 4.5 Formation of AMX-*S*-oxide obtained from AMX by photolysis under sunlight irradiation.

In particular, the product named AMX-*S*-oxide was first identified by Gozlan et al. (2010) and was obtained only under sunlight irradiation, simply as an indirect photolysis process. In addition, a significant augmentation of AMX-*S*-oxide production was obtained in the presence of natural photosensitizers (humic acids), which activate oxygen to oxidize the AMX, forming the AMX-*S*-oxide DP (Figure 4.5).

AMX-*S*-oxide is the only AMX-DP that sustains the β-lactam ring, which is the biologically active site in the AMX molecule. The possible presence of this compound in aquatic environments (e.g., wastewater, rivers, and groundwater) is of great concern, because the AMX-*S*-oxide β-lactam ring is still active and may lead to the development of resistant bacteria and even cause other possible health hazards to humans and/or wild and domestic animals. The formation of the AMX-*S*-oxide DP is accelerated in secondary effluent, probably due to the presence of natural photosensitizers, which exist in the solution. Furthermore, the fact that this product is also obtained under an ozonation process should be taken into consideration when using this technique to remove antibiotic residues (Gozlan et al., 2010).

4.3.1.3 Erythromycin

Among the antibiotics shown in Table 4.1, erythromycin, azithromycin, clarithromycin, and roxithromycin are macrolide antibiotics. These antibiotics are characterized by the property of inhibiting bacterial protein synthesis. Erythromycin prevents bacteria from growing by interfering with their protein synthesis. It binds to the 23S rRNA molecule in the 50S of the bacterial ribosome, blocking the exit of the growing peptide chain and thus inhibiting the translocation of peptides.

These antibiotics are mostly excreted in the bile. In structure, a macrocyclic compound contains a 14-membered lactone ring with 10 asymmetric centers and 2 sugars (L-cladinose and D-desoamine), making it a very difficult compound to produce by synthetic methods. Macrolides belong to the polyketide class of natural products. Erythromycin has an antimicrobial spectrum similar to or slightly wider than that of penicillin and is often used in people who have an allergy to penicillin. For respiratory tract infections, it has better coverage of atypical organisms, including mycoplasma and *Legionella*. It is also used to treat outbreaks of chlamydia, syphilis, acne, and gonorrhea. Erythromycin is produced from a strain of the actinomycete *Saccharopolyspora erythraea*, formerly known as *Streptomyces erythraeus*. Elimination of trace clarithromycin and erythromycin in WWTPs has been studied. Ternes et al. (2003) reported elimination of 0.21 µg/L clarithromycin and 0.62 µg/L erythromycin. A similar result was reported by Huber et al. (2005).

Gozlan et al. (2010) routinely detected dehydrated erythromycin in secondary efflu-ent (2008–2009) in concentrations ranging from 500 to 10,000 ng/L. Recently, they found that the dehydrated-erythromycin concentration in secondary effluents has dramatically decreased compared to azithromycin (1993 ng/L), clarithromycin (468 ng/L), and roxithromycin (13 ng/L), due to the local pharmaceutical industry stopping the production of this drug in 2010; in fact, the most consumed macrolides at the time were azithromycin, clarithromycin, and roxithromycin.

Most of the erythromycin is metabolized by demethylation in the liver. Its main elimination route is in the bile and a small portion in the urine. The elimination half-life of erythromycin is 1.5 hours.

4.3.1.4 Degradation Product of Erythromycin

Recent experiments to confirm the analytical methods for extraction (SPE) and detection of erythromycin (LC-MS) showed that erythromycin was not detected in its original form but as a DP with the apparent loss of one molecule of water. Figure 4.6 shows the suggested pathway to obtain the product due to the elim-ination of one molecule of water. It was proved that this loss already occurs in aqueous solutions at pH < 7.0 (Roth and Fenner, 1994; Hirsch et al., 1999). In pharmacologic literature, a metabolic water abstraction by the formation of an intermolecular spiroketal (Figure 4.6) is described, which accelerates at pH

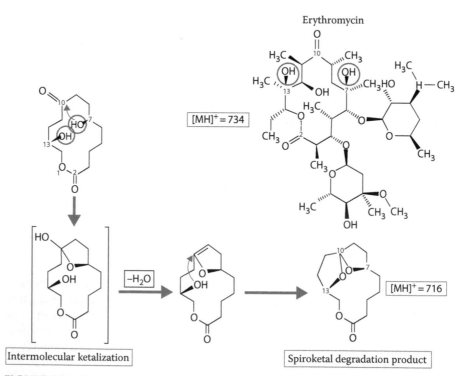

FIGURE 4.6 Suggested pathway for obtaining the spiroketal degradation product. (From Roth and Fenner, 1994.)

values below 4.0. Obviously, the same can also occur in aqueous surroundings at even higher pH values. The imperative question was to determine which form of erythromycin was present in environmental water samples. Only the degraded product was detectable, showing that the water molecule is probably already lost when the compound is present in the aquatic environment. The fact that erythromycin is applied orally and thus has to pass through strongly acidic conditions in the stomach possibly causes the degradation of erythromycin. According to the literature, the DP no longer exhibits antibiotic properties (Roth and Fenner, 1994). Other macrolide antibiotics (derivatives of erythromycin), such as roxithromycin, azithromycin, and clarithromycin, do not exhibit strong pH sensitivity. Since they are structurally modified, they are not able to react to the spiroketal form (Gozlan et al., 2010).

Gozlan et al. (2010) shows that erythromycin, which was degraded under controlled environmental conditions (under sunlight radiation), resulted in ERY-*N*-oxide and ERY-*N*-oxide-H$_2$O—spiroketal products (Figure 4.7).

The major products were isolated by preparative high-performance liquid chromatography (HPLC) and identified by liquid chromatography/mass spectrometry

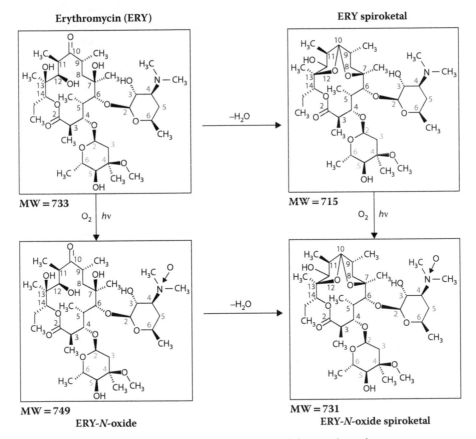

FIGURE 4.7 Suggested erythromycin degradation path in aquatic environments.

(LC/MS) and nuclear magnetic resonance (NMR) techniques. In contrast to the spiroketal products, which lost their antibiotic activity, the ERY-*N*-oxide product probably still functioned as an antibiotic drug. Consequently, Gozlan et al. (2010) detected this DP in secondary effluent (10 μg/L) and groundwater (20 ng/L) located beneath an irrigated agricultural field. This is probably the first study that proves the occurrence of the chemically stable forms of erythromycin, that is, ERY-*N*-oxide and ERY-*N*-oxide-H$_2$O, in effluent and groundwater.

4.4 ANALYTICAL METHODS

At the beginning of pharmaceutical studies (late 1990s), the main focus of most of the studies was to develop analytical methods to identify and detect pharmaceutical residues in trace concentrations (nanograms up to micrograms per liter) from environmental samples. Ternes (2001) developed an analytical method for determining the concentration levels, down to the lower nanograms per liter range, of 18 antibiotics in water. The analytes studied belonged to different groups of antibiotics such as penicillins, tetracyclines, sulfonamides, and macrolides. Nowadays, higher levels of accuracy can be achieved in measurements determining antibiotic residues and their DPs in environmental samples. Such data has been collected by Díaz-Cruz and Barceló (2005) and Díaz-Cruz et al. (2003). They developed analytical methods for determining trace antimicrobials in water, using HPLC coupled with MS. Others, such as Golet et al. (2002), were able to conduct both environmental exposure and risk assessments of fluoroquinolone antibacterial agents in wastewater and river water in Switzerland. Further analytical methods, such as preparative HPLC (separation and isolation of DPs) and NMR (identification of molecular structure of DPs), are required.

4.4.1 IDENTIFICATION OF AMOXICILLIN DEGRADATION PRODUCTS

The obtained AMX DPs were isolated from the AMX sample solutions used for degradation processes. The isolation process was carried out using a preparative HPLC system with semipreparative C$_{18}$ and phenyl columns. The preparative system consisted of a large column with a compatible flow rate and a large volume injection capability, which enables high-throughput DP separation, isolation, and collection.

To complete the identification of AMX DPs, MS and NMR analyses were performed. The principle of NMR is based on the fact that the nuclei of all elements carry a charge. When the spins of the protons and neutrons comprising a nucleus are not paired, the overall spin of the charged nucleus generates a magnetic dipole along the spin axis and the intrinsic magnitude of this dipole is a fundamental nuclear property called the nuclear magnetic moment, μ. The symmetry of charge distribution in the nucleus is a function of its internal structure, and if this is spherical (i.e., analogous to the symmetry of a 1s hydrogen orbital) it is said to have a corresponding spin angular momentum number of $I = 1/2$, examples of which are ^{1}H, ^{13}C, ^{15}N, ^{19}F, ^{31}P, and so on. The NMR spectrum output indicates the specific functional groups that a specific molecule consists of, thus allowing the characterization of known and unknown structural molecules.

For example, the structural molecular identification of AMX-*S*-oxide DP (Figure 4.8) using the NMR technique was determined by Gozlan et al. (2010). Decisive proof for the proposed AMX-*S*-oxide ([MH]$^+$ = 382, obtained by MS) structure was obtained by comparing the ^{1}H NMR spectra of AMX and of the isolated AMX-*S*-oxide. A comparison of their ^{1}H NMR spectra is presented in Table 4.2.

A lone pair of electrons resides on the AMX sulfur atom, giving it a tetrahedral molecular geometry. Subsequently, the oxygen-bonded sulfur turns into a chiral center. As a result, the bonded oxygen may appear at two positions: (1) isomer-a: *cis* to H$_{(4)}$ and CH$_{3(11)}$ and *trans* to H$_{(2)}$ and CH$_{3(10)}$; and (2) isomer-b: *trans* to H$_{(4)}$ and CH$_{3(11)}$ and *cis* to H$_{(2)}$ and CH$_{3(10)}$. To determine the position of the bonded oxygen, an nuclear overhauser effect (NOE) experiment was conducted. The NOE results point to isomer-a, due to the existence of through-space interactions between H$_{(2)}$ (shift from δ 5.44 ppm AMX to 5.25 ppm AMX-*S*-oxide) to CH$_{3(10)}$ (shift from δ 1.36 to 1.20 ppm) and H$_{(4)}$ (shift from δ 4.09 to 4.31 ppm) to CH$_{3(11)}$ (shift from δ 1.37 to 1.57 ppm) (Table 4.2). In comparison with the AMX NMR spectrum, the chemical shift of CH$_{3(11)}$ and H$_{(4)}$ of AMX-*S*-oxide, which were shifted downfield, indicate that CH$_{3(11)}$ and H$_{(4)}$ are in a *cis* position relative to the bonded oxygen. Furthermore,

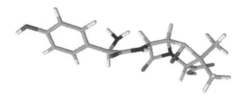

FIGURE 4.8 **(See color insert.)** Structure of AMX-*S*-oxide.

TABLE 4.2
^{1}H NMR Spectral Data for AMX and AMX-*S*-Oxide

Hydrogen	AMX (chemical shift)	AMX-*S*-oxide (chemical shift)
CH$_3$	1.36, 1.37 (s)	1.20, 1.57 (s)
H-4	4.09 (s)	4.31 (s)
H-17	5.07 (m)	4.99 (m)
H-6	5.44 (m)	5.25 (m)
H-2	5.44 (m)	5.90 (m)
H-21,23	6.92 (d)	6.94 (d)
H-20,24	7.31 (d)	7.31 (d)

$CH_{3(10)}$ and $H_{(2)}$, which were shifted upfield, signify that $CH_{(10)}$ and $H_{(2)}$ are in a *trans* position relative to the bonded oxygen (Table 4.2). The *cis* position of the bonded oxygen of isomer-a, relative to $CH_{3(11)}$ and $H_{(4)}$, is explained by the steric hindrance caused by the carbonyl group.

4.4.2 DETECTION OF AMOXICILLIN DEGRADATION PRODUCTS

In general, the analytical method consists of four steps: (1) sampling, (2) filtration, (3) extraction (solid-phase extraction [SPE]), and (4) HPLC-MS.

Of all the stages of the analytical method sample preparation is the most time-consuming procedure, and its goal is to pretreat the samples before detection and analysis. This is an essential stage in the method, and it takes place between sample collection and sample analysis by HPLC-MS. Depending on the type of sample (e.g., groundwater, raw sewage, or effluent), different procedures are preferred to prepare the sample for analytical measurement. However, the result is the same: an HPLC vial contains the concentrated analytes, ready for injection and analysis using HPLC.

Prior to extraction, a coarse and/or fine filtration of the sample is required. The main goal is to reduce the total load of suspended solids in the sample to obtain a smooth percolation without clogging the SPE cartridge.

To detect environmental trace concentrations of any targeted drug, it is obligatory to preconcentrate the analytes and to quantify them from environmental samples. SPE is an extraction method that uses a solid phase (sorbent) and a liquid phase (solvent) to isolate analytes from complex matrices. It is typically used to clean up and preconcentrate a sample before using a chromatographic or another analytical method to quantify the amount of analytes in the sample (Simpson, 2000). This technique uses cartridges that contain chemically modified sorbents to achieve high selectivity and sensitivity in the extraction of a specific type of analyte.

The general SPE procedure consists of five essential steps, as illustrated in Figure 4.9:

1. Cartridge conditioning to launch the active site of the cartridge
2. Sample addition by loading the sample onto the SPE cartridge
3. Rinsing by washing away undesired impurities
4. Eluting the desired analytes with a strong solvent into a collection tube
5. Evaporation and reconstitution by redissolving in a solvent compatible with the chromatograph

The SPE extraction of AMX DPs in aquatic environments was carried out using specific methods for the DPs (Table 4.3) (Gozlan et al., 2010). After sample preparation is completed, the extracted antibiotics are ready for HPLC-MS analysis.

Chromatography is a technique that separates a mixture into its individual components (analytes); a chromatograph is coupled with a detection unit that can characterize each type of analyte appropriately. Development of a chromatographic method is for gaining high sensitivity and selectivity in a practical time. The sensitivity can be improved by using a gradient elution program, which allows the

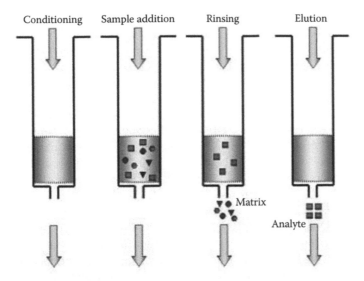

FIGURE 4.9 Essential four main steps for solid-phase extraction procedure (not including step 5 as mentioned in the text). (From the European Mycotoxin Awareness Network, http://services.leatherheadfood.com/eman/FactSheet.aspx?ID=79.)

TABLE 4.3
SPE Procedure for AMX DPs Collected from Aquatic Environments

DP	ADP1/2	ADP3,4/5	ADP6,8/9
Cartridge type	Strata X-C (500 mg/6 mL)	Oasis Max (500 mg/6 mL)	HR-P (500 mg/6 mL)
Activation	1.15 mL MeOH 2.15 mL phosphate buffer (pH 2.0)	1.6 mL MeOH 2.6 mL H$_2$O	1.6 mL MeOH 2.6 mL H$_2$O
Loading	1000 mL sample (pH adjusted to 3.0 with formic acid (FA)	1000 mL sample (pH was adjusted to 7.5 with NaOH before loading)	1000 mL sample (pH was adjusted to 4.5 before loading)
Washing	1.15 mL of FA (at pH 3.0) 2.15 mL MeOH	6 mL of H$_2$O at pH 7.5 (phosphate buffer)	6 mL of H$_2$O (adjusted to 4.5 of TFA)
Elution	15 mL phosphate buffer (pH 8.0)	1.4 mL of 1 M NaCl of MeOH:H$_2$O (15:85) at pH 4.5 2.6 mL of MeOH:H$_2$O (50:50)	30 mL of MeOH:H$_2$O (75:25) at pH 7.5 (phosphate buffer)
Evaporation	Lyophilization	N$_2$, 40°C	N$_2$, 40°C
Reconstitution	1 mL H$_2$O (pH = 4)	1 mL H$_2$O	1 mL H$_2$O

establishment of relatively narrow and high peaks, and by increasing the sample volume by dissolving the sample in a weak eluent. In addition, the chromatographic system (HPLC) is coupled with a mass spectrometer (MS) via two main sources: atmospheric pressure chemical ionization (APCI) and electrospray ionization (ESI), which is a better source for easily ionized molecules. The MS uses three modes for monitoring antibiotics to acquire sensitivity and selectivity: (1) Total ion chromatography (TIC) enables the achievement of a full MS spectrum that is characterized by low sensitivity. (2) Selected ion monitoring (SIM) is a mode that allows detection in high sensitivity but with no spectrum. (3) Multiple MS (MS^n) enables high selectivity and improves the sensitivity by increasing the signal-to-noise ratio (S/N).

The LC-MS analysis of AMX and its DPs was performed by HPLC/UV coupled to the MS using a reverse phase (RP)-phenyl column (250 × 2.1 mm, 5 μm). UV absorption was recorded at 230 and 275 nm. The HPLC mobile phase consisted of methanol (A) and water adjusted to pH 2.5 with trifluoroacetic acid (TFA) (B). The elution gradient started with 5% A, increasing to 75% over 8 and/or 15 minutes, and was then held at 75% for 2 minutes (Gozlan et al., 2010).

The results indicate that several micrograms per liter of ADP1/2 (noncalibrated results) and ADP6 (nonquantitative) were detected in wastewater secondary effluents. The detection of ADP1/2 at relatively high levels was expected, according to the AMX degradation experiments. Furthermore, it was not possible to obtain ADP1/2 as standards for calibration, due to its instability under the storage conditions and its degradation to ADP4/5 (at low pH) and ADP8/9 (at high pH). ADP6 was clearly detected in secondary effluents. In groundwater, only ADP8/9 was detected at a level of 0.03 μg/L. No ADP1/2 was detected in groundwater, possibly because it had converted to ADP8/9, which infiltrated and traveled through local groundwater (measured pH was ~7.8).

Therefore, our findings are of concern since

The sampled well is located in the vicinity of drinking water wells, and it is thus reasonable to assume that for a considerable period of time the local population was routinely consuming nanogram quantities of AMX DPs, as well as sulfa drugs (Avisar et al., 2009).

Large volumes of groundwater in the Herzlia depression (wells' area), which is located 10 km north to Tel-Aviv, where other pumping wells are located (not sampled in this study), may at present be contaminated with antibiotics and their DPs.

Wastewater effluents have been used for agricultural irrigation above large areas of the phreatic coastal plain aquifer of Israel since the early 1960s. Thus, groundwater contamination by antibiotic residues may be a widespread phenomenon.

In the search for more efficient medications, new drugs are being developed each year, adding to the environmental load. Moreover, these parent compounds are likely to degrade under natural environmental conditions, creating chemically different known and unknown DPs that may be toxic to various ecosystems and humans. Consequently, it is recommended that DPs of any further investigated drugs be detected and identified and that this be an integral part of any study, method, or technique that may be used in new studies regarding pharmaceutical residues in the aquatic environment. Chemicals derived from parent drugs are part of an unknown mixture. Chemicals do not simply disappear. Thus, we must understand what they have turned into. Therefore, environmental chemists must deal with a whole new range of contaminants.

REFERENCES

Al-Hadithi, N.N.A. 2007. "Determination of Drugs and Metabolites in Water by Use of Liquid Membrane Systems and HPLC." PhD diss. Paderborn, Germany: University of Paderborn.

Avisar, D., Lester, Y. and Ronen, D. 2009a. Sulfamethoxazole Contamination of a Deep Phreatic Aquifer. *Science of the Total Environment*. 407, 4278–4282.

Avisar, D., Lester, Y. and Ronen, D. 2009b. Sulfamethoxazole Detected in a Deep Phreatic Aquifer Beneath Effluent Irrigated Land. *The Science of the Total Environment*. 407, 4278–4282.

Ayim, J.S.K. and Rapson, H.D.C. 1972. Zinc and Copper (Ii) Ion Catalyzes of Penicillins in Alcohols. *The Journal of Pharmacy and Pharmacology*. 24, 172–175.

Batt, A.L., Snow, D.D. and Aga, D.S. 2006. Occurrence of Sulfonamide Antimicrobials in Private Water Wells in Washington County, Idaho, USA. *Chemosphere*. 64, 1963–1971.

Bentley, P.H. and Southgate, R. 1988. *Recent Advances in the Chemistry of β-Lactam Antibiotics*. The Royal Society of Chemistry, London, 108.

Blaha, J.M., Knevel, A.M., Kessler, D.P., Mincy, J.W. and Hem, S.L. 1976. Kinetic Analysisnod Penicillin Degradation in Acidic Media. *Journal of Pharmaceutical. Sciences*. 65, 1165–1170.

Cahill, J., Furlong, E., Burkhardt, M., Kolpin, D. and Anderson, L. 2004. Determination of Pharmaceutical Compounds in Surface- and Ground-Water Samples by Solid-Phase Extraction and High-Performance Liquid Chromatography–Electrospray Ionization Mass Spectrometry. *Journal of Chromatography A*. 1041, 171–180.

Calamari, D., Zuccato, E., Castiglioni, S., Bagnati, R. and Fanelli, R. 2003. Strategic Survey of Therapeutic Drugs in the Rivers Po and Lambro in Northern Italy. *Environmental Science and Technology*. 37, 1241–1248.

Chatterjee, N.R., Degani, M.S. and Singh, C.B. 1988. Degradation of Some Aqueous Semi-synthetic Penicillin Solutions in Presence of Copper (ii) Ions. *Indian Journal of Pharmaceutical Sciences*. 50(2), 128–130.

Christian, T., Schneider, R., Färber, H., Skutlarek, D., Meyer, M. and Goldbach, H. 2003. Determination of Antibiotic Residues in Manure, Soil, and Surface Waters. *Acta Hydrochimica et Hydrobiologica*. 31, 36–44.

Coyne, R., Bergh, O. and Samuelsen, O.B. 2004. One-Step Liquid Chromatographic Method for the Determination of Oxytetracycline in Fish Muscle. *Journal of Chromatography B*. 810, 325–328.

Derksen, J.G.M., Rijs, G.B.J. and Jongbloed, R.H. 2004. Diffuse Pollution of Surface Water by Pharmaceutical Products. *Water Science and Technology*. 49, 213–221.

Deshpande, A.D., Baheti, K. and Chatterjee, N.R. 2004. Degradation of β-Lactam Antibiotics. *Current Science*. 87, 12–25.

Díaz-Cruz, M., Barceló, D. 2005. LC–MS^2 Trace Analysis of Antimicrobials in Water, Sediment and Soil. *Trends in Analytical Chemistry*. 24, 645–657.

Díaz-Cruz, M., López de Alda, M. and Barceló, D. 2003. Environmental Behavior and Analysis of Veterinary and Human Drugs in Soils, Sediments and Sludge. *Trends in Analytical Chemistry*. 22, 340–351.

Garrison, A., Pope, J. and Allen, F. 1976. GC/MS analysis of organic compounds in domestic wastewaters. In: Keith, L.H. (Ed.), *Identification and Analysis of Organic Pollutants in Water*, 517–556. Ann Arbor Science Publishers, Ann Arbor, MI.

Gentili, A. 2007. Determination of Non-Steroidal Anti-Inflammatory Drugs in Environmental Samples by Chromatographic and Electrophoretic Techniques—Review. *Analytical and Bioanalytical Chemistry.* 387, 1185–1202.

Golet, E., Alder, A. and Giger, W. 2002. Environmental Exposure and Risk Assessment of Fluoroquinolone Antibacterial Agents in Wastewater and River Water of the Glatt Valley Watershed, Switzerland. *Environmental Science and Technology.* 36, 3645–3651.

Gozlan, I., Rotstein, A. and Avisar, D. 2010. Investigation of an Amoxicillin Oxidative Degradation Product Formed under Controlled Environmental Conditions. *Environmental Chemistry.* 7, 435–442.

Gozlan, I., Koren, O. and Avisar, D. 2013. Erythromycin Degradation Products Obtained Under Controlled Environmental Conditions. *Journal of Hazardous Material* (In Press).

Gutierez, N.P., Bekkouri, A. and Rodriguez, R.E. 1998. Spectrofluorimetric Study of the Degradation of α-Amino β-Lactam Antibiotics Catalyzed by Metal Ions in Methanol. *Analyst* 123, 2263–2266.

Halling-Sørensen, B., Nors Nielsen, S., Lanzky, P.F., Ingerslev, F., Holten Løtzhøft, H.C. and Jørgensen, S.E. 1998. Occurrence, Fate and Effects of Pharmaceutical Substances in the Environment—a Review. *Chemosphere.* 36, 357–393.

Halling-Sørensen, B., Sengeløv, G. and Tjørnelund, J. 2002. Toxicity of Tetracyclines and Tetracycline DP's to Environmentally Relevant Bacteria, Including Selected Tetracycline-Resistant Bacteria, *Archives of Environmental Contamination and Toxicology.* 42, 263–271.

Hamscher, G., Sczesny, S., Höper, H. and Nau, H. 2002. Determination of Persistent Tetracycline Residues in Soils Fertilized with Liquid Manure by High-Performance Liquid Chromatography with Electrospray Ionization Tandem Mass Spectrometry. *Analytical Chemistry.* 74, 1509–1518.

Heberer, T. 2002. Tracking Persistent Pharmaceutical Residues from Municipal Sewage to Drinking Water. *Journal of Hydrology.* 266, 175–189.

Heberer, T., Schmidt-Bäumler, K. and Stan H. 1998. Occurrence and Distribution of Organic Contaminants in the Aquatic System in Berlin. Part I: Drug Residues and Other Polar Contaminants in Berlin Surface and Groundwater. *Acta Hydrochimica et Hydrobiologica.* 26, 272–278.

Hirsch, R., Ternes, T., Haberer, K., Mehlich, A., Ballwanz, F. and Kratz, K. 1998. Determination of Antibiotics in Different Water Compartments via Liquid Chromatography–Electrospray Tandem Mass Spectrometry. *Journal of Chromatography A.* 815, 213–223.

Hirsch, R., Ternes, T., Haberer, K. and Kratz, K. 1999. Occurrence of Antibiotics in the Aquatic Environment. *The Science of the Total Environment.* 225, 109–118.

Hirsch, R., Ternes, T., Haberer, K. and Kratz, K.L. 1999. Occurrence of Antibiotics in the Aquatic Environment. *The Science of the Total Environment.* 225, 109–118.

Hirsch, R., Ternes, T., Haberer, K., Mehlich, A., Ballwanz, F. and Kratz, K. 1998. Determination of Antibiotics in Different Water Compartments via Liquid Chromatography–Electrospray Tandem Mass Spectrometry. *Journal of Chromatography A.* 815, 213–223.

Holm J.V., Rügge K., Bjerg, P.L. and Christensen, T.H. 1995. Occurrence and Distribution of Pharmaceutical Organic Compounds in the Groundwater Downgradient of a Landfill (Grindsted, Denmark). *Environmental Science and Technology.* 29, 1415–1420.

Huber, M.M., Gobel, A., Joss, A., Hermann, N., Loffler, D., McArdell, C.S., Ried, A., Siegrist, H., Ternes, T.A. and von Gunten, U. 2005. Oxidation of Pharmaceuticals During Ozonation of Municipal Wastewater Effluents: A Pilot Study. *Environmental Science and Technology.* 39, 4290–4299.

Isidori, M., Lavorgna, M., Nardelli, A., Parrella, A., Previtera, L. and Rubino, M. 2005. Ecotoxicity of Naproxen and Its Phototransformation Products. *Science of the Total Environment.* 348, 93–101.

Krapac, I.G., Dey, W.S., Roy, W.R., Smyth, C.A., Storment, E., Sargent, S.L. and Steele, J.D. 2002. Impact of Swine Manure Pits on Groundwater Quality. *Environmental Pollution.* 120, 475–492.

Kulshrestha, P., Rossman, F., Gies, J. and Diana, S. 2004. Investigating the Molecular Interactions of Oxytetracycline in Clay and Organic Matter: Insights on Factors Affecting its Mobility in Soil. *Environmental Science and Technology.* 38, 4097–4105.

Lalumera, G.M., Calamari, D. Galli, P., Castgilioni, S., Crosa, G. and Fanelli R. 2004. Preliminary Investigation on the Environmental Occurrence and Effects of Antibiotics Used in Aquaculture in Italy. *Chemosphere.* 54, 661–668.

Lamm, A., Rotsein, A., Gozlan, I. and Avisar, D. 2009. Detection of Amoxicillin-Diketopiperazine-2′,5′ in Wastewater Samples. *Journal of Environmental Science and Health. Part A.* 44, 1512–1517.

Lindberg, R., Jarnheimer, P., Olsen, B., Johansson, M. and Tysklind, M. 2004. Determination of Antibiotic Substances in Hospital Sewage Water using Solid Phase Extraction and Liquid Chromatography–Mass Spectrometry and Group Analogue Internal Standards. *Chemosphere.* 57, 1479–1488.

Lindberg, R., Wennberg, P., Johansson, M., Tysklind, M. and Andersson, B. 2005. Screening of Human Antibiotic Substances and Determination of Weekly Mass Flows in Five Sewage Treatment Plants in Sweden. *Environmental Science and Technology.* 39, 3421–3429.

Lindsey, M.E., Meyer, M. and Thurman, E.M. 2001. Analysis of Trace Levels of Sulfonamide and Tetracycline Antimicrobials in Groundwater and Surface Water Using Solid-Phase Extraction and Liquid Chromatography/Mass Spectrometry. *Analytical Chemistry.* 73, 4640–4646.

Martinez, J.H., Navarrom, P.G., Garcia, A.A. and de las Parras P.J. 1999. Beta-Lactam Degradation Catalyzed by Cd^{2+} Ion in Methanol. *International journal of biological macromolecules.* 25, 337–343.

Miao, X.S. and Metcalfe, C.D. 2005. Determination of Carbamazepine and Its Metabolites in Aqueous Samples Using Liquid Chromatography–Electrospray Tandem Mass Spectrometry. *Analytical Chemistry.* 75, 3731–3738.

Miao, X.S. and Metcalfe, C.D. 2005. Determination of Carbamazepine and Its Metabolites in Aqueous Samples Using Liquid Chromatography–Electrospray Tandem Mass Spectrometry. *Analytical Chemistry.* 75, 3731–3738.

Minneman, K.P., Wecker, L., Larner, J. and Brody, T.M. (Eds.). 2005. *Brody's Human Pharmacology: Molecular to Clinical.* 4th ed., Elsevier Mosby, Philadelphia, PA.

Nabil, N. and Al-Hadithi, H. 2007. Determination of Drugs and Degradation Products in Water by use of Liquid Membrane Systems and HPLC. Dean of the Faculty of Natural Sciences Department Chemistry of the University of Paderborn [*Der Fakultät für Naturwissenschaften Department Chemie Der Universität Paderborn*].

Nägele, E. and Moritz, R. 2005. Structure Elucidation of DP's of the Antibiotic Amoxicillin with Ion Trap MS^n and Accurate Mass Determination by ESI TOF. *Journal of the American Society for Mass Spectrometry.* 16, 1670–1676.

Navarrom, P.G., Blázquez, H.I., Osso, B.Q., Martínez de las Parras, P.J., Puentedura, M.I., Garcia, A.A. 2003. Penicillin Degradation Catalyzed by Zn(II) Ions in Methanol. *International Journal of Biological Macromolecules.* 23, 159–166.

Perez-Parada, A., Aguera Del Mar, A., Gomez-Ramos, M., Garcia-Reyes, J.F., Heinzen, H. and Fernandez-Alba, A.R. 2011. Behavior of Amoxicillin in Wastewater and River Water: Identification of Its Main Transformation Products by Liquid Chromatography/Electrospray Quadrupole Time-of-Fight Mass Spectrometry. *Rapid Communications in Mass Spectrometry.* 25, 731–742.

Petrovic, M., Hernando, M.D., Diaz-Cruz, M.S. and Barcelo, D. 2005. Liquid Chromatography–Tandem Mass Spectrometry for the Analysis of Pharmaceutical Residues in Environmental Samples: A Review. *Journal of Chromatography. A.* 1067, 1–14.

Richardson, S.D. 2004. Environmental Mass Spectroscopy: Emerging Contaminates and Current Issues. *Analytical Chemistry*. 76, 3337–3364.

Roth, H.J. and Fenner. H. 1994. *Struktur-Bioreaktivität-Wirkungsbezogene Eigenschaften*. 2nd ed. Georg Thieme Verlag, Stuttgart, Germany, 65–68.

Sacher, F., Lange, F., Brauch, H. and Blankenhorn, I. 2001. Pharmaceuticals in Groundwater Analytical Methods and Results of a Monitoring Program in Baden-Württemberg, Germany. *Journal of Chromatography A*. 938, 199–210.

Simpson, N. 2000. Solid-Phase Extraction: Principles, Techniques and Applications. CRC Press, New York.

Soulet, B., Tauxe, A. and Tarradellas, J. 2002. Analysis of Acidic Drugs in Swiss Wastewaters. *International Journal of Environmental Analytical Chemistry*. 82, 659–667.

Tauxe-Wuersch, A., De Alencastro, L., Grandjean, D., and Tarradellas J. 2005. Occurrence of Several Acidic Drugs in Sewage Treatment Plants in Switzerland and Risk Assessment. *Water Research*. 39, 1761–1772.

Ternes, T. 2001. Analytical Methods for the Determination of Pharmaceuticals in Aqueous Environmental Samples. *Trends in Analytical Chemistry*. 20, 419–434.

Tilles, S.A. 2001. Practical Issues in the Management of Hypersensitivity Reactions: Sulfonamides. *Southern Medical Journal*. 94, 817–824.

Tutt, D.E. and Schwartz, M.A. 1971. Spectrophotometric Assay of Ampicillin (α-Aminobenzylpenicillin) Involving Initial Benzoylation of the Side Chain α-Amino Group. *Analytical chemistry*. 43, 338–342.

Valvo, L., Alimonti, S., Alimenti, R., De Sena, C., Signoretti, E.C., Draisci, R. and Giannetti, L. 1997. Investigation of a New Amoxicillin Sodium Impurity Unstable in Solution, *Journal of Pharmaceutical and Biomedical Analysis*. 15, 487–493.

Vogel, J.R., Verstraeten, I.M., Coplen, T.B., Furlong, E.T., Meyer, M.T. and Barber, L.B. 2003. *Occurrence of Selected Pharmaceutical and Non-Pharmaceutical Compounds, and Stable Hydrogen and Oxygen Isotope Ratios, in a Riverbank Filtration Study, Platte River, Nebraska, 2001 to 2003*. U.S. Geological Survey Data Series 117, Volume 1, 1–39.

Xu, W., Zhang, G., Zou, S., Li, X., and Liu, Y. 2007. Determination of Selected Antibiotics in the Victoria Harbour and the Pearl River, South China Using High-Performance Liquid Chromatography–Electrospray Ionization Tandem MS. *Environmental Pollution*. 145, 672–679.

Zhang, G., Zou, S., Li, X. and Liu, Y. 2007. Determination of Selected Antibiotics in the Victoria Harbour and the Pearl River, South China using High-Performance Liquid Chromatography-Electrospray Ionization Tandem MS. *Environmental Pollution*. 145, 672–679.

Zhang, Y., Geien, S.U. and Gal, C. 2008. Carbamazepine and Diclofenac: Removal in Wastewater Treatment Plants and Occurrence in Water Bodies. *Chemosphere*. 73, 1151–1161.

Zhu, J., Snow, D.D., Cassade, D.A., Monson, S.J. and Spalding, R.F. 2001. Analysis of Oxytetracycline, Tetracycline, and Chlortetracycline in Water Using Solid-Phase Extraction and Liquid Chromatography-Tandem Mass Spectrometry. *Journal of Chromatography A*. 928, 177–186.

Zuccato, E., Castiglioni, S. and Fanelli, R. 2005. Identification of the Pharmaceuticals for Human Use Contaminating the Italian Aquatic Environment. *Journal of Hazardous Materials*. 122, 205–209.

Zuehlke, S., Duennbier, U. and Herberer, T. 2004. Determination of Polar Drug Residues in Sewage and Surface Water Applying Liquid Chromatography–Tandem Mass Spectrometer. *Analytical Chemistry*. 54, 311–320.

5 Sample Preparation for Chromatographic Analysis

Antonio Martín-Esteban

CONTENTS

5.1 INTRODUCTION

The determination of organic compounds in food, environmental, and biosamples at low concentrations is always a challenge. In recent years, great advances in analytical instrumentation have occurred, eventually allowing the determination of any compound in such samples. Typically, target analytes are determined by chromatographic or electrophoretic techniques coupled to a convenient detector such as ultraviolet (UV), fluorescence or, more recently, mass spectrometry (MS), or tandem MS detectors. However, even when using the powerful and selective MS detection, direct injections of crude sample extracts are not recommended since matrix components can inhibit or enhance analyte ionization, hampering accurate determination. Thus, a clean sample is generally convenient to improve separation and detection, whereas a poorly treated sample may invalidate the whole analysis. Another additional and valuable aspect of the use of cleaned samples is the reduction of time to maintain instruments, thereby reducing associated costs. Besides, new objectives in sample preparation have recently appeared, such as using smaller initial sample sizes, improvement of selectivity in extraction, facilitating automation, and minimizing the amount of glassware and organic solvents to be used. Consequently, the last decade has led to the development of new microextraction techniques, mostly based on miniaturization of traditional sample preparation methods.

In this chapter, the different sample treatment techniques that are currently available and most commonly used in analytical laboratories as well as anticipated future trends for the analysis of organic compounds in different kinds of samples by chromatographic techniques are described.

5.2 SAMPLE PRETREATMENT

Generally, sampling techniques provide amounts of samples much higher (2–10 L of liquid samples and 1–2 kg of solid samples) than those needed for the final chromatographic analysis (just a few milligrams). Thus, it is always necessary to carry out some pretreatments to get a homogeneous and representative subsample. Even apparently simple and homogeneous samples, that is, aqueous samples, will need to be filtered to remove suspended particles that could affect the final determination of target analytes. However, some hydrophobic analytes (i.e., organochlorine pesticides) could be adsorbed onto particle surfaces and thus, depending on the objective of the analysis, might be necessary to analyze such particles.

Usually, environmental water samples require just filtration, whereas liquid food samples might be subjected to other kinds of pretreatments depending on the objective of the analysis. However, solid samples (both environmental and food samples) need to be more extensively pretreated to get a homogeneous subsample. The wide variety of solid samples prevents an exhaustive description of the different procedures in this chapter; however, some general common procedures will be described in Section 5.2.2.

5.2.1 DRYING

The presence of water or moisture in solid samples might produce alterations (i.e., hydrolysis) of the matrix and/or analytes, which will obviously affect the final analytical results. Besides, water content varies depending on atmospheric conditions and thus it is recommended to refer the contents of target analytes to the mass of the dry sample.

Sample drying is carried out before crushing and sieving steps, although it is recommended to dry the sample again before final determination since rehydration might occur. Typically, the sample is dried inside an oven at a temperature of about 100°C. Higher temperatures can be used to decrease the drying time, but losses of volatile analytes might occur. In this sense, it is important to know a priori the physicochemical properties of target analytes to preserve the integrity of the sample. A more conservative approach, using low temperatures, can be followed, but it will unnecessarily increase the drying time. Alternatively, lyophilization is recommended if a high risk of analyte loss exists, and it is an appropriate procedure for drying food, biological material, and plant samples. However, even when following this procedure, losses of analytes might occur depending on their physical properties (i.e., solubility and volatility).

It is not possible to establish a general rule on how to perform sample drying. Thus, studies on the stability of target analytes in spiked samples should be carried out to guarantee the integrity of the samples before final determination of the analytes.

5.2.2 HOMOGENIZATION

As mentioned in Section 5.2, samples are heterogeneous in nature and thus they must be treated to get a homogeneous distribution of target analytes.

Generally, soil samples are crushed, grinded, and sieved through a 2 mm mesh. Grinding can be done manually or automatically using specially designed equipments (i.e., ball mills). This procedure might provoke the local heating of a sample, and thus thermolabile or volatile compounds might be affected. Also, due to heating water content may vary, making it necessary to recalculate sample moisture.

Food samples are cut down into small pieces with a laboratory knife before further homogenization with automatic instruments (i.e., a blender). Sample freezing is a general practice to ease blending, which is especially recommended for samples with high fat content (i.e., cheese) and soft samples with high risk of phase separation during blending (i.e., liver, citrus fruits, etc.).

Finally, it is important to point out that in most of the cases samples need to be stored for certain periods of time before performing the analysis. Although sample storage cannot be considered a sample pretreatment, establishment of the right conditions of storage (i.e., at room temperature or in the fridge) and addition of preservatives to minimize analyte/sample degradation are typical procedures carried out at this stage of the analytical process and need to be taken into account to guarantee the accuracy of the final results.

5.3 EXTRACTION AND PURIFICATION

5.3.1 Solid–Liquid Extraction

Solid–liquid extraction is probably the most widely used procedure in the analysis of organic compounds in solid samples. Solid–liquid extraction includes various extraction techniques based on the contact of a certain amount of sample with an appropriated solvent. Figure 5.1 shows a scheme of the different steps that take place in a solid–liquid extraction procedure and will influence the final extraction efficiency. In the first stage (step 1), the solvent must penetrate the pores of the sample particulates to achieve desorption of the analytes bound to matrix active sites (step 2). Subsequently, analytes have to diffuse through the matrix (step 3) to be dissolved in the extracting solvent (step 4). Again, the analytes must diffuse though the solvent to leave the sample pores (step 5) and be finally swept away by the external solvent (step 6). Obviously, the proper selection of the solvent to be used is a key factor in a solid–liquid extraction procedure. However, other parameters such as pressure and temperature have an important influence on extraction efficiency. Working at high pressure facilitates the solvent to penetrate sample pores (step 1) and, in general, increasing the temperature increases the solubility of the analytes. Moreover, high temperatures increase diffusion coefficients (steps 3 and 5) and the capacity of the solvent to disrupt matrix–analyte interactions (step 2). Depending on the strength of the interaction between the analyte and the sample matrix, the extraction will be performed in soft, mild, or aggressive conditions. Table 5.1 shows a summary and a comparison of the drawbacks and advantages of the different solid–liquid extraction techniques (which are described in Sections 5.3.1.1 through 5.3.1.4) most commonly used in the analysis of organic compounds in food, environmental, and biosamples.

5.3.1.1 Shaking

Shaking is the simplest procedure to extract target analytes weakly bound to a sample such as, for instance, pesticides from fruits and vegetables. It consists of

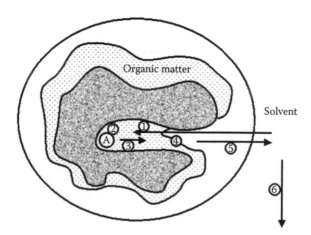

FIGURE 5.1   Scheme of the different steps involved in the extraction of a target analyte A from a solid particle.

TABLE 5.1
Solid-Liquid Extraction Techniques

Technique	Description	Advantages	Drawbacks
Shaking	Samples and solvent are placed in a glass vessel. Shaking can be done manually or mechanically.	• Simple • Fast (15–30 minutes) • Low cost	• Filtration of the extract is necessary • Matrix dependent • Moderate solvent consumption (25–100 mL)
Soxhlet	Sample is placed in a porous cartridge and solvent recirculates continuously by distillation–condensation cycles.	• Standard method • No further filtration of the extract necessary • Independent of kind of matrix • Low cost	• Time-consuming (12–48 hours) • High solvent volumes (300–500 mL) • Solvent evaporation needed
USE	Samples and the solvent are placed in a glass vessel and introduced in an ultrasonic bath.	• Fast (15–30 minutes) • Low solvent consumption (5–30 mL) • Bath temperature can be adjusted • Low cost	• Filtration of the extract is necessary • Matrix dependent
MAE	Sample and solvent are placed in a reaction vessel. Microwave energy is used to heat the mixture.	• Fast (~15 minutes) • Low solvent consumption (15–40 mL) • Easily programmable	• Filtration of the extract is necessary • Addition of a polar solvent is required • Moderate cost
PSE	Sample is placed in a cartridge and pressurized with a high-temperature solvent	• Fast (20–30 minutes) • Low solvent consumption (30 mL) • Easy control of extraction parameters (temperature, pressure) • High temperatures achieved • High sample processing	• Initial high cost • Matrix dependent

Note: Ultrasound-assisted extraction (USE); microwave-assisted extraction (MAE); pressurized solvent extraction (PSE).

(manually or automatically) shaking the sample in the presence of an appropriated solvent for a certain period of time. The most commonly used solvents are acetone and acetonitrile due to their miscibility with water, making the diffusion of analytes from the solid sample to the solution easy, although immiscible solvents such as dichloromethane or hexane can also be used for the extraction depending on the properties of target analytes. In a similar manner, the use of mixtures of solvents is a common practice when analytes of different polarity are extracted in multiresidue analysis. Once the analytes are extracted, the mixture needs to be filtered before further treatment. Besides, since the volume of organic solvents used following this procedure is relatively large, it is usually necessary to evaporate the solvent before final determination.

However, shaking might not be effective enough to extract analytes that are strongly bound to a sample. To achieve more effective shaking, the use of ultrasound-assisted extraction (USE) is recommended. Ultrasound radiation provokes the vibration of molecules and eases the diffusion of the solvent into the sample, favoring the contact between both phases. Thanks to this improvement, both time and the amount of solvents are considerably reduced.

An interesting and useful modification for reducing the amounts of both sample and organic solvents is the so-called USE in small columns, proposed by Sánchez-Brunete et al. [1,2] for the extraction of pesticides from soils. Briefly, this procedure just consists of placing the sample (~5 g) in a glass column equipped with a polyethylene frit. Subsequently, samples are extracted with around 5–10 mL of an appropriate organic solvent in an ultrasonic water bath. After extraction, columns are placed on a multiport vacuum manifold where the solvent is filtered and collected for further analysis.

5.3.1.2   Soxhlet Extraction

As indicated in Section 5.3.1.1, in some cases shaking is not enough for disrupting the interactions between analytes and matrix components. To this end, an increase in the temperature of extraction is recommended. A simpler approach to isolate analytes bound to solid matrices at high temperatures is Soxhlet extraction, introduced by F. Soxhlet in 1879, which is still nowadays the more used technique. Besides, the performance of new extraction techniques uses to be assessed by comparison to that provided by Soxhlet extraction.

The sample is placed in an apparatus (Soxhlet extractor), and the extraction of analytes is achieved by means of a hot condensate of a solvent distilling in a closed circuit. Distillation in a closed circuit allows the sample to be extracted many times with fresh portions of the solvent, and exhaustive extraction can be performed. Its weak points are the long time required for the extraction and the large amount of organic solvents used.

To minimize the aforementioned drawbacks, several attempts toward the automation of the process have been proposed. Among them, Soxtec systems (Foss, Hillerød, Denmark), the most extensively accepted and used ones in analytical laboratories, allow the reduction of extraction times by about five times compared with classical Soxhlet extraction.

Table 5.2 shows a comparison of the recoveries obtained for several pesticides in soils after extraction using different techniques [3]. In this case, it is clear that USE allows the isolation of target analytes, whereas simple shaking is not effective enough to quantitatively extract the selected pesticides. It is important to stress that recoveries after Soxhlet extraction were too high, which means that a large amount of matrix components were coextracted with target analytes. In this regard, it is clear that exhaustive extraction is not always required and a balance between the recoveries of target analytes obtained and the amount of matrix components coextracted needs to be established.

TABLE 5.2

Recoveries in Percentage of Pesticides in Soils Obtained by Different Extraction Techniques

Pesticide	Concentration (mg·mL^{-1})	USE	Soxhlet Extraction	Shaking
Atrazine	0.04	103.5 ± 2.8	201.9 ± 14.6	108.3 ± 6.2
Pyropham	0.05	79.7 ± 6.3	143.0 ± 18.6	65.1 ± 9.3
Chlorpropham	0.05	93.6 ± 7.9	155.6 ± 20.4	88.1 ± 10.0
α-Cypermethrin	0.12	97.2 ± 4.4	128.4 ± 16.4	90.1 ± 9.1
Tetramethrin	0.26	83.4 ± 4.2	64.3 ± 16.0	52.0 ± 8.3
Diflubenzuron	0.02	92.8 ± 4.0	182.5 ± 17.4	98.1 ± 8.9

Source: Babic et al., *J. Chromatogr. A*, 823, 3, 1998.

Note: Experimental conditions: 10 g of soil sample spiked at indicated concentration level. USE: 20 mL of acetone, 15 minutes; Soxhlet extraction: 250 mL of acetone, 4 hours; shaking: 20 mL of acetone, 2 hours.

5.3.1.3 Microwave-Assisted Extraction

Microwave-assisted extraction (MAE) has appeared during the past few years as a clear alternative to Soxhlet extraction due to the ability of microwave radiation to heat the sample–solvent mixture in a fast and efficient manner. Besides, the existence of several commercially available instruments that are able to perform the simultaneous extraction of several samples (up to 14 samples in some instruments), which allows extraction parameters (pressure, temperature, and power) to be perfectly controlled, has made MAE a very popular technique.

Microwave energy is absorbed by molecules with high dielectric constants. In this regard, hexane, a solvent with a very low dielectric constant, is transparent to microwave radiation and thus does not get heated, whereas acetone is heated in a few seconds due to its high dielectric constant. However, solvents with low dielectric constants can be used if the compounds contained in the sample (i.e., water) absorb microwave energy.

The use of solvent mixtures (especially for the extraction of compounds of different polarity), combining the ability of heating of one of the components (i.e., acetone) with the solubility of the more hydrophobic compounds in the other solvent of the mixture (i.e., hexane), is a typical practice.

In summary, the recoveries typically obtained are quite similar to those obtained by Soxhlet extraction, but the significant decrease in extraction time (~15 minutes) and that in the volume of organic solvents (25–50 mL) have caused MAE to be extensively used in analytical laboratories.

5.3.1.4 Pressurized Solvent Extraction

Pressurized solvent extraction (PSE), also known as accelerated solvent extraction, pressurized liquid extraction, and pressurized fluid extraction, uses solvents at high temperatures and pressures to accelerate the extraction process. The higher temperature increases extraction kinetics, whereas the elevated pressure keeps the solvent in liquid phase above its boiling point, leading to rapid and safe extractions [4].

Figure 5.2 shows a scheme of the instrumentation and the procedure used in PSE. Experimentally, the sample (~10 g) is placed in an extraction cell and filled with an appropriate solvent (15–40 mL). Subsequently, the cell is heated in a furnace up to temperatures below 200°C while the pressure of the system is increased (up to 20 MPa) to perform the extraction. After a certain period of time (10–15 minutes), the extract is directly transferred to a vial without the necessity of subsequent filtration of the obtained extract. Then the sample is rinsed with a portion of pure solvent and, finally, the remaining solvent is transferred to the vial with a stream of nitrogen (N_2). The whole process is automated and each step can be programmed, allowing the sequential unattended extraction of up to 24 samples.

This technique is easily applicable for the extraction of organic compounds from any kind of sample, and the high temperature used allows performing a very efficient extraction in a short time. In addition, the considerable reduction of the amount of organic solvents used makes PSE a very attractive technique. The main limitations of this technique are the high cost of the apparatus and the unavoidable necessity of purifying obtained extracts, which is common to other efficient extraction techniques based on the aforementioned use of organic solvents.

Superheated water extraction (SHWE) (also called subcritical water or pressurized water extraction), although typically treated independently, can be considered a variant of PSE since in SHWE also extraction is carried out at a high temperature (above 100°C up to 250°C) under pressure to keep water in its liquid state. In such conditions, water solubility properties can be tuned and thus its ability to extract not only polar analytes but also hydrophobic compounds can be successfully achieved. By 200°C, water has a similar relative permittivity to common extraction solvents, such as methanol ($\varepsilon = 32$) or acetonitrile ($\varepsilon = 37$) and thus should be a good solvent for many organic analytes. Both static and dynamic or combined extraction systems, collecting the sample extracts for further chromatographic analysis, have been used, but the coupling of the extraction step directly to an analytical separation method has

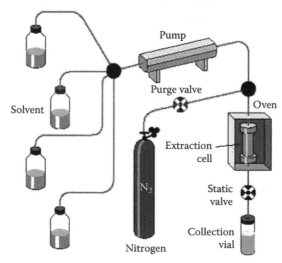

FIGURE 5.2 Pressurized solvent extraction equipment. (Courtesy of Dionex Corporation.)

also been proposed for the extraction of a great variety of compounds (i.e., polycyclic aromatic hydrocarbons [PAHs], pesticides, and essential oils) in environmental, food, plant, and pharmaceutical samples. In general, quantitative recoveries with excellent reproducibility have been achieved since cleaner extracts were obtained. Besides, the use of water as the extraction solvent makes SHWE an environmentally friendly technique, avoiding the necessity of waste disposal and treatment associated with the extraction of toxic and flammable organic solvents. Unfortunately, as yet there is no equipment on the market dedicated to SHWE and the current PSE commercial systems have only a limited temperature range. As a result, SHWE has been carried out using homemade systems, limiting its widespread use in analytical laboratories. However, its simplicity, solvent-free properties, and excellent extraction capabilities might make SHWE one of the breakthrough technologies in the near future [5].

5.3.2 SUPERCRITICAL FLUID EXTRACTION

Supercritical fluid extraction (SFE) has been widely used for the isolation of a great variety of organic compounds from almost any kind of solid samples. Supercritical fluids can be considered as a hybrid between liquids and gases, and they possess properties ideal for the extraction of target analytes from solid samples. Supercritical fluids have in common with gases the ability to diffuse through a sample, which facilitates the extraction of analytes that are located in pores that are not easily accessible. In addition, the solvation power of supercritical fluids is similar to that of liquids, allowing the release of target analytes from the sample to the fluid.

Carbon dioxide (CO_2) has been widely used in SFE because it can be obtained with high purity, it is chemically inert, and its critical point (31.1°C and 71.8 atm) is easily accessible. Its main drawback is its apolar character, limiting its applicability to the extraction of hydrophobic compounds. To overcome, at least to some extent, this drawback, the addition of a small amount of an organic solvent modifier has been proposed and it permits varying the polarity of the fluid, thus increasing the range of extractable compounds. Table 5.3 shows the recoveries in the extraction of PAHs in

TABLE 5.3
Recoveries in Percentage of PAHs in Sludge by CO_2-SFE with and without Modifier (Toluene)

	Supercritical Fluid	
Compound	CO_2	CO_2 + 5% toluene (v/v)
Pyrene	68	76
Benzo(a)anthracene	69	79
Crisene	67	85
Dibenzo[ah]anthracene	32	68
Benzo[ghi]perylene	20	71
Indeno[1, 2, 3-cd]pyrene	27	75

Source: Miege et al., *J. Chromatogr. A*, 823, 219, 1998.

sludge by SFE using CO_2 with or without a modifier (toluene) [6]. It is clear that the presence of toluene favors the extraction of target analytes. Toluene molecules are able to establish π–π* interactions with the aromatic rings of PAHs, releasing them from the sample or with the matrix active sites displacing target analytes.

Once target analytes are in the supercritical fluid phase, they have to be isolated for further analysis, which is accomplished by the decompression of the fluid through a restrictor with analytes trapped on a liquid trap or a solid surface. With a liquid trap, the restrictor is immersed in a suitable liquid and, thus, the analyte is gradually dissolved in the solvent while CO_2 is discharged to the atmosphere. In the solid surface method, analytes are trapped on a solid surface (i.e., glass vial, glass beads, or solid-phase sorbents) and cryogenically cooled directly by the expansion of the supercritical fluid or with the aid of liquid N_2. Alternatively, SFE can be directly coupled to gas chromatography (GC) or to supercritical fluid chromatography being the success of such online coupling dependent of the interface used, which determines the quantitative transfer of target analytes to the analytical column [7].

SFE has been widely used for the extraction of organic compounds from solid samples thanks to the effectiveness and selectivity of the extraction and the possibility of online coupling to chromatographic techniques. However, the costs of instrumentation and apparatus in the market of new, less-sophisticated extraction instruments is causing SFE to be displaced by other extraction techniques, especially PSE.

5.3.3 Liquid–Liquid (Micro) Extraction

Liquid–liquid extraction (LLE) has been widely used for the extraction of organic compounds from aqueous liquid samples and, although to a lesser extent, for the purification of organic extracts. LLE is based on the partitioning of the target analyte between two immiscible liquids. The efficiency of the process depends on the affinity of the analyte for the solvents, the ratio of volumes of each phase, and the number of successive extractions.

Hexane and cyclohexane are typical organic solvents used for extracting nonpolar compounds such as organochlorine and organophosphorous pesticides, and dichoromethane and chloroform are used for medium-polarity organic compounds such as triazines or phenylurea herbicides. However, quantitative recoveries for relatively polar compounds by LLE are difficult to achieve. To increase the efficiency and thus the range of application, the partition coefficients may be increased by using mixtures of solvents, changing the pH (preventing the ionization of acids or bases), or adding salts ("salting-out" effect).

The high number of possible combinations of solvents and pHs ideally makes possible the isolation of any compound from water samples by LLE, which has been traditionally considered a great advantage of LLE. However, LLE is not exempt from important drawbacks. One of the most important ones is the toxicity of the organic solvents used, leading to a large amount of toxic residues. In this sense, the costs of disposal of toxic solvents are rather high. However, it is important to mention that this problem is minimized when LLE is used for cleanup steps where low volumes are usually employed. Besides, the risk of exposure of the chemist to toxic solvents and vapors always exists. From a practical point of view, the formation of emulsions, which are sometimes difficult to break up; the handling of large water samples; and the difficulties in the automation of the whole process cause LLE to be considered a tedious, time-consuming, and costly technique.

To minimize the drawbacks associated with traditional LLE, research has been conducted to miniaturize it into so-called liquid-phase microextraction (LPME) techniques. In LPME, extraction normally takes place between a small amount of a water-immiscible solvent and an aqueous phase containing the analytes of interest. The volume of the receiving phase is in the microliter or submicroliter range. In this way, high enrichment factors can be obtained owing to the high sample volume to acceptor phase volume ratio. Depending on the format, several LPME techniques have been developed in the past few years.

5.3.3.1 Single-Drop Microextraction

Single-drop microextraction (SDME) is based on the extraction of analytes from aqueous samples into a small drop of an organic solvent (toluene and mixtures of CH_2Cl_2 and CCl_4 are the solvents most often used); it was introduced in 1996 by Liu and Dasgupta [8] and Jeannot and Cantwell [9]. Typically, the small drop of organic solvent is suspended from the tip of a GC syringe and immersed (direct immersion SDME) into or held above (headspace [HS] SDME) the aqueous solution for extraction. During extraction, target analytes are extracted from the aqueous sample into the hanging drop based on passive diffusion, and extraction recoveries are essentially determined by the organic solvent to water partition coefficients. When finished, the drop of organic solvent is withdrawn into the syringe and subsequently injected in GC. Although hanging-drop LPME is very simple and efficient and reduces the consumption of organic solvents per sample to a few microliters, it is still used only in a limited number of research laboratories. One reason for this may be attributed to the low stability of the hanging drop, which is easily lost into the sample during extraction. For a more in-depth discussion on SDME, we recommend the readers to the work of Xu et al. [10].

5.3.3.2 Liquid Membrane Extraction

Liquid membrane extraction techniques (supported liquid membrane extraction [SLME] and microporous membrane liquid–liquid extraction [MMLLE]) are based on the use of a hydrophobic membrane, containing an organic solvent, which separates two immiscible phases. These extraction techniques are a combination of three simultaneous processes: (1) extraction of the analyte into organic phase, (2) membrane transport, and (3) re-extraction in an acceptor phase. The chemical gradient existing between the two sides of the liquid membrane causes permeation of solutes. The compounds present in the donor phase diffuse across the organic liquid membrane to the acceptor phase, where they accumulate at a concentration generally greater than that in the donor phase. Depending on the sample volume, different membrane unit formats for liquid membrane extraction are applied, as shown in Figure 5.3 [11]. The effects of membrane type and solvents used; flow rate of donor phase; as well as donor, membrane, and acceptor phase composition on analyte transport across the liquid membrane are parameters to be investigated to

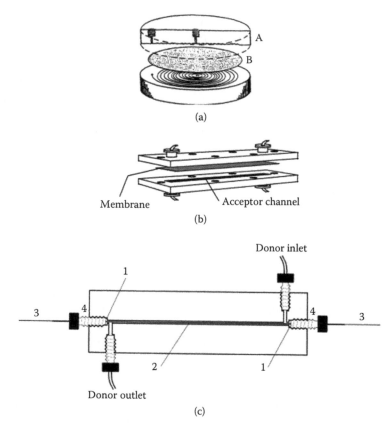

FIGURE 5.3 (a) Membrane unit with 1 mL channel volume (A = blocks of inert material, B = membrane). (b) Membrane unit with 10 mL channel volume. (c) Hollow fiber membrane unit with 1.3 mL acceptor channel. (Reproduced from *Journal of Chromatography A*, 902, Jönsson, J.A., and Mathiasson, L., Membrane-based techniques for sample enrichment, 205, 2000, with permission from Elsevier.)

optimize the extraction efficiency. The main advantages of liquid membrane extraction over traditional separation methods are small amounts of organic phases used; mass transfer is performed in one step, and it is possible to achieve high separation and concentration factors.

The distinguishing factor of the use of SLME or MMLLE is the possibility of connecting them online with an analytical system. MMLLE is easily interfaced to GC and normal-phase high-performance liquid chromatography (HPLC), whereas SLME is compatible with reversed-phase HPLC. These online connections result in an improvement of the overall reliability of the analysis, since the number of steps involved in sample preparation is decreased and allows method automation. Additionally, significant reduction of analysis time is achieved. Till now, SLME and MMLLE have been successfully applied for the enrichment of phenoxy acid, sulfonylurea, and triazine herbicides from environmental water samples. In those examples, similar or even better results were obtained in comparison with conventional sample preparation methods.

Thanks to their flexibility, SLME and MMLLE have proved to be interesting techniques that can be combined with a second pretreatment technique (e.g., solid-phase extraction [SPE]). In this regard, detection limits as low as 30 μg·L⁻¹ have been achieved by the combination of SLME and SPE for the determination of atrazine in fruit juices (orange, apple, black currant, and grape) [12].

In spite of such advantages, the technical configuration, including at least two peristaltic pumps to provide a flow system, makes SLME and MMLLE less attractive than the hollow-fiber LPME (HF-LPME) technique, which is discussed in Section 5.3.3.3.

5.3.3.3 HF-LPME

Pedersen-Bjergaard and Rasmussen introduced HF-LPME [13] with the aim of improving the stability and reliability of LPME techniques. In this concept, the extracting phase is placed inside the lumen of a porous polypropylene hollow fiber. The fiber had a porosity of 70%; the pore size was 0.2 μm, wall thickness was 200 μm, and internal diameter was 600 μm. A supported liquid membrane was formed by dipping the hollow fiber into the organic solvent (like n-octanol, dihexyl ether, or toluene) for a few seconds. The solvent penetrated the pores of the hollow fiber and was bound by capillary forces to the polypropylene network comprising the fiber wall. The high porosity enabled the immobilization of a considerable volume of solvent as a thin film. A 1 cm length of the fiber was able to immobilize approximately 8 μL of solvent as a 200 μm film within the polypropylene network. The extracting phase (acceptor solution), which was placed in the lumen of the fiber, was mechanically protected inside the hollow fiber and it was separated from the sample by the supported liquid membrane (organic solvent). This prevented dissolution of the extracting phase (acceptor solution) into the sample. In HF-LPME, analytes are extracted from an aqueous sample into the organic solvent immobilized as an SLM and into the acceptor solution placed inside the lumen of the hollow fiber. Subsequently, the acceptor solution is removed by a microsyringe and transferred to final chemical analysis. The chemical principle of HF-LPME does not differ from that of other liquid membrane extraction techniques (see Section 3.3), but the technical setup, on the other hand, differs considerably since SLM extractions are performed with a small piece of hollow fiber in a stagnant system without the use of pumps to deliver the sample and the acceptor phase.

HF-LPME can be performed in either the two-phase or the three-phase mode. Prior to extraction, the sample is filled into a sample vial and for acidic or basic compounds the pH of the sample is adjusted to suppress the ionization of the target analytes. A short piece (~1 cm) of a porous hollow fiber is used, and this may have either a rod configuration with a closed bottom or a U-configuration where both ends of the hollow fiber are connected to guiding tubes. After loading the SLM, the acceptor solution is filled into the lumen of the hollow fiber. This acceptor solution can be an organic solvent (same as that used for SLM) resulting in a two-phase extraction system, or it may be an acidic or alkaline aqueous solution resulting in a three-phase extraction system.

HF-LPME is an equilibrium extraction technique where the concentration of analytes in the acceptor solution increases to a certain level, and subsequently the system enters equilibrium and the analyte concentration in the acceptor phase remains constant over time. Thus, LPME is not an exhaustive extraction technique like LLE and SPE. The extraction recovery in HF-LPME is determined by actual partition coefficients, the sample volume, the volume of SLM, and the volume of the acceptor phase. Typically, HF-LPME recoveries reported in the literature range between 10% and 90%. As analytes are extracted from relatively large sample volumes into a very small volume of acceptor solution, most HF-LPME applications provide substantial analyte enrichment. High enrichments are obtained directly without the need for solvent evaporation and reconstitution, as required for high enrichment by LLE and SPE, and this is one of the major advantages of HF-LPME. Besides, especially in the three-phase mode, HF-LPME provides very clean extracts from a variety of samples like environmental waters and biological samples [14]. An example of the latter is illustrated in Figure 5.4, where citalopram and methamphetamine are spiked to human urine, plasma, and whole blood [14]. In spite of the complex matrices of these samples, almost no other components than the two basic drugs were recovered in the acceptor solution.

5.3.4 SPE

SPE, as LLE, is based on the different affinity of target analytes for two different phases. In SPE, a liquid phase (liquid sample or liquid sample extracts obtained by following other techniques) is loaded onto a solid sorbent, which is packed with disposable cartridges or enmeshed in the inert matrix of an extraction disk. Those compounds with a higher affinity for the sorbent will be retained on it, whereas others will pass through it unaltered. Subsequently, if target analytes are retained, they can be eluted using a suitable solvent with a certain degree of selectivity.

The typical SPE sequence involving several steps is depicted in Figure 5.5. First, the sorbent needs to be prepared by activation with a suitable solvent and by conditioning with the same solvent in which analytes are dissolved. Then, the liquid sample or a liquid sample extract is loaded onto the cartridge. Usually, target analytes are retained together with other components of the sample matrix. Some of these compounds can be removed by the application of a washing solvent. Finally, analytes are eluted with a small volume of an appropriate solvent. In this sense, it is possible to obtain final, very clean sample extracts by SPE, thanks to the cleanup performed, with high enrichment factors due to the low volume of solvent used for eluting target analytes. These aspects together with the simplicity of operation and the ease of automation (see Section 5.3.4.5.4) have made SPE a very popular technique that is widely used in analytical laboratories.

The success of an SPE procedure depends on the knowledge about the properties of target analytes and the kind of sample, which will help in the proper selection of the sorbent to be used. Understanding the mechanism of interaction between the sorbent and the analyte is a key factor in the development of an SPE method since it will make choosing the right sorbent from the wide variety of sorbents available in the market easy.

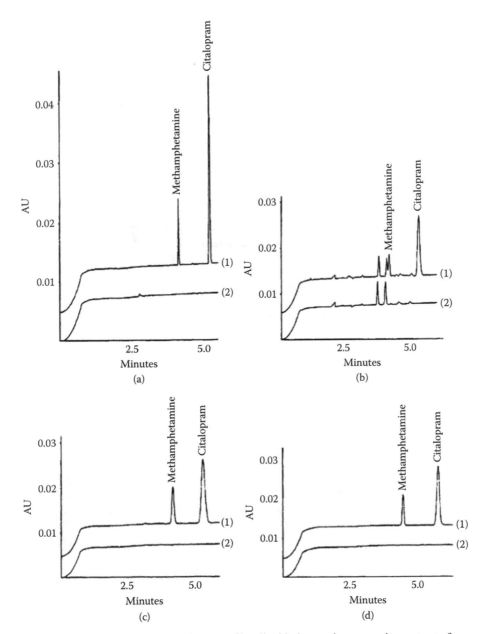

FIGURE 5.4 CE-UV analysis of hollow-fiber liquid-phase microextraction extracts from (1) unspiked and (2) spiked (a) water, (b) urine, (c) plasma, and (d) whole blood. Samples are spiked with 100 ng/mL methamphetamine and citalopram. (Reproduced from *Journal of Chromatography A*, 1184, Pedersen-Bjergaard, S., and Rasmussen, K.E., Liquid-phase microextraction with porous hollow fibers, a miniaturized and highly flexible format for liquid–liquid extraction, 132, 2008, with permission from Elsevier.)

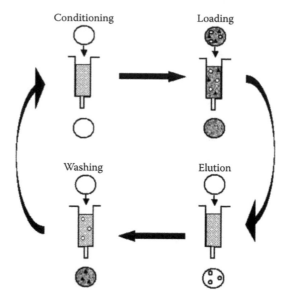

FIGURE 5.5   Solid-phase extraction (SPE) steps.

5.3.4.1   Polar Sorbents

The purification of organic sample extracts is usually performed by SPE onto polar sorbents. The most used sorbent is silica, which possesses active silanol groups in its surface that are able to interact with target analytes. This interaction is stronger for organic compounds with base properties due to the slightly acidic character of silanol groups. Other common polar sorbents are alumina (commercially available in its acid, neutral, and base forms) and Florisil.

In the loading step analytes compete with the solvent for the active adsorption sites of the sorbent, and elution is performed by displacing the analytes from the active sites using an appropriate solvent. In this sense, the more polar the solvent, the higher its elution power. The elution power is established by the eluotropic strength ($\varepsilon°$), which is a measure of the adsorption energy of a solvent in a given sorbent. The eluotropic series of different common solvents in alumina and silica are shown in Table 5.4. By a careful selection of solvents (or their mixtures), analytes (or interferences) will be retained on a sorbent by loading in a nonpolar solvent that is subsequently eluted using a second solvent with a higher eluotropic strength. Obviously, the selection of these solvents will be determined by the polarity of the analytes. For instance, in pesticide multiresidue analysis, hydrophobic pesticides such as pyrethroids can be eluted with a mixture of hexane and diethyl ether after loading, whereas for eluting carbamates a more polar mixture such as hexane:acetone is necessary.

5.3.4.2   Nonpolar Sorbents

Nonpolar sorbents are appropriate for the trace enrichment and cleanup of organic compounds in polar liquid samples (i.e., environmental waters). Traditionally, octyl- and octadecyl-silica, both in cartridges or disks, have been used due to their ability to

TABLE 5.4
Eluotropic Series

Solvent	ε° SiOH	ε° Al$_2$O$_3$
Pentane	0.00	0.00
Hexane	0.00–0.01	0.00–0.01
Iso-octane	0.01	0.01
Cyclohexane	0.03	0.04
Carbon tetrachloride	0.11	0.17–0.18
Xylene	—	0.26
Toluene	0.22	0.20–0.30
Chlorobenzene	0.23	0.30–0.31
Benzene	0.25	0.32
Ethyl ether	0.38–0.43	0.38
Dichloromethane	0.32–0.32	0.36–0.42
Chloroform	0.26	0.36–0.40
1,2-Dichloroethane	—	0.44–0.49
Methyl ethyl ketone	—	0.51
Acetone	0.47–0.53	0.56–0.58
Dioxane	0.49–0.51	0.56–0.61
Tetrahydrofuran	0.53	0.45–0.62
Methyl *t*-butyl ether	0.48	0.3–0.62
Ethyl acetate	0.38–0.48	0.58–0.62
Dimethyl sulfoxide	—	0.62–0.75
Acetonitrile	0.50–0.52	0.52–0.65
1-Butanol	—	0.7
n-Propyl alcohol	—	0.78–0.82
Isopropyl alcohol	0.6	0.78–0.82
Ethanol	—	0.88
Methanol	0.70–0.73	0.95

retain nonpolar and moderate-polar compounds from liquid samples. The retention mechanism is based on van der Waals forces and hydrophobic interactions, which allow the handling of large sample volumes and the subsequent elution of target analytes in a small volume of a suitable organic solvent (i.e., methanol, acetonitrile, and ethyl acetate) getting high enrichment factors. However, for more polar compounds, the strength of the interaction is not high enough and low recoveries are obtained because the corresponding breakthrough volume is easily reached.

An easy manner of increasing breakthrough volumes is to increase the amount of sorbent used, which will increase the number of interactions that take place, or to add salts to the sample, diminishing the solubility of target analytes (salting-out effect) and thus favoring their interactions with the sorbent. However, these approaches do not always provide satisfactory results. In this case, the most direct way of increasing breakthrough volumes of polar compounds is the use of sorbents with higher affinity for target analytes. These sorbents include styrene–divinylbenzene-based polymers with a high specific surface (~1000 m^2·g^{-1}), which are commercialized by several companies under different trademarks (i.e., Lichrolut, Oasis, Envichrom, etc.). The interaction of analytes with these sorbents is also based on hydrophobic interactions, but the presence of aromatic rings within the polymeric network leads to strong π–π* interactions with the aromatic rings present in the chemical structure of many organic compounds. Another alternative is the use of graphitized carbon cartridges or disks, which have a great capacity for the preconcentration of highly polar compounds thanks to the presence of various functional groups, including positively charged active centers on its surface.

5.3.4.3 Ion-Exchange Sorbents

Ionic or easily ionizable compounds can be extracted by these sorbents. Sorption occurs at a pH in which the analyte is in its ionic form, and then it is eluted by a change in the pH value with a suitable buffer. The mechanism involved provides a certain degree of selectivity. Phenols, carboxylic acids, or phenoxyacid herbicides can be extracted using anion exchangers, whereas anilines, amines, and *n*-heterocycles can be extracted using cation exchangers. However, its use is rather limited due to the presence of high amounts of inorganic ions in the samples, which overload the capacity of the sorbent, leading to low recoveries of target analytes.

5.3.4.4 Affinity Sorbents

The retention mechanisms taking place between target analytes and the sorbents described in Sections 5.3.4.1, 5.3.4.2 and 5.3.4.3 are not selective (hydrophobic or ionic interactions), leading to the simultaneous extraction of matrix compounds, which can negatively affect subsequent chromatographic analysis. Even when using selective detectors (i.e., MS), the presence of matrix compounds can suppress or enhance analyte ionization, hampering accurate quantification.

The use of antibodies immobilized on a suitable support, the so-called immunosorbent (IS), for the selective extraction of organic compounds (mainly for pesticides and mycotoxins) from different samples appeared some years ago as a clear alternative to traditional sorbents [15,16]. In this approach, only the antigen that produced the immune response, or very closely related molecules, will be able to bind the antibody. Thus, when the sample is run through the IS, the analytes are selectively retained and subsequently eluted free of coextractives. However, this methodology is not free from important drawbacks. Obtaining antibodies is expensive and time-consuming, and few antibodies are commercially available. Besides, once obtained antibodies have to be immobilized on an adequate support, which may result in poor antibody orientation or even complete denaturation. Because of these limitations, the preparation and use of molecularly imprinted polymers (MIPs) has been proposed as a promising alternative.

To overcome the drawbacks associated with the preparation and use of ISs, MIPs, tailor-made macroporous materials with selective binding sites that are able to recognize a particular molecule have been proposed as an alternative [17]. Their synthesis, depicted in Figure 5.6, is based on the formation of defined (covalent or noncovalent) interactions between a template molecule and functional monomers during a polymerization process in the presence of a cross-linking agent. After polymerization, the

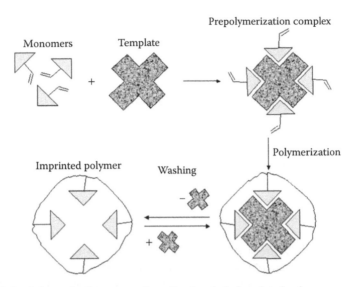

FIGURE 5.6 Scheme for the preparation of molecularly imprinted polymers.

template molecule is removed, leaving cavities that are complementary in size and shape to the analyte. Thus, if a sample is loaded on it, in an SPE procedure, the analyte (the template) or closely related compounds will be able to rebind selectively the polymer being subsequently eluted free of coextractives. This methodology, that is, molecularly imprinted solid-phase extraction (MISPE), has been successfully used in the determination of a great variety of organic compounds in environmental, food, and biological samples. Because of their ease of preparation and excellent physical stability and chemical characteristics (high affinity and selectivity for the target analyte), MIPs have received special attention from the scientific community in several fields. Besides, there already are MISPE cartridges commercially available for the extraction of certain analytes and some companies offer custom synthesis of MIPs for SPE, which will make the implementation of MISPE in analytical laboratories easy [18].

5.3.4.5 Other Sorbents and Modes

5.3.4.5.1 Multimodal and Mixed-Phase Separations

Multimodal SPE refers to the intentional use of two separation modes by using two cartridges each packed with a different sorbent with the cartridges connected in series or using one cartridge packed with multifunctional particles or two different sorbents blended in a given proportion. Also, the use of layered packing has been reported, where two or more different packings are used to isolate differing molecular species.

5.3.4.5.2 Restricted Access Materials

Restricted access materials (RAMs) are specially designed phases for SPE for the isolation of small analytes in biological fluids [19,20]. The packing consists of small pore particles with a C_8 or C_{18} layer inside the pores, whereas a hydrophilic layer covers the external surface of the particle. After loading, big molecules (i.e., proteins) are unable to enter the pores and are not retained by the hydrophilic particle surface passing through the column. However, small analytes enter the pores and are retained inside, being eluted after the proteins.

5.3.4.5.3 QuEChERS: Quick, Easy, Cheap, Effective, Rugged, and Safe Extraction

Quick, easy, cheap, effective, rugged, and safe extraction (QuEChERS) was proposed by Anastassiades et al. [21] in 2003 for the extraction of pesticides (multiresidue) in fruits and vegetables. The QuEChERS procedure involves an initial extraction with acetonitrile or acetone to a homogenized portion of the sample. Then, salts (i.e., $MgSO_4$) are added, leading to the separation of organic solvents from water present in the sample and promotes the extraction of target analytes into the organic solvent. After centrifugation, an aliquot of the raw extract is then cleaned up by dispersive SPE, which involves the addition of small amounts of bulk SPE packing (i.e., C_{18}, amino, etc.) to remove interferences from the organic solvent. After centrifugation, the final organic extract is directly amenable to analysis by liquid chromatography (LC)-MS and/or GC-MS.

5.3.4.5.4 Online SPE

The coupling of SPE to LC is especially simple to perform in any laboratory and has been extensively described for the online preconcentration of organic compounds in environmental water samples [22]. The simplest way of SPE-LC coupling is shown in Figure 5.7, where a precolumn (1–2 cm × 1–4.6 mm i.d.) filled with an appropriated sorbent is inserted in the loop of a six-port injection valve. After sorbent conditioning, the sample is loaded by a low-cost pump and the analytes are retained in the precolumn. Then, the precolumn is connected online to the analytical column by switching the valve so that the mobile phase can desorb the analytes prior to their separation in the chromatographic column. Alkyl-bonded silica (mainly C_{18} silica) has been widely used as a precolumn sorbent, although it is being replaced by styrene–divinylbenzene copolymers, which offer higher affinity for polar analytes and permits the utilization of larger sample volumes without exceeding the

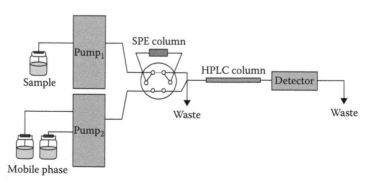

FIGURE 5.7 SPE–liquid chromatography coupling setup.

breakthrough volumes of analytes. Other materials that have been successfully employed are small extraction disks, graphitized carbons, ISs, and MIPs.

The coupling of SPE to GC is also possible thanks to the ability of injecting large volumes into a gas chromatograph using a column of deactivated silica (retention gap) located between the injector and the analytical column. SPE-GC uses the same sorbents as those used in SPE-LC; but in this case, after the preconcentration step the analytes are desorbed with a small volume (50–100 μL) of an appropriate organic solvent, which is directly introduced into the chromatograph. In general, using only 10 mL of a water sample, it is possible to reach detection limits at the microgram per liter level by using common detectors.

5.3.5 SOLID-PHASE MICROEXTRACTION

As stated previously, SPE is a very useful procedure for the extraction of organic compounds from any kind of sample. However, it still requires the use of toxic organic solvents (although to a lower extent than LLE) and its applicability is restricted to liquid samples. With the aim of eliminating these drawbacks, Arthur and Pawliszyn [23] introduced solid-phase microextraction (SPME) in 1989. Its simplicity of operation, its solventless nature, and the availability of commercial fibers have caused SPME to be rapidly implemented in analytical laboratories.

The SPME device is quite simple and just consists of a silica fiber coated with a polymeric stationary phase similar to the stationary phases used in GC columns, as depicted in Figure 5.8. The fiber is located inside the protecting needle of a

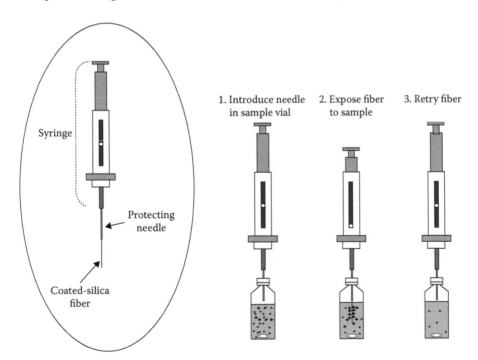

FIGURE 5.8 Solid-phase microextraction device and typical mode of operation.

syringe that is specially designed to allow exposure of the fiber during sample analysis. As in any SPE procedure, SPME is based on the partitioning of target analytes between the sample and the stationary phase and consists of two consecutive steps, extraction and desorption. An intermediate washing step can also be performed.

5.3.5.1 Extraction

The extraction step can be performed both by exposure of the fiber to the HS (restricted to volatile compounds in liquid or solid samples) and by direct immersion of the fiber into the sample (aqueous-based liquid samples). The experimental procedure (see Figure 5.8) is very simple. First, the fiber is inside the protecting needle, which is introduced into the sample vial. Then, the fiber is exposed to the sample to perform extraction by sorption of the analytes to the stationary phase. Finally, the fiber is retried inside the needle for further desorption and the whole device is removed.

Obviously, a proper selection of SPME sorbent is a key factor in the success of the analysis. In general, the polarity of the fiber should be as similar as possible to that of the analyte of interest. In this sense, nowadays there are a great variety of commercially available fibers that cover a wide range of polarities (i.e., carbowax/ DVB for polar compounds or polydimethylsiloxane [PDMS] for hydrophobic compounds). Also, both the fiber thickness and the porosity of the sorbent will influence final extraction efficiency. Besides, other physical and chemical parameters such as temperature, exposition time, agitation, pH, and ionic strength (salting-out effect) of the sample can be optimized.

It is important to mention that from the mathematical model governing SPME, it can be concluded that when the sample volume is much higher than the fiber volume the extraction efficiency becomes independent of the sample volume. Although this is not applicable for laboratory samples (low volumes), this fact makes SPME a very interesting tool for in-field sampling procedures, since the fiber can be exposed to the air or directly immersed into a lake or a river regardless of the sample volume.

5.3.5.2 Desorption

Desorption can be performed thermally in the injection port of a gas chromatograph, or it is possible by the elution of analytes by means of a suitable solvent. In the latter case, desorption can be carried out in a vial containing a small volume of the solvent to be further analyzed by chromatographic techniques or eluted with the mobile phase on a specially designed SPME-HPLC interface.

The thermal desorption of analytes in the injector port of the GC instrument is based on an increase in gas-fiber the partition coefficient with increasing temperature. In addition, a constant flow of carrier gas inside the injector facilitates the removal of the analytes from the fiber. The main advantage of thermal desorption is that the total amount of extracted analytes is introduced in the chromatographic system and analyzed, thus compensating the low recoveries usually obtained in the extraction step. However, unfortunately thermal desorption cannot be used for nonvolatile or thermolabile compounds; thus it is necessary to use desorption with solvents. The procedure is similar to SPE elution, but in this case the fiber is immersed in a small volume of elution solvent and agitated or heated to favor the transfer of analytes to the solvent solution. Today, there are commercially available interfaces that allow the direct coupling of SPME to liquid chromatography, the coupling being similar to that described in Figure 5.7 for SPE-HPLC apart from placing a specially designed little chamber instead of a precolumn in the loop of a six-port injection valve.

Traditionally, some drawbacks have been identified in SPME. First, the available fused-silica fibers were rather fragile and possessed short lifetimes. However, the durability and robustness of SPME fibers has been addressed by the introduction of new-generation metal (superelastic) fibers. Besides, the amount of coatings available has grown during recent years, extending the field of application of SPME. Other developments, such needle trap devices [24], high-surface-area samplers (thin-film microextraction) [25], *in vivo* SPME samplers [26], or internally cooled SPME devices [27], have made SPME one of the most used and promising extraction technique.

A technique derived from SPME is stir-bar sorptive extraction (SBSE), which is based on the partitioning of target analytes between the sample (mostly aqueous-based liquid samples) and a stationary phase–coated magnetic stir bar [28]. To date, only PDMS-coated stir bars are commercially available, restricting the range of applications to the extraction of hydrophobic compounds due to the apolar character of PDMS. The experimental procedure followed in SBSE is quite simple. The liquid sample and the PDMS-coated magnetic stir bar are placed in a container. Then the sample is stirred for a certain period of time (30–240 minutes) until no additional recovery for target analytes is observed. Finally, the stir bar is removed and placed in a specially designed unit in which thermal desorption and transfer of target analytes to the head of the GC column take place. The simplicity of this operation and its solventless nature make SBSE a very attractive technique, and the development of new stir bars coated with more polar and selective sorbents are expected in the near future.

5.3.6 Solid–Solid Extraction: Matrix Solid-Phase Dispersion

Matrix solid-phase dispersion (MSPD), introduced by Barker et al. in 1989 [29], is based on the complete disruption of the sample (liquid, viscous, semisolid, or solid) while the sample components are dispersed into a solid sorbent. Most methods use C_8- and C_{18}-bonded silica as solid supports. Other sorbents such as Florisil and silica have also been used, although to a lesser extent. Experimentally, the sample is placed in glass mortar and blended with the sorbent until a complete disruption and dispersion of the sample on the sorbent is obtained. Then the mixture is directly packed into an empty cartridge such as that used in SPE. Finally, analytes are eluted after a washing step for removing interfering compounds. The main difference between MSPD and SPE is that the sample is dispersed through the column instead of only onto the first layers of the sorbent, which typically allows one to obtain rather clean final extracts. MSPD has been successfully applied for the extraction of pesticides in fruit juices, honey, oranges, cereals and soil, veterinary drugs in tissue, and antibiotics in milk, among others, and the achieved performance, compared to other classical extraction methods, has been found to be superior in most cases [30]. The main advantages of MSPD are the short extraction times needed; the small amounts of sample, sorbent, and solvents required; and the possibility of performing extraction and cleanup in one single step. However, the technique is not easily automated and could be time-consuming for a large number of samples.

5.3.7 EXTRACTION FROM THE VAPOR PHASE: HEADSPACE ANALYSIS

Some target analytes are volatile and can be directly extracted from the vapor phase above the sample (HS), thus avoiding coextraction of the nonvolatile matrix components. HS analysis has existed since the late 1950s and is still actively used. HS is rather simple, and it is easily coupled to GC for the analysis of volatile analytes. Both static and dynamic systems can be used, as described in Sections 5.3.7.1 and 5.3.7.2, and the HS can also be sampled using other techniques such as SPME or SDME.

5.3.7.1 Static Headspace

Static HS involves heating an aliquot of a liquid or solid sample in a sealed vial at a given temperature for a given amount of time. The increase in temperature leads to an increase in vapor phase concentration and, thus, volatile analytes begin to partition between the sample and the gas phase. Eventually, an equilibrium state is reached and an aliquot of the gas phase is sampled with a gastight syringe, which is further analyzed by GC. The effects of sample volume, temperature, and modifications to the sample matrix must be considered [31] as reproducible analysis requires the exact replication of analytical conditions. Its main drawback is its limited sensitivity, since only a small aliquot of the whole gas phase is sampled, although this is enough for the determination of solvents in biological fluids and pharmaceuticals, monomers in polymers, flavors in food and volatiles in drinking water and wastewater, and so on [32].

5.3.7.2 Dynamic Headspace

Dynamic HS (or continuous gas extraction) involves the continual sampling of the gas phase above the sample by flushing with an inert gas, and the volatiles are transferred to an absorbent trap. As the gas is removed, the equilibrium will reestablish itself and an exhaustive extraction of volatiles can be obtained. Then, the trap is heated and the analytes are released and transferred to the gas chromatograph. In this manner, the low sensitivity associated with a static HS is overcome, reaching detection limits of several parts per trillion.

A variant of dynamic HS is the so-called "purge-and-trap" technique. In this case, the inert gas is bubbled through the sample, which favors the extraction of target analytes. However, a purge-and-trap system requires more maintenance and is subject to problems such as foaming of the sample.

5.4 CONCLUSIONS

In this chapter, a description of the different techniques developed during the past few years for the extraction and cleanup of organic compounds from environmental, food, and bio samples prior to further chromatographic analysis has been made. It is evident that a great effort has been made during the past years to improve the techniques and procedures used for sample preparation, and the analyst has at his or her disposal a wide variety of sample preparation techniques. Obviously, depending on the required information, different extraction techniques should be chosen. However, in spite of such efforts, sample preparation is still the limiting step of the analysis.

Thus, since sample preparation cannot be avoided, further studies toward its simplification are expected in the near future. In this regard, environmentally friendly, cost-effective, and selective procedures are required. In parallel, advances in miniaturization and automation will ease the integration of sample preparation and instrumental analysis, leading to faster procedures with improved performance in terms of accuracy, precision, and traceability.

REFERENCES

1. Sánchez-Brunete, C., Pérez, R.A., Miguel, E. and Tadeo, J.L., Multiresidue herbicide analysis in soil samples by means of extraction in small columns and gas chromatography with nitrogen-phosphorus and mass spectrometric detection, *J. Chromatogr. A*, 823 (1–2), 17, 1998.
2. Castro, J., Sánchez-Brunete, C. and Tadeo, J.L., Multiresidue analysis of insecticides in soil by gas chromatography with electron-capture detection and confirmation by gas chromatography-mass spectrometry, *J. Chromatogr. A*, 918 (2), 371, 2001.
3. Babic, S., Petrovic, M. and Kastelan, M., Ultrasonic solvent extraction of pesticides from soil, *J. Chromatogr. A*, 823, 3, 1998.
4. Björklund, E., Nilsson, T. and Bøwadt, S., Pressurised liquid extraction of persistent organic pollutants in environmental analysis, *Trends Anal. Chem.*, 19 (7), 434, 2000.
5. Smith, R.M., Superheated water extraction, in *Handbook of Sample Preparation*, Pawliszyn, J. and Lord, H., Eds., John Wiley & Sons, Inc., Hoboken, New Jersey, 2010, chap. 10.
6. Miege, C., Dugay, J. and Hennion, M.-C., Optimization and validation of solvent and supercritical-fluid extractions for the trace-determination of polycyclic aromatic hydrocarbons in sewage sludges by liquid chromatography coupled to diode-array and fluorescence detection, *J. Chromatogr. A*, 823, 219, 1998.
7. Zougagh, M., Valcarcel, M. and Rios, A., Supercritical fluid extraction: A critical review of its analytical usefulness, *Trends Anal. Chem.*, 23 (5), 399, 2004.
8. Liu, H. and Dasgupta, P.K., Analytical chemistry in a drop. Solvent extraction in a microdrop, *Anal. Chem.*, 68, 1817, 1996.
9. Jeannot, M.A. and Cantwell, F.F., Solvent microextraction into a single drop, *Anal. Chem.*, 68, 2236, 1996.
10. Xu, L., Basheer, C. and Lee, H.K., Developments in single-drop microextraction, *J. Chromatogr. A*, 1152, 184, 2007.
11. Jönsson, J.A. and Mathiasson, L., Membrane-based techniques for sample enrichment, *J. Chromatogr. A*, 902, 205, 2000.
12. Khrolenko, M., Dzygiel, P. and Wieczorek, P., Combination of supported liquid membrane and solid-phase extraction for sample pretreatment of triazine herbicides in juice prior to capillary electrophoresis determination, *J. Chromatogr. A*, 975, 219, 2002.
13. Pedersen-Bjergaard, S. and Rasmussen, K.E., Liquid–liquid–liquid microextraction for sample preparation of biological fluids prior to capillary electrophoresis, *Anal. Chem.*, 71, 2650, 1999.
14. Pedersen-Bjergaard, S. and Rasmussen, K.E., Liquid-phase microextraction with porous hollow fibers, a miniaturized and highly flexible format for liquid–liquid extraction, *J. Chromatogr. A*, 1184, 132, 2008.

15. Pichon, V., Chen, L., Hennion, M.-C., Daniel, R., Martel, A., Le Goffic, F., Abian, J. and Barceló, D., Preparation and evaluation of immunosorbents for selective trace enrichment of phenylurea and triazine herbicides in environmental waters, *Anal. Chem.*, 67, 2451, 1995.

16. Martín-Esteban, A., Fernández, P. and Cámara C., Immunosorbents: A new tool for pesticide sample handling in environmental analysis, *Fresenius' J. Anal. Chem.*, 357, 927, 1997.

17. Sellergren, B., *Molecularly Imprinted Polymers: Man-Made Mimics of Antibodies and their Applications in Analytical Chemistry*, 1st ed., Elsevier Science BV, Amsterdam, 2001.

18. Sellergren, B. and Martin-Esteban, A., The use of molecularly imprinted polymers for sampling and sample preparation, in *Handbook of Sample Preparation*, Pawliszyn, J. and Lord, H., Eds., John Wiley & Sons, Inc., Hoboken, New Jersey, 2010, chap. 23.

19. Boos, K.-S. and Rudolphi, A., The use of restricted access media in HPLC, part I. Classification and review, *LC GC Int.*, 15, 602, 1997.

20. Rudolphi, A. and Boos, K.-S., The use of restricted access media in HPLC, part II. Applications, *LC GC Int.*, 15, 814, 1997.

21. Anastassiades, M., Lehotay, S.J., Stajnbaher, D. and Schenk, F.J., Fast and easy mutiresidue method employing acetonitrile extraction/partitioning and "dispersive solid-phase extraction" for the determination of pesticides residues in produce, *J. AOAC Int.*, 86, 412, 2003.

22. Hennion, M.-C. and Scribe, P., Sample handling strategies for the analysis of organic compounds from environmental water samples, in *Environmental Analysis: Techniques, Applications and Quality Assurance*, Barceló, D., Ed., Elsevier Science Publishers BV, Amsterdam, 1993, chap. 2.

23. Arthur, C.L. and Pawliszyn, J., Solid phase microextraction with thermal desorption using fused silica optical fibers, *Anal. Chem.*, 62, 2145, 1990.

24. Wnag, A.-P., Fang, F. and Pawliszyn, J., Sampling and determination of volatile organic compounds with needle trap devices, *J. Chromatogr. A*, 1072, 127, 2005.

25. Bruheim, I., Liu, X. and Pawliszyn, J., Thin-film microextraction, *Anal. Chem.*, 75, 1002, 2003.

26. Musteata, F.M. and Pawliszyn, J., In vivo sampling with solid phase microextraction, *J. Biochem. Biophys. Methods*, 70, 181, 2007.

27. Zhang, Z. and Pawliszyn, J., Quantitative extraction using and internally cooled solid phase microextraction device, *Anal. Chem.*, 67, 34, 1995.

28. Baltussen, E., Sandra, P., David, F. and Cramers, C., Stir bar sorptive extraction (SBSE), a novel extraction technique for aqueous samples: Theory and principles, *J. Microcolumn. Sep.*, 11, 737, 1999.

29. Barker, S.A., Long, A.R. and Short, C.R., Isolation of drug residues from tissues by solid phase dispersion, *J. Chromatogr.*, 475, 353, 1989.

30. Kristenson, E.M., Ramos, L. and Brinkman, U.A.Th., Recent advances in matrix solid-phase dispersion, *Trends Anal. Chem.*, 25, 96, 2006.

31. Kolb, B.and Ettre, L.S., *Static Headspace–Gas Chromatography Theory and Practice*, Wiley-VCH, Hoboken, New Jersey, 1997.

32. Penton, Z.E., Headspace gas chromatography, in *Handbook of Sample Preparation*, Pawliszyn, J. and Lord, H., Eds., John Wiley & Sons, Inc., Hoboken, New Jersey, 2010, chap. 2.

L. Mark Hall, Dennis W. Hill, Lowell H. Hall,
Tzipporah M. Kormos, and David F. Grant

CONTENTS

6.1 INTRODUCTION

Metabolomics is the study of low-molecular-weight compounds (50–1000 Da) that are the products of cellular chemical processes. The presence, absence, and concentration of these metabolites in a tissue or biofluid may be considered a profile that carries significant information about the state of the cellular system. Metabolomics uses high-resolution spectrometric techniques to separate and identify compounds and was initially based on data collected from proton nuclear magnetic resonance (^{1}H-NMR) analyses of urine, blood, saliva, tears, cerebrospinal fluid, cells, and tissue samples [1]. These studies showed differences in metabolites and metabolite concentrations between treated and control samples that could be statistically associated with toxin exposure or specific diseases [2–5]. More recent metabolomic studies have added computational methods to facilitate high-throughput analyses of samples. The output from these combined methods consists of experimental "features" that have statistically significant associations with phenotypes of interest. Subset "omics" have also recently emerged, including glycomics, [6] lipidomics [7–9], and peptidomics [10]. These "targeted" subset omic approaches relate to specific metabolites, or classes of metabolites, that are expected to be important in specific disease states [9]. Thus, a distinction can be made between targeted studies and "nontargeted" studies where there is no expectation of focusing on a specific metabolite or class of metabolites.

Metabolomic studies have progressed from concept development to methods development and more recently to the identification of disease biomarkers [11–18], differentiating gender and species [19–23], and toxicity studies [24–28]. An increased interest in metabolomics has led to the development of chemometric tools that extract chromatographic data and apply multivariate statistical tools to determine significant analytical features that are differentially expressed between healthy and disease states.

Although it is possible to use a reproducible analytical feature as a surrogate for chemical structure, metabolite identification is the ultimate goal. Structure identification serves to explain the biological significance of a biomarker [1] and is important in suggesting reaction mechanisms and druggable therapeutic targets for the related pathology. Structure identification also aids in the development of high-throughput biomarker assays for use in the clinical setting. In addition, agencies such as the U.S. Food and Drug Administration require thorough validation of a putative biomarker. The validation procedure generally requires an explanation of the mechanism of action, and thus structure identification is an important component of this procedure [29,30]. Although NMR spectroscopy, gas chromatography–mass spectrometry (GC-MS), and high-performance liquid chromatography–mass spectrometry (HPLC-MS) are all commonly used for metabolomics, the use of HPLC-MS techniques is becoming increasingly more prevalent.

The identification of an unknown small molecule generally requires analysis with multiple analytical techniques. In the case of HPLC-MS-based nontargeted metabolomics studies, current methods focus on exact mass, retention time, and mass spectral fragmentation. Each of these techniques can be quickly performed on HPLC-MS instrumentation without extensive sample preparation. Exact mass is used to determine potential molecular formulas for unknown compounds. Retention times and fragmentation patterns are then used to help distinguish a particular unknown compound from the multitude of candidate structures with the same molecular formula. Unfortunately, retention times and collision-induced dissociation (CID)-based mass spectral fragmentation are not typically standardized or reproducible between instruments. Therefore, a database of fragmentation patterns and retention times created on one instrument will not necessarily be useful in another laboratory. This has led some laboratories to create their own databases of experimental variables [31]. However, creating an internal database large enough for the identification of a significant number of metabolites is time consuming, expensive, and not feasible for most laboratories.

Currently, the majority of nontargeted metabolomic structure identification methods are based on external database searching. As mentioned earlier, the initial analysis of biofluids using HPLC-MS yields a list of unknowns defined by an m/z value, retention time, and CID spectra. The mass of the unknown ($\pm$instrument mass accuracy) can be used as a query to extract potential structures from a database by filtering out compounds with mass values outside the query range. Currently, there are a number of freely accessible online databases that can be used for this purpose. They range from general chemical structure databases such as PubChem [32], ChemSpider [33], and ZINC [34] to more biologically oriented databases such as the Human Metabolome Database (HMDB) [35], KEGG COMPOUND [36], METLIN [37], and Chemical Entities of Biological Interest [38]. Databases are also being developed for specific diseases [39] or classes of metabolites [40,41]. These databases can be searched by name, mass, or structure. Some databases have additional information (predicted log P values, mass spectrometry [MS] fragmentation patterns, or metabolic pathways) that may assist the structure identification process.

Biological databases, such as HMDB, include data useful for metabolomic analysis of biofluids and tissues in which each metabolite has been detected, as well as known concentrations of metabolites in control and disease samples [42]. Many laboratories limit their searches to biological databases in the belief that filtering a general chemical database such as PubChem will result in an excessive number of nonendogenous candidate compounds (i.e., false positives). There is an understandable logic to this approach, but there are notable drawbacks to the exclusive use of biological databases. The use of general chemical databases tends to produce a larger number of potential matches to an experimentally observed unknown. However, it must be kept in mind that there is no species for which the entire metabolome has been completely identified. Although the practice of searching biological databases tends to minimize the number of false positives, such databases cannot be used to identify the structure of previously unidentified metabolites. The likelihood of the correct structure being present in the filtered results is greatly increased by the use of general chemical databases, since these databases contain the vast majority of known chemicals in addition to those found in biological databases. There is a clear need to create a balance between a filtered candidate list that is too large to efficiently process and a smaller candidate list that may not contain a match to the unknown structure. One possible solution involves the application of multiple independent filters to a large general database. The use of multiple descriptors in filtering can minimize the impact of a large initial database and result in a filtered list of manageable size.

In addition to m/z, matching chromatographic retention times is a technique commonly used to help confirm structure identity. Analytes partition between stationary and mobile phases in a chromatographic column due to noncovalent interactions between the analyte and both phases. Since all such interactions fundamentally arise from the structure of the analyte, and nearly every atom participates to some degree, the retention characteristic is useful in characterizing structure with a high level of discrimination. Unfortunately, slight variations in the composition of the mobile phase, the status of the column, or the temperature of the room can shift the retention time, making retention time a difficult descriptor to reproduce over long periods. This is the case even when measurements are taken on the same instrument. Matching the retention time of an analyte between different instruments is even more difficult. Retention index (RI) is a normalized version of retention time that is more robust and has a greater potential for use in structure identification [43–45]. RI determination uses a series of standards, such as a homologous series of n-nitroalkanes, as a reference index for retention time. The retention times of the analytes are then converted to an index value relative to the number of carbons in the standards that bracket its retention. Additional internal standards can be used to adjust for any slight changes between the analysis of the reference standards and the unknown to make the data even more reproducible.

Database filtering for structure identification depends on access to a database of structures for which many of the experimental endpoints necessary to confirm structure have been measured. Features such as exact mass can be calculated for any structure, but it is necessary to make measurements for other descriptors such as RI and CID spectra. As stated previously, it is not feasible to acquire and measure these values for a sufficient number of compounds to facilitate identification of previously unknown metabolites. The possibility of using computationally derived versions of descriptors such as RI and CID spectra has also been explored as an alternative to the use of measured values [46,47]. Unlike experimentally determined descriptors, values calculated by predictive models can be generated for any compound in a database. These compounds do not need to exist, except in the virtual sense, and only need to be obtained and measured if there is a high probability that they are a match for an experimentally observed unknown. In this way, predicted values could be used to facilitate high-throughput identification through database searching.

Computational models of reasonable accuracy have been created to predict RI based on structure. Since there is no reason to expect a linear correlation between structural descriptors and RI, modeling has turned to artificial neural networks (ANNs) for their superior predictive power. Guo et al. [48] used the RI values of 25 alcohols analyzed on six different GC stationary phases to create a model that takes into account structure and stationary phase. For these studies, Randić, as well as Kier and Hall structural descriptors were evaluated; statistically significant multiple linear regression (MLR) and ANN models were created using four variables related to structure and one related to stationary phase. The ANN models showed superior predictive power. Acevedo-Martínez et al. [49] determined the RI values for 49 imines on a GC and used MLR and ANN to predict RI values for imines. The MODEST program was used to determine Randić, topographic, and valence descriptors, and the best model with six variables was chosen. Both modeling methods yielded statistically significant correlations, with the ANN model providing superior predictive power. Konoz et al. [50] modeled RI values from 35 aliphatic ketones and aldehydes that were analyzed on four different GC stationary phases, each at four different temperatures to form one predictive model. The CODESSA program was used to calculate constitutional, topological, geometrical, electrostatic, and quantum chemical descriptors. The MLR model contained three descriptors in addition to a variable for the polarity of the column and one for the column temperature. The ANN model used the same three descriptors and temperature to predict the RI value on each of the four different stationary phases. Although both models were statistically significant, the ANN model had far better predictive power. A limitation of the use of predictive chromatographic retention models is that each model is defined by the specific chromatographic parameters used to develop the model. All of these models were based on datasets of chemical subclasses and focused on a limited set of functional groups such as imines or aliphatic ketones.

RI models can also be created from HPLC-MS data. Such models could be useful for predicting RI values for candidate compounds from a large database and comparing these predicted values with the RI values of experimentally observed unknowns. RI values are experimentally determined at the same time as exact mass, as long as retention standards are included in the sample set. Since RI values are quantitative, the accuracy and precision of a predictive model can be calculated. Therefore, potential matches extracted from the database using exact mass that are not within the error range of the RI predictive model can be excluded from the candidate list.

Previous models of RI were based on small datasets of chemically related compounds and had a limited structural applicability domain. To be of use for database filtering, an RI model needs to be based on a large dataset of structurally diverse compounds so that it will have the ability to make reasonable predictions for the broadest possible range of structures. In this chapter, we describe our efforts to develop computational models of RI. The intent is to use these models for nontargeted metabolomics applications where structure identification is the ultimate goal.

6.2   PRELIMINARY RETENTION INDEX MODEL

To evaluate the concept of using computationally predicted values of experimental features such as RI for use in filtering a virtual database, it was necessary to determine whether or not it was reasonable to model RI using descriptors that may be rapidly calculated for any molecular structure. As proof of concept, a preliminary ANN ensemble model was created using a previously measured dataset of RI values. This diverse dataset of compounds was measured by a modification of the HPLC method reported by Hill and Kind [41] and was judged to be large and diverse enough to provide a reasonable measure of the potential of the proposed methods. For a model to be effective in database filtering as described earlier, the ± 2 standard error (SE) of prediction range needs to be narrow enough to filter out a substantial number of candidates without being so narrow as to risk inadvertently discarding the compound matching the unknown. To address this issue, an initial goal was set to maximize the number of predicted values within 100 RIU of the experimental value. It is unlikely that a model that does not make a substantial percentage of predictions at this accuracy level would prove effective in the filtering task. The methodology and results of this modeling project are given in sections 6.2.1 through 6.2.7.

6.2.1 RETENTION INDEX DETERMINATIONS FOR THE PRELIMINARY DATASET

The retention times of the compounds in the dataset were measured on an HP1090 high-performance liquid chromatography–diode array detector (HPLC-DAD) system. The compounds were eluted from a Zorbax C8, 4.6 mm × 250 mm (5 μm particle), column that was maintained at 30°C using a solvent-programmed mobile phase consisting of a 2.0 minute isocratic hold at 100% solvent A (0.15 M phosphoric acid and 0.05 M triethylamine) followed by a linear gradient to 100% solvent B (0.15 M phosphoric acid, 0.05 M triethylamine, and 20% water in acetonitrile) in 20 minutes with a 5 minute isocratic hold at 100% solvent B. The flow rate was 2.0 mL/min. The RI scale for each batch analysis was established by running n-nitroalkanes C through C_{10}. Acetophenone was coanalyzed with all samples, including n-nitroalkanes.

Retention times were measured for each sample test compound, and the relative retention time (RR_t) was calculated as the ratio of the test compound retention time to that of the acetophenone. The RI was then calculated by the formula given in Equation 6.1 [41], where RR_{tx} is the relative retention time of the test compound, RR_{tz} is the relative retention time of the n-nitroalkane eluting just before the test compound, RR_{tz+1} is the relative retention time of the n-nitroalkane eluting just after the test compound, and z is the number of carbons in the n-nitroalkane eluting just before the test compound.

$$RI = \left(\frac{(RR_{tx} - RR_{tz})100}{RR_{tz+1} - RR_{tz}} \right) + 100z \qquad (6.1)$$

The diversity of RI values for the dataset was significant, covering a range of nearly 1100 RIU. The minimum RI value in the dataset was 55 (thiamine), whereas the largest was 1140 (hexahydrocannabinol). The average value was 354 RIU with a standard deviation (SD) of 156. The distribution of RI values was not normal, being skewed to the low end. Less than 5% of the data was found with RI greater than 600 RIU, and only 10 values fell between 800 RIU and the maximum at 1140 RIU. A histogram of the RI values is given in Figure 6.1.

6.2.2 PRELIMINARY DATASET STRUCTURAL CHARACTERISTICS

The overall structural diversity of the preliminary dataset is considerable. The data contain simple aromatic hydrocarbons (biphenyl and naphthalene) and also very complex structures such as α-amanitin (a 25-heteroatom cyclic polyamide, with a dual-fused 25-membered ring system). More than 99% of structures contain one or more rings, and 82% contain an aromatic ring. Compounds have an average of 1.4 hydrogen bond donors, 4 hydrogen bond acceptors, and 5.7 rotatable bonds. Nearly 40% of the structures contain a molecular configuration amenable to the formation of an internal hydrogen bond. The average formula weight and number of hydrogen bond donors and acceptors are consistent with those of drug-like compounds.

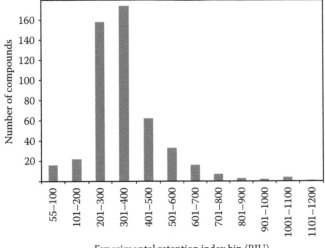

FIGURE 6.1 Experiential retention index histogram for the preliminary dataset.

Table 6.1 gives an outline of the structural characteristics of the dataset, including the functional groups present.

Many compounds are found in the preliminary dataset that contain features with low population. Examples are carbamate, guanidine, sulfate, and nitrogen atoms with a permanent charge. These functional groups occur in ≤3% of the overall data and are listed in Table 6.2 along with their populations. Underpopulated structural features are often ignored when creating descriptors for a dataset, but because of the physical nature of the RI endpoint these structural features likely make a significant contribution to the RI value for compounds containing the feature. For this reason, it will be necessary to account for these features in the structure description system.

6.2.3 PRELIMINARY DATASET MODEL DESCRIPTORS

Structure descriptors for compounds in the preliminary dataset were generated from an SD file using winMolconn v1.4 [52]. The winMolconn software package calculates structure descriptors in three broad categories: E-State indices (electrotopological state), chi indices (molecular connectivity), and elementary structure information indices. This overall method of structure descriptor development has come to be known as "structure information representation" (SIR) [53], and it is also referred to as topological structure description.

The E-State is based on an atom-level representation of electron accessibility. An atom-level E-State is calculated for each atom in a molecule, and the magnitude of the value arises from a combination of electron distribution and local atom topology. These calculations are derived from graph connectivity and are not based on an optimized three-dimensional (3D) geometry or molecular orbital calculations. Each atom is assigned an intrinsic state value based on valance state and Kier–Hall

TABLE 6.1
HPLC-RI Preliminary Model Dataset Structural Characteristics, Including Features with >3% Population

Global Attribute	Average[a]	Count[b]	Percentage[c]
MW	262.7	—	—
Ring	2.38	494	99.2
Aromatic ring	1.08	408	81.9
Heteroaromatic ring	0.23	106	21.3
Nonaromatic ring	1.06	266	53.4
Fused ring system	0.44	221	44.3
Rotatable bonds	5.7	477	95.8
Heteroatom	4.3	495	99.4
Hydrogen bond acceptor	4.01	495	99.3
Hydrogen bond donor	1.4	388	77.9
Internal hydrogen bond	0.95	197	39.6
TPSA[d]	58.9	495	99.4
Feature Attribute			
Amine	0.42	202	40.6
Aniline	0.08	38	7.6
Pyridine nitrogen	0.20	77	15.5
Pyrrole nitrogen	0.09	47	09.4
Ether	0.38	109	21.9
Ester	0.11	52	10.4
Ketone	0.17	62	12.4
Alcohol	0.35	108	21.7
Carboxylic acid	0.12	56	11.2
Phenol	0.24	93	18.7
Amide	0.25	108	21.7
Urea	0.09	44	8.8
Sulfonamide	0.09	42	8.4
Chlorine	0.15	62	12.4
Fluorine	0.09	24	4.8

[a] Average count/value in the dataset for the specified feature; ring = 2.38 means that compounds had an average of 2.38 rings.

[b] Number of compounds with a specified attribute; ring = 494 means that 494 compounds have at least 1 ring.

[c] Percentage of compounds in the dataset with at least one example of the specified attribute.

[d] TPSA = static surface area of O, N, P, and S along with associated H atoms as calculated by the Ertl method [51].

TABLE 6.2
HPLC-RI Preliminary Model Dataset Structure Features
with <3% Population

Feature	Number[a]	Percentage[b]
Nitro	14	2.8
Vinyl CH	13	2.6
Carbamate	9	1.8
Oxazole	5	1.0
Guanadine	4	0.8
N-oxide	4	0.8
Thiazole	4	0.8
Cyano	4	0.8
N+ (permanent)	4	0.8
Bromine	4	0.8
Furan	3	0.6
Benzamidine	3	0.6
Thiourea	3	0.6
Thiophene	2	0.4
Thioamide	2	0.4
Acetylene CH	2	0.4
Sulfate	1	0.2
Thiol	1	0.2

[a] Number of compounds with at least one example of the feature.
[b] Percentage of the dataset with at least one example of the feature.

electronegativity (a variation of Mulliken–Jaffe electronegativity). The intrinsic state value is modified by a contribution from every other atom in the molecule to reflect molecular redistribution of electron density. The magnitude of this effect is based on the square of the graph distance between the atoms and their intrinsic state values. The resulting atom-level E-State is a quantification of the buildup of electron density at an atom and the topological accessibility of the atom to participate in intermolecular interactions. The largest atom-level E-State values occur for highly electronegative terminal atoms such as $=O$ and $-F$, whereas the smallest values occur for buried atoms with highly electronegative alpha neighbors, such as the phosphorous atom in a phosphate group. In addition to the basic atom-level index, E-State information has been extended to additional structure descriptors including atom types, bond types, functional groups, and internal hydrogen bonding descriptors. Atom-level contributions from multiple atoms or bonds are summed into aggregate indices to form descriptors with a broader application than the atom-level indices.

The molecular connectivity (chi) descriptors encode features of molecular skeletal structure including the influence of heteroatoms in the skeletal structure, ring structure and degree of ring substitution, degree and pattern of branching, volume/surface area, skeletal shape, and molecular size. The different chi indices represent skeletal variation independent of molecular size. In contrast to the largely electronic information in E-State descriptors, chi indices primarily encode the topology of the molecular skeleton.

In addition to the molecular connectivity and E-State descriptors, winMolconn calculates basic structure information descriptors such as molecular weight, feature counts, and simple indices of molecular graph characteristics such as the number of terminal vertices. Examples of indices from all three classes were used in creating models of HPLC RI.

Based on the structure features of the compounds in the preliminary RI dataset, a group of 38 SIR structure descriptors was selected for the modeling process. As opposed to the use of an automated feature selection algorithm, descriptors were chosen in an attempt to explicitly describe every feature of every molecule that would likely influence the retention time under known measurement conditions.

Significant challenges exist when modeling a dataset that is small compared to its level of diversity, the most significant being the need to encode all of the chemical information relative to the modeled endpoint in a statistically acceptable number of inputs. RI is often used to aid in characterization of compounds. Since nearly every atom in a given molecule can influence the relative molecular distribution between the mobile and stationary phases, small changes in structure result in measurable changes in the RI value. For this reason, it is ideal that the set of descriptors used to create a model of RI should encode information about each atom in the molecule. For a structurally diverse dataset such as that used in this project, statistical limitations on the number of input descriptors made it impossible to encode every feature in every compound in an independent index.

To address this issue, the Interaction Group (IGroup) E-State family of descriptors [54] was utilized. The IGroup methodology attempts to explicitly represent every atom in a molecule by combining atom-level contributions from atoms that would be expected to undergo similar noncovalent interactions in solution. Functional subgraphs are categorized into groups, and a separate descriptor is created for each non-carbon atom in a group. These individual IGroup indices are based on the atom-level E-States of atoms assigned to the group. A list of the IGroup descriptors used for the preliminary model and the functional groups assigned to each index is given in Table 6.3.

In addition to the IGroup indices, connectivity and structure information descriptors were added to encode bulk structure features that likely influence partitioning characteristics during liquid chromatography. These indices include information about size, shape, branch points, rings, flexibility, hydrophilicity, lipophilicity, and internal hydrogen bonding and are listed in Table 6.4.

6.2.4 Ensemble Model Dataset Partitioning

In preparation for creating a computational model, a list of compound names and RI values was used to create an electronic database of the compound structures. A search of the PubChem database was made on each compound name to verify each structure. When a compound matching the name was found, the structure was

TABLE 6.3
IGroup E-State Descriptors Used in Neural Network and MLR
Models of the Preliminary Dataset

Index	Types[a]	Functional Groups[b]
SAMdO	=O	Amide, sulfonamide, nitro, N-oxide
SAMdS	=S	Thioamide-like
SAMsssN	>N–, =N–	Amide-like, guanadine
SHAMsNH	–H	On any amide-like nitrogen
ACDdO	=O	Any acid group
SACDsO	–O–	Any acid group
SAniN	>N–, =N–	Aniline groups
SHAniNH	–H	On any aniline nitrogen
SAromN	aNa, aNa–	Pyridine/pyrrole groups
SHAromNH	–H	Pyrrole aNa–H
SalphN[c]	>N–, =N–	Amine, imine
SHalphNH[c]	–H	On any aliphatic nitrogen
SAlphO[d]	–O–, =O	Ether, ester, ketone, alcohol
SHAlphOH[d]	–H	On any aliphatic oxygen
SalphS[e]	–S–, =S	Thioether, thione, thiol
SCspN	≡N	Cyano, azide
SallNp	>N+<, aaN+–	Quaternary amine, pyridinium
SsBrI	–Br, –I	Bromine and iodine
Ssp3C	All sp^3 carbon atoms	
Ssp2C	All sp^2 carbon atoms	
SspC	All sp carbon atoms	
SaromC	All aromatic carbon atoms	
SHacCH	–H on acidic CH (vinyl, acetylene)	

[a] Valence state atom types included in the index.
[b] The index comprises atoms of the prescribed type from these functional groups.
[c] Nitrogen atoms not assigned to another IGroup are added to this group by default.
[d] Oxygen atoms not assigned to another IGroup are added to this group by default.
[e] Sulfur atoms not assigned to another IGroup are added to this group by default.

encoded into an MDL format Molfile and the data for the compound (compound name, PubChem CID number, molecular weight, and experimental RI value) were entered into a spreadsheet, along with the name of the Molfile containing the corresponding structure. A total of 73 compounds were left out of the study because the name could not be matched to a structure in the PubChem database. The remaining 498 compounds were used for the modeling project and were partitioned into the appropriate subsets for an ensemble model.

The generation of a cross-validated ensemble model requires the dataset to be divided into two subsets: model fitting and model external validation. Initially,

TABLE 6.4
Global and non-IGroup Feature Descriptors Used in Neural Network and MLR Models of the Preliminary Dataset

Index	Description and Information Encoded
xv0	**chi valence 0** Molecular volume, surface area
dx2	**difference simple chi 2** Branching independent of molecular size
xch5	**chi simple chain 5** Number, substitution degree of 5-membered rings
xch6	**chi simple chain 6** Number, substitution degree of 6-membered rings
xch7	**chi simple chain 7** Number, substitution degree of 7-membered rings
ncirc	**number of graph circuits** Distinguishes between naphthalene and diphenyl
EPSA	**E-State polar surface area** Sum of E-state values for N, O, S, and P E-State version of the Ertl TPSA index [51]
nrbond	**number of rotatable bonds** Molecular flexibility
SaaO	**aromatic oxygen atom type E-State** Electron accessibility of aromatic oxygen atoms
SaaS	**aromatic sulfur atom type E-State** Electron accessibility of aromatic sulfur atoms
SsF	**fluorine atom type E-State** Electron accessibility of fluorine atoms
SsCl	**chlorine atom type E-State** Electron accessibility of chlorine atoms
Hmax	**largest hydrogen atom-level HE-State** Site of the most polar hydrogen atom
ka3	**kappa alpha 3 shape index** Shape index, largest values for linear structures
THBdInt[a]	**E-State total internal hydrogen bonding** Increases with number and strength of potential internal H-bonds of path 3, 4, and 5 skeletal bonds

[a] For each atom pair in a configuration where an internal hydrogen bond may form, the product of the acceptor atom-level E-State and donor hydrogen atom-level HE-State is added to the index sum.

a percentage of the data is set aside for external validation and is not used in the learning process. The remainder of the data is used for the fit set. Typically, the external-validation subset comprises between 10% and 50% of the overall data. For this preliminary model, 102 compounds were used, constituting approximately 20% of the overall data. The model fitting portion of the data is used to fit descriptors of molecular structure to the experimental target endpoint during training or learning. This is the data on which the algorithm learns the problem. To enhance the learning process, model fitting data are partitioned into 10 training folds, in which each fold is a variation of the fit data with approximately 10% left out as a test set. A total of 10 folds are created, each with a unique 10% of the data as a test set. This allows each compound in the fit set to be used in exactly one test set. The result is 10 train sets, each of which contains approximately 90% of the fit data, and 10 corresponding test sets, each of which contains a unique approximately 10% of the fit data.

Machine learning algorithms process the training data for each fold, attempting to derive the endpoint value for each structure based in the input descriptors. Statistics for this fitting process improve as the machine goes through repeated iterations of learning. At the end of each iteration, statistics are calculated for the 10% leave out test set, and model learning continues as long as the quality of the test set statistics continues to improve. Learning stops when the test set statistics begin to degrade and the learning algorithm proceeds to the next fold in the series. The final result is 10 distinct models of the experimental endpoint based on 10 different versions of the data. Finally, each of the 10 models is used to make predictions for the external-validation data and the average of the 10 models is used as the final predicted value.

The subsets described in Section 6.2.3 (train, test, and validate) have often been generated by picking compounds at random. The use of completely random external-validation and test sets can lead to unintended bias, particularly when the diversity of both molecular structure (parameter space) and experimental endpoint (activity space) is not well balanced. For this study, Ward's hierarchical clustering was used to create model subsets that were as balanced as possible in both structure and activity space. This was done because the dataset was small compared to its level of diversity and there was a significant danger that random partitioning could result in situations where all compounds with a given structural feature are assigned to the test set of a fold. In this situation, all train compounds would have a value of 0 for the input, which means the feature and its impact would be zeroed out in the model. Test set predictions would be made in the complete absence of information about the feature, even though there are test set compounds that contain the feature.

To create data splits that were balanced in both structure and activity space, the dataset was analyzed by Ward's hierarchical clustering using the MDL quantitative structure–activity relationship software [55]. The winMolconn v1.4 software package was used to generate a comprehensive set of structure descriptors encoding every atom, bond, and functional group, along with a significant quantity of bulk structure information. This set of molecular descriptors was used as the input to the Ward's algorithm. An effort was made to keep the cluster size to 15–30 compounds. This cluster size is small enough for the compounds to be closely related but large enough that removal of compounds to the external-validation subset will not overly deplete the train set of information on the kind of compound contained in the cluster. A total of 27 clusters were produced, ranging in size from 11 to 35 compounds with an average size of 21. After the Ward's algorithm had processed the data, compounds with an underpopulated structural feature were manually removed to a separate cluster for each feature. These features included thiazole/thiophene, permanently charged nitrogen, cyano, bromine, furan/oxazole, and thioamide/thiourea. This procedure resulted in a total of 33 clusters for use to partition the data into subsets for model fitting and external validation.

The compounds in the 33 clusters were rank ordered on their RI values from low to high, and every fifth row was assigned to the external validation set. This method evenly samples each identified structural subclass and takes compounds from the full range of observed activity for the cluster. The external validation compounds were removed to a separate data file, and the remaining compounds were partitioned into the 10 training folds by iteratively assigning every tenth row to a test subset. This procedure creates validation and test sets that are representative of the overall data and are computationally expedient to generate. The intent is to reduce or eliminate bias in the external-validation statistics associated with random phenomena. Since all types of compounds in the overall data are represented in the validation set, a model created from data partitioned by this method is likely to show if there are specific kinds of structures that the method cannot accurately fit or predict.

6.2.5 ARTIFICIAL NEURAL NET TECHNIQUES

After the data were partitioned and descriptors selected, the Emergent ANN software package [57,58] was used to create an ensemble model of the dataset by generating a separate model for each of the 10 train–test folds. A three-layer back-propagation network with 38 input neurons and 20 hidden neurons was used with online weight updates, 0.25 learning rate, and 0.9 momentum. For each fold, learning was conducted from a total of 50 different random initial-weight starting points and the network was allowed to iterate through learning epochs until the mean absolute error (MAE) for the test set reached a minimum. At that point, the learning rate was dropped to 0.01 and the training continued as long as the test set MAE continued to improve. Training was stopped when the test MAE began to deteriorate. Statistics were calculated for the train and test subsets. After the learning from all 50 initial-weight sets was completed, the run that produced the smallest test MAE was kept as the model for that fold. This process was repeated for each of the 10 folds. When all folds were finished, each of the 10 models was used to predict the compounds in the validation set and the nonweighted average of all 10 models was used as the final predicted value for these compounds.

6.2.6 PRELIMINARY MODEL RESULTS

At the conclusion of model learning, statistics were generated for train set fitting, cross-validation of the training data by tenfold leave-10%-out, and external valida-tion on a set of compounds not used for modeling. The statistical results for both the neural net and MLR learning algorithms are summarized in Table 6.5.

TABLE 6.5
Statistical Results for MLR and ANN Models of the Preliminary Dataset

Model	Train Set					Cross-Validation Set					External-Validation Set				
	n^a	r^{2b}	MAE	SE	$<100^c$	n	q^2	MAE	SE	<100	n	v^2	MAE	SE	<100
MLR	396	0.65	84	87	65%	396	0.44	84	94	68%	102	0.49	80	81	73%
ANN	396	0.93	30	39	98%	396	0.76	54	75	86%	102	0.83	41	61	91%
DIFF	—	+0.28	−53	−48	+33%	—	+0.32	−30	−20	+20%	—	+0.34	−39	−20	+18%

MAE, mean absolute error; SE, standard error.

[a] Number of compounds in a given subset.

[b] Square of the correlation coefficient.

[c] Percentage of compounds with an absolute residual less than 100 RIU.

The ensemble model generated using MLR gave marginally acceptable correlation statistics (training $r^2 = 0.65$, cross-validation $q^2 = 0.44$, and external-validation $v^2 = 0.49$). The MLR MAE constituted approximately 7%–8% of the data range (train MAE = 84 RIU, cross-validate MAE = 84 RIU, and external-validate MAE = 80 RIU). Whereas the MLR ensemble model captures an observable correlation between the structure descriptors and the experimental RI value, the train set fit values are within 100 RIU of the experimental values for only 65% of the data rows. Figure 6.2 shows a plot of predicted versus experimental values for the MLR data. The erratic distribution of residuals for the training and validation datasets suggests a nonlinear relationship [58] between the RI value and molecular structure.

The ANN ensemble model achieved a significantly higher correlation with the experimental data (training $r^2 = 0.93$, cross-validation $q^2 = 0.76$, and external-validation $v^2 = 0.83$). The ANN MAE constitutes only 2.8%–4.9% of the data range (train MAE = 30 RIU, cross-validate MAE = 54 RIU, and external-validate MAE = 41 RIU) and is approximately half the relative error of the MLR model. A total of 91% of ANN validation predictions were within 100 RIU of the experimental value. A plot of the training and cross-validation data points for the ANN model is given in Figure 6.3. An examination of the plot in Figure 6.3 shows performance that is similar across the RI data range for both train and validate subsets. There are several predicted values with large residuals from the ANN model, but it is clear that the ANN algorithm provides substantially better results than MLR.

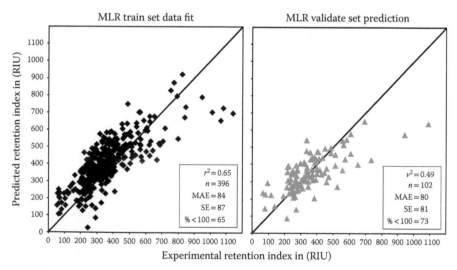

FIGURE 6.2 Multiple linear regression (MLR) results for the preliminary model showing predicted values versus experimental values. Model statistics given are the squared correlation coefficient, mean absolute error (MAE), standard error (SE), and percentage of compounds with residuals less than 100 RIU.

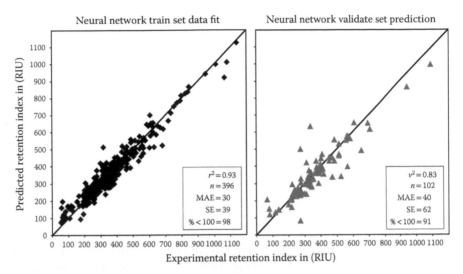

FIGURE 6.3 Artificial neural network (ANN) results for the preliminary model showing predicted values versus experimental values. Model statistics given are the squared correlation coefficient, MAE, SE, and percentage of compounds with residuals less than 100 RIU.

6.2.7 COMMENTS ON THE PRELIMINARY MODEL

The preliminary model of RI was made with the specific goal of maximizing the number of absolute residuals less than 100 RIU. This was deemed to be the widest absolute error range that would eliminate a significant number of candidate compounds with a high probability that the matching compound would not be inadvertently filtered out. This model achieved 91% of prediction values within 100 RIU of the experimental value. The range of ± 2 SEs of prediction was 124 RIU, and this value corresponds to a 95% confidence interval. These values give a reasonable indication that the model is of sufficient quality for use in database filtering.

These results also demonstrate the utility of the combination of IGroup, atom type, internal hydrogen bonding, and global E-State, along with molecular connectivity, kappa shape, and elementary structure information descriptors in the prediction of RIs. External-validation prediction statistics indicate that the nonlinear ANN algorithm is superior to MLR for modeling this endpoint with these descriptors. It is reasonable to assert that a model developed by these methods can be used to support the identification of unknown substances in cases where experimental HPLC-RI data are not available for candidate structures. The proposed database filtering method depends on MS-based experimental endpoints such as exact mass and CID spectra. This means that data for a useful RI model must be measured on an HPLC-MS system where all of the relevant endpoints can be measured. Since the mobile phase used to make the measurements for the preliminary dataset is not compatible with electrospray ionization MS, it will not be possible to reuse the original dataset for further exploration of the proposed method.

This data included a number of elements and functional groups that are not found in endogenous human substances or their metabolites. The dataset also contains a number of drugs, dyes, and other compounds not related to biological systems. For this reason, a new dataset was assembled based on known biological compounds, metabolites, and some drugs that were consistent with the endogenous structure space.

6.3   REVISED RETENTION INDEX MODEL

The relative success of the preliminary model led to the development of a new database of compounds and RI data, along with a revised model based on this data. This new dataset was analyzed by HPLC/MS to obtain multiple experimental endpoints to assist in the identification process. A new dataset was compiled with as many endogenous human substances and metabolites as could be reasonably obtained. A number of drugs, and drug metabolites, were also included because of their similarity to the metabolite-like structure space. A total of 411 compounds were collected for the study. RI determinations were made for all compounds in the new dataset in preparation for the development of a new model.

6.3.1   Retention Index Determination for the Metabolite-Like Dataset

Both HPLC and MS measurements for the metabolite-like compounds required different instrumentation and experimental protocols from those used to mea-sure the preliminary dataset. The phosphoric acid and triethylamine used in the previous HPLC mobile phase are not compatible with positive-ion electrospray analysis. The retention times of the compounds in the metabolite-like dataset were measured on an Agilent 1100 HPLC system using an Agilent DAD for the detection of n-nitroalkane reference compounds and a Micromass Q-TOF 2 mass spectrometer for the detection of test compounds. Compounds were eluted from a Zorbax C8, 4.6 mm × 250 mm, column that was maintained at 30°C using a solvent-programmed mobile phase consisting of a linear gradient from 100% solvent A (0.01% heptafluorobutyric acid) to 100% solvent B (0.01% heptafluorobutyric acid, and 10% water in acetonitrile) in 17 minutes with a 5 minute isocratic hold at 100% solvent B. The flow rate was 75 µL/min.

The RI scale for each batch analysis was established by analyses of 1 µL of a homologous mixture of C through C_{10} n-nitroalkane reference standards (1.12 µM in mobile solvent B), run before and after each batch of test samples. Ethisterone (0.3 µM) was coanalyzed with the n-nitroalkanes and with all test samples. The retention times of the C through C_{10} n-nitroalkane reference stan-dards analyzed at the beginning and end of a set of test sample analyses were averaged for use as the calibration references for determining test compound RI values.

The effluent tubing from the HPLC column was attached to either the DAD or the electrospray source of the Q-TOF 2 mass spectrometer depending on whether the n-nitroalkanes or the test samples were being analyzed. The retention time for test compounds was calculated from the respective reconstructed m/z molecular ion chromatographic peak. The use of this method allowed for the association of a specific molecular ion mass with each observed HPLC peak. For each test batch, the difference between the average ethisterone retention time at the DAD and the average ethisterone retention time at the mass spectrometer was used to calculate an offset that was added to the retention time to correct for the void time difference between the two detectors. All test compounds were run in duplicate, and the final, determined RI value was the average of the duplicate measurements.

Initially, the mobile phase in the HPLC column is exclusively solvent A; thus the composition of the mobile phase remains constant at the initial composition. The reason for this is that there is a volume of solvent that must pass through the tubing from the mixing chamber at the pumps before the first change in mobile-phase composition reaches the column. The period of time required for solvent B to traverse the dwell volume is dependent on the mobile phase flow rate and is referred to as the dwell time. Once the dwell time elapses, the mixture of the two mobile phases traverses the dwell volume and the mobile phase starts becoming more hydrophobic from the increased proportion of solvent B. The mobile phase continues to increase in hydrophobicity until the end of the run. Compounds eluted before the end of the dwell time are eluted under isocratic conditions. Compounds that elute after the dwell time are eluted under gradient conditions; thus the dwell time delineates the end of isocratic conditions. RI is calculated differently for the isocratic and gradient portions of an HPLC separation because the composition of the mobile phase is different during the two phases. The dwell time marks the point at which a switch is made from the isocratic calculation method to the gradient calculation method. The methods used for determining the RI value during the respective phases are given here.

During the isocratic portion of an analysis, a linear relationship exists between the logarithm of the retention times of the n-nitroalkane series and the number of carbon atoms of each specific n-nitroalkane [59]. The RI of a given n-nitroalkane was defined as 100 times the number of carbon atoms. Based on this relationship, the RI values of test compounds that eluted during the isocratic portion of the mobile phase were calculated according to Equation 6.2:

$$\text{RI} = \left(\frac{\left(\log T_x - \log T_z \right) 100}{\log T_{z+1} - \log T_z} \right) + 100z \tag{6.2}$$

where T_x is the corrected retention time of the analyte, T_z is the retention time of the nitroalkane eluting just before the analyte, T_{z+1} is the retention time of the nitroalkane eluting just after the analyte, and z is the number of carbons in the nitroalkane eluting just before the analyte.

To establish an RI scale for compounds that elute during the gradient phase, it is first necessary to establish the dwell time for the measurement system. When a change in gradient time does not affect the retention time of a compound, the compound is eluting under isocratic conditions. Based on the test compound measurements, the longest observed retention time for which a 17 minute and a 34 minute gradient had the same value was for 1-methylpyrrolidine with an R_t value of 5.54 minutes for the 17 minute solvent gradient and 5.53 minutes for the 35 minute solvent gradient. Based on these values, a dwell time of 5.54 minutes was used to delineate the end of the isocratic conditions. Using Equation 6.2, this dwell time corresponds to an RI value of 221.8. The 221.8 RI value of the dwell time is used as an offset to create the first point in the gradient RI calculation, ensuring that RI remains a continuous scale through both conditions of an analysis.

As a result of the changing composition of the mobile phase, the relationship between the logarithm of the retention times of the n-nitroalkane series and the number of carbon atoms of each specific n-nitroalkane is no longer linear during the gradient portion of the analysis. Because of this nonlinearity, a trend function was used to determine the fourth-order polynomial relationship between RT and RI for n-nitroalkanes C_3 through C_{10}, which were found to elute during the gradient portion of the analysis. This equation was recalculated for each run using the average of the RT values for the n-nitroalkanes that were run both before and after each run of test samples. This equation was used to calculate the RI value for test compounds based on the offset-corrected RT.

Reproducibility of the RI determinations was evaluated based on 12 replicate measurements of a set of 15 test compounds. A two-factor mixed effects model [60] was calculated from the resulting data, suggesting a total measurement error of 2.24 RIU. This measurement error corresponds to an experimental reproducibility of ±6.7 RIU for an error range of three SEs (99.7% confidence interval).

A histogram of the resulting RI values is given in Figure 6.4. The measurement range was 578 RIU where phosphoserine had the minimum value of 15 RIU and carbazole had the maximum of 593 RIU.

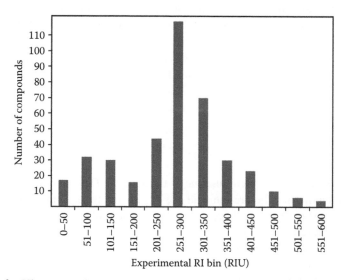

FIGURE 6.4   Histogram of experimental RI values for the metabolite-like dataset.

The overall RIU range is almost half the observed range for the preliminary dataset. The RI distribution is not normal in the statistical sense, but the majority of the data points cluster around the mean at 264 RIU (SD = 114 RIU) and the median value of 268 RIU is close to the mean. The observed RI values ranged from −2.2 to +2.9 SDs from the mean. As a result, compounds were not removed from the study as experimental outliers ($> \pm 3$ SDs from the mean).

6.3.2 METABOLITE-LIKE DATASET STRUCTURAL CHARACTERISTICS

Because the revised model was created specifically to aide in characterizing endogenous human metabolites, compounds in the dataset were limited to the set of biological elements (carbon, hydrogen, oxygen, nitrogen, sulfur, and phosphorous). Compounds selected for the study contained at least one protonatable atom to facilitate detection by MS in positive-ion mode. The protonatable atom was nitrogen for the majority of compounds, but 12 steroid-like compounds with α,β-unsaturated ketone groups were included because of the prevalence of the scaffold in biological systems. Compounds have been detected by Q-TOF mass spectrometer in positive-ion mode even when an α,β-unsaturated ketone oxygen atom was the only heteroatom in the molecule [54]. It is therefore asserted that the ketone group protonates at pH levels associated with electrospray ionization.

Molecular weights for the metabolite-like compounds ranged from 79 to 609 Da with an average value of 218 Da ($s = 94$). More than 40 common organic functional groups are present, along with multiple analogs of common functional groups. The dataset displays a number of carbon skeletal features such as phenyl, naphthyl, cyclohexyl, and t-butyl groups. A summary of the dataset characteristics and a list of the highly populated functional groups ($>3\%$ population) are given in Table 6.6.

6.3.3 METABOLITE-LIKE DATASET MODEL DESCRIPTORS

In a manner similar to that used to select descriptors for the preliminary model, a group of 33 SIR descriptors was selected for use in creating an ANN ensemble model of the metabolite-like data. The set of descriptors used for the revised model was similar to that used for the model of the preliminary data. Again, descriptors were chosen in an attempt to explicitly describe every feature of every molecule that would likely influence the retention time under known measurement conditions. The main difference in the descriptors results from elements and functional groups that were present in the preliminary data being absent from the metabolite-like dataset. Descriptor values were calculated using the winMolconn v2.1 software [61]. As with the preliminary dataset, the IGroup family of descriptors was utilized, along with additional global and feature descriptors to encode bulk structural features that likely influence partitioning characteristics during liquid chromatography.

Low-population functional subgraphs exist in the metabolite-like dataset that are chemically related to more prevalent groups, such as amide. Some of these subgraphs, such as urea, imide, guanine, xanthine, cytosine, uracil, barbiturate, uric acid, carbamate, hydantoin, pyrazolidinone, and flavin, can be considered as

TABLE 6.6
HPLC-RI Metabolite-Like Model Dataset Structural Characteristics, Including Features with >3% Population

Structural Feature	Average[a]	Count[b]	Percentage[c]
MW	218.2	400	100
Ring	1.7	319	80
Aromatic ring	0.6	186	47
Heteroaromatic ring	0.3	95	24
Nonaromatic ring	0.8	180	45
Fused ring system	0.3	128	32
Rotatable bonds	6.1	381	95
Heteroatom	4.4	400	100
Hydrogen bond acceptor	4.1	396	99
Hydrogen bond donor	1.9	343	86
Internal hydrogen bond	1.8	232	58
TPSA[d]	68.2	399	99
Amine	0.57	212	53.0
Aniline	0.12	44	11.0
Pyridine	0.11	40	10.0
Pyrrole	0.09	33	8.3
Ether	0.32	80	20.0
Ester	0.08	27	6.8
Ketone	0.11	33	8.3
Alcohol	0.42	107	26.8
Carboxylic acid	0.32	113	28.3
Phenol	0.12	41	10.3
Amide	0.20	73	18.3

[a] Average count/value in the dataset for the specified feature; rings = 1.7 means that compounds had an average of 1.7 rings.

[b] Number of compounds with a specified attribute; ring = 319 means that 319 compounds have at least 1 ring.

[c] Percentage of compounds in the dataset with at least one example of the specified feature.

[d] TPSA = static surface area of O, N, P, and S along with associated H atoms calculated by the Ertl method [51].

"amide-like" in that they contain at least one amide subgraph that is part of an extended conjugated system of oxygen, carbon, and nitrogen atoms. Description of such systems presents a particular challenge in that there are a large number of functional groups in this class, along with a vast number of substructure variations that are similar but do not fall into a specific group. This is especially true of metabolites, which are enzymatically modified versions of the parent structure. Fully describing the variation in oxygen, nitrogen, and hydrogen atoms in just the amide-like features in this dataset would require more than 100 fragment descriptors. Such a large number of descriptors would exceed the number of inputs from which a statistically valid model can be produced. Even if these descriptors were included in a pool from which the total number of inputs could be reduced and optimized by a selection algorithm, the very low population of each individual descriptor would so limit the leverage over the dataset that they would be passed over during selection. It is highly unlikely that reasonable predictions can be made for compounds containing these features in the absence of information about them in the model descriptors.

The IGroup methodology addresses this situation by combining the atom-level E-State values of atoms in underpopulated functional groups into a set of IGroup indices. The amide-like IGroup consists of three indices: one for double bond oxygen (SamideO), one for nitrogen atoms (SamideN), and one for hydrogen bonded to amide-like nitrogen (SHamideNH). The SamideO index is the sum of the atom-level E-State values for all =O atoms found in amide-like groups. The SamideN index is the sum of the atom-level E-State values for all >N– and =N– atoms found in amide-like groups, and the SHamideNH index is the sum of the hydrogen atom–level HE-State value for all –H atoms bonded to amide-like nitrogen. The use of the IGroup method condenses this information from more than 100 fragment indices into three descriptors that can be used to encode information for any subgraph of the amide-like type. This method also has the advantage of forcing the modeling algorithm to treat chemically similar functional groups in the same way. The revised model IGroup indices are described in Table 6.7.

There are many underpopulated structural features in the dataset other than the amide-like features discussed in the preceding paragraph. These groups were managed in the same way by combining their atom-level E-State values with chemically similar features of higher populations. For example, the atom-level E-State values for all oxygen atoms in phosphate, sulfonic acid, sulfinic acid, and phosphonate were combined with the oxygen values for carboxylate to form the acid-like descriptor SacidO. In some cases, the atom-level E-State values were distributed to more than one IGroup index. For example, the E-State values for the nitrogen atoms in variations of guanine were divided among the amide-like, pyridine, and pyrrole IGroups. The five-membered ring of guanine is aromatic, so the nitrogen atoms in the ring were considered as pyridine and pyrrole and not amide-like. These atoms have a greater proton affinity than the atoms in the six-membered ring, and so they are more likely to protonate at the low pH of the HPLC mobile phase. The contribution of the aromatic nitrogen atoms to the retention time is likely different from that of the nitrogen atoms conjugated with carbonyl oxygen, so they are summed in a different descriptor. If nitrogen atoms in the conjugated system have associated hydrogen atoms, they likely participate in multiple tautomers with carbonyl oxygen and other nitrogen atoms. This is another reason to include these nitrogen atoms in a different descriptor than the aromatic nitrogen atoms.

Using this method, the atom-level E-State value for every heavy atom in every molecule in the dataset was explicitly assigned to a descriptor. The hydrogen atom–level HE-State was also encoded in cases where it was not too highly correlated with the E-State for the associated heavy atom. This is very important when modeling an endpoint–like RI in which every atom in the molecule contributes to the retention

TABLE 6.7
IGroup E-State Descriptors Used in Metabolite-Like Neural Network Model

Index	Atom Types[a]	Functional Groups[b]
SaliphaticN[c]	>N–, =N	Amine, guanidine, imine
SHaliphaticNH	–H	On any aliphatic nitrogen
SanilineN	>N–, =N–	Aniline, benzamidine
SHanilneNH	–H	On any aniline nitrogen
SpyridineN	aNa	Pyridine, diazole
SpyrroleN[d]	aNa–	Pyrrole, indole, diazole
SHpyrroleNH	–H	On any pyrrole nitrogen
SamideO	=O	Amide-like groups
SamideN	>N–, =N–	Amide-like groups
SHamideNH	–H	On any amide-like nitrogen
SaliphaticO[e]	–O–, =O	Ether, ester, ketone, aldehyde
SHaliphaticOH	–H	On any aliphatic oxygen
SacidO	=O, –OH	COOH, SO_2H, SO_3, PO_3, PO_4
SphenolO	–OH	Phenol
SalphaticS[f]	–S–, =S<	Thioether, disulfide, sulfoxide
SprotN	>N+<, aaN+–	Quaternary amine, pyridinium
SaromS	aSa	Thiophene, thiazole
Ssp3C	All sp^3 carbon atoms	
Ssp2C	All sp^2 carbon atoms	
SaromC	All aromatic carbon atoms	
SHacCH	–H on acidic CH (vinyl, acetylene)	

[a] Valence-state atom types included in the index.
[b] The index comprises atoms of the prescribed type from these functional groups.
[c] Nitrogen atoms not assigned to another IGroup are added to this group by default.
[d] Diphenylamine nitrogen atoms are included in this group.
[e] Oxygen atoms not assigned to another IGroup are added to this group by default.
[f] Sulfur atoms not assigned to another IGroup are added to this group by default.

time. A list of all the underpopulated features ($\leq 3\%$ population) is given in Table 6.8 along with the IGroup descriptor where information for that subgraph was assigned.

Some changes were made to the IGroup assignments for the model of the metabolite-like dataset. Some of these changes relate to differences in the datasets, and some relate to an evolution of the IGroup method. Nitrogen atoms in guanidine groups were added to the aliphatic nitrogen IGroup index instead of the amide-like nitrogen IGroup index. The aromatic nitrogen IGroup used for the first dataset was split into pyridine and pyrrole groups because of the significant difference in pK_a

TABLE 6.8
HPLC-RI Metabolite-Like Dataset Low-Population Structure Features and Corresponding IGroup Descriptors

Feature	Count[a]	Percentage[b]	IGroup[c]
Aldehyde	3	0.8	Aliphatic O
Diazole	12	3.0	Pyridine, pyrrole
Purine	12	3.0	Pyridine, aniline, pyrrole
Guanidine	6	1.5	Aliphatic N
Pyrimidine	4	1.0	Pyridine, aniline
Diphenylamine	1	0.3	Pyrrole
Carbamate	8	2.0	Amide, aliphatic O
Oxime	1	0.3	Aliphatic N, aliphatic O
Guanine	9	2.3	Amide, pyridine, pyrrole
Xanthine	9	2.3	Amide, pyridine, pyrrole
Cytosine	8	2.0	Amide
Uracil	6	1.5	Amide
Urea	6	1.5	Amide
Barbiturate	4	1.0	Amide
Uric acid	4	1.0	Amide
Imide	3	0.8	Amide
Hydantoin	3	0.8	Amide
Flavin	2	0.5	Amide
Pyrazolidinone	2	0.5	Amide
Phosphate	16	4.0	Acid O
Sulfonic acid	7	1.8	Acid O
Sulfinic acid	1	0.3	Acid O
Phosphonate	1	0.3	Acid O
Thioether	9	2.3	Aliphatic S
Thioanisole	6	1.5	Aliphatic S
Disulfide	3	0.8	Aliphatic S
Sulfoxide	1	0.3	Aliphatic S, aliphatic O
Vinyl	13	3.3	Acidic CH
Quaternary N+	9	2.3	Protonated N
Pyridinium N+	5	1.3	Protonated N
Thiazole	3	0.8	Aromatic S, pyridine
Thiophene	2	0.5	Aromatic S

[a] Count of compounds in the dataset with at least one example of the feature.
[b] Percentage of compounds in the dataset with at least one example of the feature.
[c] IGroup descriptor or descriptors where structure information for atoms in the group was assigned.

values. Similarly, diphenylamine nitrogen was removed from the aniline IGroup and added to the pyrrole IGroup because the pK_a value for diphenylamine is closer to that of pyrrole than aniline. The preliminary dataset used two oxygen descriptors for acid groups: one for =O and another for –O–. All acid oxygen atoms were combined into a single descriptor for the model of metabolite-like data. Groups such as sulfonamide, N-oxide, nitro, thioamide, cyano, azide, chlorine, fluorine, bromine, iodine, and sp carbon are absent from the metabolite-like dataset, resulting in a small number of inputs for the dataset. Most of these groups are not found in human endogenous compounds, and so the compounds containing these groups were not acquired for measurement.

As with the preliminary dataset, the IGroup descriptors for the metabolite-like data were augmented with descriptors to encode bulk structural features that likely influence partitioning characteristics during liquid chromatography. These descriptors contain information about size, shape, branch points, rings, flexibility, hydrophilicity, lipophilicity, and internal hydrogen bonding. All of these structural characteristics influence solvation with the mobile phase, the stationary phase, or both, but they are not specifically related to functional groups or atom-level intermolecular interactions such as electrostatic dipole, dispersion, or hydrogen bonding. A list of these descriptors, their definitions, and the encoded information is given Table 6.9.

The values for the 33 input descriptors were normalized and then scaled from 0.1 to 0.6 to accommodate the input requirements of the neural net software, which limits input values to the scale of 0–1. The practice of using 0.6 as the scale maximum is intended to leave headroom in the scale in the event that a compound submitted for prediction has a larger value for the input than the largest value in the model dataset. During normalization, only rows with a nonzero value for the input descriptor were used to calculate the mean and SD. For inputs that encode a specific molecular feature or functional group, including the IGroups, an input value of zero indicates that the feature is not present in the molecule. The information is preserved through scaling and normalization by retaining the value of 0.0. This is important in that the descriptor values for some features can run from positive to negative numbers. If care is not taken during the normal-scale process, input values of zero can be transformed to nonzero values. Inputs with a 0.0 value are set aside during the normal-scale process and assigned a final input value of 0.0.

After normalization, any compound with a normalized descriptor value of 5 or greater was excluded from the study. These compounds have a descriptor value that is greater than 5 SDs above or below the mean value for all compounds with a nonzero value for the index. These structures are considered outliers in parameter space and tend to have undue influence on the outcome of the model. Eleven compounds were excluded for this reason, leaving 400 compounds in the dataset used to create the ensemble model.

6.3.4 FULL DATASET ENSEMBLE MODEL PARTITIONING

After compounds with large normalized values were removed, the remaining 400 compounds were partitioned into a 4 × 10 × 10 full dataset ensemble. Unlike the

TABLE 6.9
Global and Non-IGroup Feature Descriptors Used in the Metabolite-Like Neural Network RI Model

Global Descriptors[a]		Feature Descriptors	
Index	Description and Information Encoded	Index	Description and Information Encoded
xv0	chi valence 0 Molecular volume, surface area	nrings	number of independent rings Total independent rings of all sizes
ka3	kappa alpha 3 shape index Shape, largest for linear structures	ncirc	number of graph circuits Distinguishes between naphthalene and diphenyl
phia	kappa flexibility index Rotatable bonds increase the value rings and conjugation decreases the value	xch5,6,7	chi simple chain 5, 6, 7 Presence, number, and degree of substitution for 5-, 6-, and 7-membered rings
rvalHyd[b]	valence hydrophilic E-State ratio Hydrophilicity (molecule size independent)	xpc4	chi simple path cluster 4 Each branch point increases the index value
sumLip[c]	total lipophilic E-State Lipophilicity (molecule size dependent)	THBdInt[d]	E-State total internal hydrogen bonding Increases with number and strength of potential internal H bonds (path 3, 4, 5)

[a] Every compound in the dataset has a nonzero value for these descriptors.

[b] Percentage taking quotient of atom-level E-State value for hydrophilic atoms divided by atom-level E-State values for all atoms, multiplied by 100. Negative atom-level E-State values can result in values exceeding 100.

[c] Sum of the atom-level E-States for all lipophilic atoms in the molecule.

[d] For each atom pair in a configuration where an internal hydrogen bond may form, the product of the acceptor atom-level E-State and donor hydrogen atom-level HE-State is added to the index sum.

$1 \times 10 \times 10$ ensemble used for the model of the preliminary dataset, the use of a $4 \times 10 \times 10$ full dataset ensemble allows each compound to be included in model training, cross-validation, and external validation. The full dataset ensemble configuration is created in the same manner as the $1 \times 10 \times 10$ ensemble; but the procedure is repeated four times, creating four separate splits of the overall data. For each of the four splits, approximately 25% of the compounds were assigned to an external-validation subset. The result was four splits of the data, each with approximately 75% of the data reserved for model fitting and 25% reserved for external validation. Each compound was used in exactly one of the four validation subsets. As with the preliminary dataset, clusters were created using the Ward's algorithm and then rank ordered on the experimental RI values. The validation compounds were sampled from the rank-ordered clusters so that the structural diversity and experimental activity were as similar as possible across all four sets. For each split, the 75% fit set was further subdivided into 10 train–test folds as was done with the preliminary dataset. For each fold, a unique 10% of the 75% is set aside for cross-validation and the remainder is used for neural net training.

After model learning, a validation prediction was made for each compound in the dataset using the 10×10 ensemble model from one of the four data splits. Each compound was used for model fitting in the other three splits. All the validate subsets were created by the same method; therefore, it is unlikely that the statistics from any one set will lead to an estimate of prediction capability that is intrinsically more reliable than the others. It is suggested that the use of all the available data spread across the four validate sets gives a more complete picture of prediction accuracy than the use of any one of the four sets by itself and limits the likelihood of bias associated with validation compound selection.

6.3.5 Artificial Neural Net Techniques

As with the preliminary model, the Emergent ANN software package [be, bf] was used to develop neural net models. A separate model was developed for each of the 10 individual train–test folds in each data split, resulting in four tenfold leave-10%-out ensemble models. The 40 total models constitute the full dataset ensemble model.

For model learning, a three-layer network with 33 input neurons and 16 hidden neurons was trained by back-propagation with online weight updates, 0.25 learning rate, and 0.9 momentum. For each fold, learning was started from 50 different random initial weight sets and the network was allowed to iterate through learning epochs until the MAE for the test set reached a minimum. As with the preliminary model, when the test MAE minimum was reached, the learning rate was reduced to 0.01 and training continued as long as the test set MAE continued to improve. At the point when additional training caused degradation of the test set MAE, training was stopped and statistics were calculated for train and test subsets. This process resulted in 50 different models of the fold data. These models were rank-ordered on the test set MAE and the model that produced the smallest test MAE was kept as the final model for the fold. The other 49 models were discarded. This process was repeated for all 10 folds.

When all 10 folds were finished, each of the 10 models was used to make RI predictions for the compounds in the validate. The nonweighted average of the predicted RI values from all 10 models was used as the final predicted value from the model based on the current data split. This process was repeated for all four data splits. Statistical results from the four ensemble models were used to create average statistics for the train, test, and validate subsets across the full dataset ensemble. The final training statistics are the average of 36 data-fit calculated values for each compound, the final test statistics are the average of 3 predicted values for each compound, and the final validation statistics are the average of 10 predicted values per compound. When the model is used to make predictions for new compounds, the predictions are the average of the predicted values from all 40 models in the $4 \times 10 \times 10$ full dataset ensemble model.

6.3.6 REVISED MODEL RESULTS

Because the results of the preliminary model indicated that linear modeling by MLR produced substantially inferior results to neural net models of the same data [62], the metabolite-like dataset was only modeled using an ANN ensemble. As discussed in Section 6.3.5, statistics were generated for train set fit, cross-validation prediction of training data by tenfold leave-10%-out, and external-validation prediction on a set of compounds not used for modeling. With the ensemble model partitioning technique, each compound was used for training in 9 of the 10 folds of each of the three splits where the compound was not in the validate set. This resulted in 36 training fit values for every compound, each calculated by a different ANN model. The 36 training fit values for each compound were averaged to calculate a train set fit value for the full dataset $4 \times 10 \times 10$ ensemble model. These $4 \times 10 \times 10$ training fit values were used to generate train set statistics for the $4 \times 10 \times 10$ full dataset ensemble model. The train set fit values show a very high correlation with measured data with $r^2 = 0.95$, MAE = 19 RIU, SE = 25 RIU, and 99% of fit values within 75 RIU of the experimental value.

Every compound in the dataset was used for validation in one of the four splits of the data. When a compound was in the validate subset, its predicted RI value was calculated as the average of the predicted values from the 10 models from the split where the compound was left out. The average value predictions for all 400 compounds were used to generate validate set statistics for the full dataset $4 \times 10 \times 10$ ensemble model. Validation predictions are also well correlated with experimental values with $v^2 = 0.87$, MAE = 30 RIU, SE = 38 RIU, and 93% of predictions within 75 RIU of the experimental value. The largest observed validation error was for bitolterol at 161 RIU.

Every compound in the dataset was used in the cross-validation test set for 1 of the 10 folds of each of the three splits in which the compound was not used for validation. The resulting three predictions were averaged to generate the cross-validation predicted RI value for the full dataset $4 \times 10 \times 10$ ensemble model. The cross-validation predictions show a lower correlation than training or validation data with $q^2 = 0.83$, MAE = 36 RIU, SE = 46 RIU, and 90% of cross-validation predictions within 75 RIU of the experimental value. It is likely that the small number of averaged values is in part responsible for the lower correlations compared with the external-validation subset, but the cross-validation statistics are still quite reasonable. A summary of the model statistics is given in Table 6.10. This table gives the individual statistics for each of the four splits of the data, as well as the averages for the full dataset $4 \times 10 \times 10$ ensemble model.

Plots of the training and validation data shown in Figure 6.5 show similar predictive ability across the RI range. The performance of validation predictions is somewhat diminished above 400 RIU with an MAE of 40 RIU, compared with 30 RIU for the overall dataset. The majority of compounds with experimental RI values > 400 RIU are underpredicted. This systematic underprediction could result from the paucity of experimental data in the upper end of the RI range, neural net architecture or transfer function that is not sufficiently nonlinear, or limitations in lipophilic feature descriptors. The validation statistics for compounds with RI values over 400 RIU are

TABLE 6.10
Statistical Results of ANN Models on All Four Data Splits and Final Average

Data Split	Train Set				Cross-Validation Set					External-Validation Set				
	n^a	r^{2b}	MAE	SE	<75[c]	n	q^2	MAE	SE	<75	n	v^2	MAE	SE	<75
A	300	0.95	18	24	98%	300	0.79	39	52	85%	100	0.88	29	39	95%
B	299	0.96	18	23	100%	299	0.81	37	50	89%	101	0.86	31	40	91%
C	302	0.95	19	24	98%	302	0.74	41	58	84%	98	0.89	29	35	94%
D	299	0.90	27	32	97%	299	0.81	37	46	88%	101	0.87	33	39	93%
Final	400	0.95	19	25	99%	400	0.83	36	46	90%	400	0.87	30	38	93%

MAE, mean absolute error; SE, standard error.

[a] Number of compounds in a given subset.

[b] Square of the correlation coefficient.

[c] Percentage of compounds with an absolute residual less than 75 RIU.

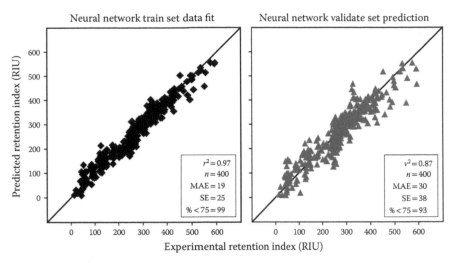

FIGURE 6.5 ANN results for revised model on metabolite-like data showing predicted values versus experimental values. Model statistics given are the squared correlation coefficient, MAE, SE, and percentage of compounds with residuals less than 75 RIU.

still quite usable for database filtering since the majority of data points fall within 75 RIU of the experimental value.

Since more than 40% of compounds in the dataset have at least one underpopulated feature, an analysis of prediction accuracy was performed to estimate the success of the model in predicting these compounds. The external-validation MAE for compounds with at least one low-population feature is only slightly higher than the MAE for that overall dataset (32 RIU compared with 30 RIU). Predictions for compounds with two to four underpopulated features are of similar quality to compounds with only one underpopulated feature, and also to the overall dataset. It does not appear that the number of underpopulated features is a significant factor in prediction accuracy. The MAE values for all subgroups of underpopulated features are within two SEs, but there are some subsets with MAE values much larger than those of the overall data. These occur in subsets in which only one or two compounds with the feature are present in the data. There are also a number of subsets with only one or two examples in the dataset that are well predicted, so the picture remains somewhat unclear. Overall, the performance when predicting compounds with a low population is reasonable and similar to predictions for the overall dataset. This implies that the IGroup methodology of combining structural features with similar solution interaction characteristics into group descriptors has been at least partially successful in mitigating the problems associated with underpopulated structural features. Statistics for all subsets of underpopulated features are given in Table 6.11.

6.3.7 COMMENTS ON THE REVISED MODEL

A significant goal of the revised model project was to generate a new model from a dataset heavily populated with human endogenous compounds while still retaining a high percentage of external-validation absolute residuals less than 100 RIU. In fact,

TABLE 6.11

MAE of Prediction for Compounds with Low Population Features

Group	Count[a]	MAE[b]
Overall dataset	400	30.4
1–4 Low population features	165	32.3
1 Low population feature	140	32.2
2 Low population features	23	33.4
3–4 Low population features	2	26.5
Feature		
Phosphate	16	30.7
Vinyl	13	22.9
Diazole	12	48.9
Purine	12	36.3
Xanthine	9	20.9
Thioether	9	40.7
Quaternary N^+	9	33.6
Cytosine	8	22.3
Carbamate	8	33.6
Sulfonic acid	7	33.4
Guanosine	6	31.9
Guanidine	6	33.7
Uracil	6	28.9
Urea	6	52.9
Thioanisole	6	24.9
Pyridinium N^+	5	25.2
Barbiturate	4	25.4
Pyrimidine	4	35.9
Methyluric acid	4	24.0
Aldehyde	3	15.4
Disulfide	3	21.9
Guanine	3	37.8
Hydantoin	3	38.0
Imide	3	19.8
Thiazole	3	31.3
Flavin	2	26.6
Pyrazolidine	2	56.5
Thiophene	2	35.2
Phosphonate	1	19.0
Diphenylamine	1	40.1
Oxime	1	65.1
Sulfinic acid	1	49.3
Sulfoxide	1	8.4

[a] Count of compounds with at least one example of the feature.

[b] Mean absolute error for all compounds with at least one example of the indicated feature.

the revised model achieved 93% of prediction values within 75 RIU of the experimental value and a ± 2 SE of prediction range of 78 RIU. This 95% confidence error range is nearly 40% smaller than the range for the preliminary model and represents a significant improvement in prediction statistics. The use of the revised model for database filtering will likely result in a smaller number of final candidates, as a greater number of compounds from the list of exact mass matches will fall outside the filter range and be excluded.

Additionally, the revised model used every compound in the dataset for validation in one of the four ensemble models. This gives a broader picture of prediction potential for the model than would be obtained from a single validate set. As is seen in the prediction MAE values for underpopulated subsets, the prediction accuracy varies for the different classes of compounds in the dataset. This information can be used as a guide for acquiring additional compounds to add to the dataset and improve the model. This information would have been unavailable if a smaller validate set was used and some classes of compounds did not appear in the validation data.

Several changes were made in the IGroup descriptor methodology for the revised model. These changes involved altering the IGroup assignment of atoms in some specific functional groups. The reasonable external-validation statistics from the revised model would appear to indicate that these changes did not have a detrimental effect on the model outcome and may be in part responsible for some of the improvement. These results also give a reasonable indication that the model is of sufficient quality for use in database filtering and an improvement over the preliminary model.

6.4 DISCUSSION

The results of both models demonstrate the utility of the combination of ANN techniques and SIR descriptors in creating a model of RI. The SIR descriptors used for this model can be calculated for any valid molecular structure and are computationally efficient to generate at 1,500–10,000 compounds per minute depending on the desktop system used. This makes the methodology well suited to filtering large virtual databases.

A common concern when creating computational models involves the issue of overfitting. Overfitting can occur in a neural net when back-propagation is allowed to iterate through an excessive number of learning epochs. The use of too many input descriptors can also result in an overfit model. This type of overfitting occurs with other learning algorithms as well as with neural networks. Overfitting from prolonged learning can be avoided by tracking the cross-validation statistics from the test set at the end of each epoch. Learning can be stopped at the point where cross-validation statistics begin to deteriorate, using a method referred to as "early stopping" [63]. Determining the maximum number of statistically acceptable inputs is a more difficult matter.

When the ratio of measured data rows to column descriptor inputs is too small, the learning algorithm tends to "memorize" a solution for the dataset as opposed to generalizing a solution. Such overfit models tend to have excellent training statistics but comparatively poor statistics for external validation and cross-validation. In MLR learning, when the number of inputs is one less than the number of training rows ($n - 1$), the training correlation will be 1.0 (a perfect fit of the data). MLR is able to exactly fit each data point by using one of the input values or the intercept constant as a correction for each row. In this situation, the information content of the input columns is no longer relevant and even the use of $n - 1$ random number inputs will result in an r^2 of 1.0. Such a model would not be able to make predictions for compounds other than the training data. Neural net models will not necessarily reach a training correlation of 1.0 because of the interconnectedness of the neurons, but the issue still persists and must be accounted for. Poor validation is certainly observed in cases where the number of inputs is much smaller than $n - 1$, but there is no empirically validated minimum row-to-column ratio for neural network models. A ratio of 10:1 is the minimum that is commonly deemed acceptable to obtain quality validation statistics.

In the case of both preliminary and revised models, the external-validation and cross-validation statistics are reasonable despite a row-to-column ratio of approximately 8:1. The quality of both models is sufficient to question whether or not these traditionally accepted ratios are overly conservative, or if some other effect is involved. It is possible that there are some datasets for which the use of a row-to-column ratio smaller than 10:1 will not lead to degradation in prediction statistics. Another possibility is that the population of input descriptors should be considered as well as the number of descriptors. In other words, it may be more informative to evaluate the count of input descriptors with nonzero values instead of the total count of inputs. Because of the normalization and scaling method used to prepare the descriptor inputs to the neural network, an input value of 0.0 was conserved for all inputs representing structural features that were absent in the molecule. In the ANN back-propagation algorithm, each input is multiplied by the weight of the connection between the input neuron and each hidden neuron. The result is that there is no contribution to the final predicted value from any input with a value of 0.0. It is reasonable to assert that it is not informative to include inputs with a value of 0.0 when evaluating whether or not a model has a statistically excessive number of inputs. It may be that the count of nonzero inputs provides a more useful measure of the actual structure information content of the descriptor inputs for a given molecule.

Looking at the nonzero population for the revised dataset, on average, compounds have a nonzero input value for 18.6 of the 33 descriptors ($s = 2.98$, min = 10, and max = 26). This represents a row-to-column ratio close to 15:1 if only descriptors with nonzero values are counted. A 15:1 ratio exceeds the minimal ratio and passes the much more conservative ($\sqrt{n}$) $- 1$ rule. Eriksson et al. [64] suggest that models do not have a significant number of irrelevant inputs when the difference between training and validation correlation coefficients does not exceed 0.254, which is well satisfied for both models. The quality validation results for both models suggest that the population of input descriptors should be taken into account when evaluating the ratio of inputs to training rows.

The data used for the revised RI model are missing structures with retention times that represent the full range observed for endogenous small molecules. RI values of 1000 RIU and larger have been observed for some endogenous metabolites, but the largest RI value in the current metabolite-like dataset is 593 RIU.

Since an RI model used in filtering a virtual database for the purpose of structure identification would need to make a reasonable prediction for any candidate compound, use of the current model will be limited until compounds with longer retention times can be added to the dataset. The current revised model would likely be useful for structure identification of unknowns with observed RI values between 10 and 600 RIU.

The current data also lack a significant number of simple structures with single functional groups or a small number of functional groups. Since a complex structure can be approximated as a combination of simple structures, the addition of a wider variety and number of simple compounds may expand the accessibility domain for the model. It is also reasonable to suggest that the neural net could better generalize the overall problem with access to a larger and more continuous range of structures and their corresponding RI values.

There are a large number of compounds in the metabolite-like dataset that have at least one feature with low population in the dataset. The quality of external-validation statistics for the revised model would appear to indicate that the IGroup method was reasonably successful in mitigating the issue of underpopulated features, but it is also likely that increasing the population of these features would enhance overall predictive performance.

Overall, future improvements in the RI model will arise largely from the addition of newly measured data that will normalize the distribution of RI values and fill in gaps in the structure space. New experimental data will also provide invaluable independent validation statistics for the current model. Even though the current model was created using extensive validation procedures, all of the data were used for both model training and model validation. Though we believe that the validation criteria used in this study were well constructed, validation with data on which the model is partially based will never reveal its predictive capability in the way that newly measured data can. Independent validation data could assist in defining the model applicability domain by unambiguously identifying compounds for which the model fails to make accurate predictions. Revealing the strengths and weaknesses of the current ensemble model will assist in tuning the model dataset, input descriptors, and the neural net learning process.

ACKNOWLEDGMENTS

This work was funded by the National Institute of Health (1R01GM087714); Pfizer, Inc., Groton, Connecticut; the University of Connecticut Foundation; and by an American Foundation for Pharmaceutical Education predoctoral fellowship awarded to Tzipporah M. Kormos.

REFERENCES

1. J. K. Nicholson, J. C. Lindon, E. Holmes. 'Metabolomics': Understanding the metabolic responses of living systems to pathophysiological stimuli via multivariate statistical analysis of biological NMR spectroscopic data. *Xenobiotica* 29(11) (1999) 1181–1189.
2. K. P. Gartland, F. W. Bonner, J. K. Nicholson. Investigations into the biochemical effects of region-specific nephrotoxins. *Mol. Pharmacol.* 35(2) (1989) 242–250.

3. R. A. Iles, A. J. Hind, R. A. Chalmers. Use of proton nuclear magnetic resonance spectroscopy in detection and study of organic acidurias. *Clin. Chem.* 31(11) (1985) 1795–1801.
4. J. K. Nicholson, D. P. Higham, J. A. Timbrell, P. J. Sadler. Quantitative high resolution ^{1}H NMR urinalysis studies on the biochemical effects of cadmium in the rat. *Mol. Pharmacol.* 36(3) (1989) 398–404.
5. J. K. Nicholson, M. P. O'Flynn, P. J. Sadler. Proton-nuclear-magnetic-resonance studies of serum, plasma and urine from fasting normal and diabetic subjects. *Biochem. J.* 217(2) (1984) 365–375.
6. K. F. Aoki-Kinoshita. An introduction to bioinformatics for glycomics research. *PLoS Comput. Biol.* 4(5) (2008) e1000075.
7. E. Fahy, D. Cotter, R. Byrnes, M. Sud, A. Maer, J. Li, D. Nadeau, Y. Zhau, S. Subramaniam. Bioinformatics for lipidomics. *Methods Enzymol.* 432 (2007) 247–273.
8. C. Wolf, P. J. Quinn. Lipidomics: Practical aspects and applications. *Prog. Lipid. Res.* 47(1) (2008) 15–36.
9. C. Wolf, P. J. Quinn. Lipidomics in diagnosis of lipidoses. *Subcell. Biochem.* 49 (2008) 567–588.
10. P. Schulz-Knappe, M. Schrader, H. D. Zucht. The peptidomics concept. *Comb. Chem. High Throughput Screen.* 8(8) (2005) 697–704.
11. J. T. Brindle, H. Antti, E. Holmes, G. Tranter, J. K. Nicholson, H. W. L. Bethell, S. Clarke, et al. Rapid and noninvasive diagnosis of the presence and severity of coronary heart disease using ^{1}H-NMR-based metabolomics. *Nat. Med.* 8(12) (2002) 1439–1444.
12. A. Sreekumar, L. M. Poisson, T. M. Rajendiran, A. P. Khan, Q. Cao, J. Yu, B. Laxman, R, et al. Metabolomic profiles delineate potential role for sarcosine in prostate cancer progression. *Nature* 457(7231) (2009) 910–914.
13. T. M. Tsang, J. T. Huang, E. Holmes, S. Bahn. Metabolic profiling of plasma from discordant schizophrenia twins: Correlation between lipid signals and global functioning in female schizophrenia patients. *J. Proteome Res.* 5(4) (2006) 756–760.
14. R. P. Woolas, M. R. Conaway, F. J. Xu, I. J. Jacobs, Y. J. Yu, L. Daley, A. P. Davies, et al. Combinations of multiple serum markers are superior to individual assays for discriminating malignant from benign pelvic masses. *Gynecol. Oncol.* 59(1) (1995) 111–116.
15. P. Yin, X. Zhao, Q. Li, J. Wang, J. Li, G. Xu. Metabolomics study of intestinal fistulas based on ultraperformance liquid chromatography coupled with Q-TOF mass spectrometry (UPLC/Q-TOF MS). *J. Proteome Res.* 5(9) (2006) 2135–2143.
16. S. Zhang, G. A. Nagana Gowda, V. Asiago, N. Shanaiah, C. Barbas, D. Raftery. Correlative and quantitative ^{1}H NMR-based metabolomics reveals specific metabolic pathway disturbances in diabetic rats. *Anal. Biochem.* 383(1) (2008) 76–84.
17. Z. Zhang, S. D. Barnhill, H. Zhang, F. Xu, Y. Yu, I. Jacobs, R. P. Woolas, A. Berchuck, K. R. Madyastha, R. C. Bast Jr. Combination of multiple serum markers using an artificial neural network to improve specificity in discriminating malignant from benign pelvic masses. *Gynecol. Oncol.* 73(1) (1999) 56–61.
18. X. Zhao, W. Wang, J. Wang, J. Yang, G. Xu. Urinary profiling investigation of metabolites with cis-diol structure from cancer patients based on UPLC-MS and HPLC-MS as well as multivariate statistical analysis. *J. Sep Sci.* 29(16) (2006) 2444–2451.
19. M. P. Hodson, G. J. Dear, A. D. Roberts, C. L. Haylock, R. J. Ball, R. S. Plumb, C. L. Stumpf, J. L. Griffin, J. N. Haselden. A gender-specific discriminator in Sprague-Dawley rat urine: The deployment of a metabolic profiling strategy for biomarker discovery and identification. *Anal. Biochem.* 362(2) (2007) 182–192.
20. R. S. Plumb, J. H. Granger, C. L. Stumpf, K. A. Johnson, B. W. Smith, S. Gaulitz, I. D. Wilson, J. Castro-Perez. A rapid screening approach to metabolomics using UPLC and oa-TOF mass spectrometry: Application to age, gender and diurnal variation in normal/Zucker obese rats and black, white and nude mice. *Analyst.* 130(6) (2005) 844–849.

21. R. S. Plumb, K. A. Johnson, P. Rainville, J. P. Shockcor, R. Williams, J. H. Granger, I. D. Wilson. The detection of phenotypic differences in the metabolic plasma profile of three strains of Zucker rats at 20 weeks of age using ultra-performance liquid chromatography/ orthogonal acceleration time-of-flight mass spectrometry. *Rapid Commun. Mass. Spectrom.* 20(19) (2006) 2800–2806.

22. S. Wagner, K. Scholz, M. Donegan, L. Burton, J. Wingate, W. Volkel. Metabolomics and biomarker discovery: LC-MS metabolic profiling and constant neutral loss scanning combined with multivariate data analysis for mercapturic acid analysis. *Anal. Chem.* 78(4) (2006) 1296–1305.

23. R. E. Williams, E. M. Lenz, J. S. Lowden, M. Rantalainen, I. D. Wilson. The metabolomics of aging and development in the rat: An investigation into the effect of age on the profile of endogenous metabolites in the urine of male rats using ^{1}H NMR and HPLC-TOF MS. *Mol. Biosyst.* 1(2) (2005) 166–175.

24. A. D. Delinsky, D. C. Delinsky, S. Muralidhara, J. W. Fisher, J. V. Bruckner, M. G. Bartlett. Analysis of dichloroacetic acid in rat blood and tissues by hydrophilic interaction liquid chromatography with tandem mass spectrometry. *Rapid Commun. Mass. Spectrom.* 19(8) (2005) 1075–1083.

25. J. L. Griffin. Metabolomics: NMR spectroscopy and pattern recognition analysis of body fluids and tissues for characterisation of xenobiotic toxicity and disease diagnosis. *Curr. Opin. Chem. Biol.* 7(5) (2003) 648–654.

26. E. M. Lenz, J. Bright, R. Knight, F. R. Westwood, D. Davies, H. Major, I. D. Wilson. Metabolomics with ^{1}H-NMR spectroscopy and liquid chromatography-mass spectrometry applied to the investigation of metabolic changes caused by gentamicin-induced nephrotoxicity in the rat. *Biomarkers.* 10(2–3) (2005) 173–187.

27. E. M. Lenz, J. Bright, R. Knight, I. D. Wilson, H. Major. A metabolomic investigation of the biochemical effects of mercuric chloride in the rat using ^{1}H-NMR and HPLC-TOF/MS: Time dependent changes in the urinary profile of endogenous metabolites as a result of nephrotoxicity. *Analyst.* 129(6) (2004) 535–541.

28. E. M. Lenz, J. Bright, R. Knight, I. D. Wilson, H. Major. Cyclosporin-a-induced changes in endogenous metabolites in rat urine: A metabolomic investigation using high field ^{1}H-NMR spectroscopy, HPLC-TOF/MS and chemometrics. *J. Pharm. Biomed. Anal.* 35(3) (2004) 599–608.

29. F. Goodsaid, F. Frueh. Biomarker qualification pilot process at the U.S. Food and Drug Administration. *AAPS J.* 9(1) (2007) E105–E108.

30. R. S. Niedbala, C. Mauck, P. Harrison, G. F. Doncel. Biomarker discovery: Validation and decision making in product development. *Sex Transm. Dis.* 36(3 Suppl) (2009) S76–S80.

31. J. Ryals, K. Lawton, D. Stevens, M. Milburn. Metabolon, Inc. *Pharmacogenomics.* 8(7) (2007) 863–866.

32. The PubChem Project. http://pubchem.ncbi.nlm.nih.gov/ (accessed Feb 17, 2009).

33. http://www.chemspider.com/.

34. J. J. Irwin, B. K. Shoichet. ZINC: A free database of commercially available compounds for virtual screening. *J. Chem. Inf. Model.* 45(1) (2005) 177–182.

35. D. S. Wishart, D Tzur, C. Knox, R. Eisner, A. C. Guo, N. Young, D. Cheng, et al. HMDB: The human metabolome database. *Nucleic Acids Res.* 35(Database issue) (2007) D521–D526.

36. K. F. Aoki, M. Kanehisa. Using the KEGG database resource. *Curr. Protoc. Bioinformatics.* Oct, Chapter 1 (2005) Unit 1.12.

37. C. A. Smith, G. O'Maille, E. J. Want, C. Qin, S. A. Trauger, T. R. Brandon, D. E. Custodio, R. Abagyan, G. Siuzdak. Metlin: A metabolite mass spectral database. *Ther. Drug. Monit.* 27(6) (2005) 747–751.

38. K. Degtyarenko, P. De Matos, M. Ennis, J. Hastings, M. Zbinden, A. McNaught, R. Alcántara, M. Darsow, M. Guedj, M. Ashburner. ChEBI: A database and ontology for chemical entities of biological interest. *Nucleic. Acids. Res.* 36(Database issue) (2008) D344–D350.

39. S. Mazurek, E. Eigenbrodt. The tumor metabolome. *Anticancer Res.* 23(2A) (2003) 1149–1154.

40. A. H. Merrill Jr. SphinGOMAP:– A web-based biosynthetic pathway map of sphingo-lipids and glycosphingolipids. *Glycobiology* 15(6) (2005) 15G.

41. E. Yasugi, K. Watanabe. LIPIDBANK for web, the newly developed lipid database. *Tanpakushitsu Kakusan Koso.* 47(7) (2002) 837–841.

42. D. S. Wishart. Current progress in computational metabolomics. *Brief. Bioinform.* 8(5) (2007) 279–293.

43. D. W. Hill, T. R. Kelley, K. J. Langner, K. W. Miller. Determination of mycotoxins by gradient high-performance liquid chromatography using an alkylphenone retention index system. *Anal. Chem.* 56(13) (1984) 2576–2579.

44. D. W. Hill, A. J. Kind. Reversed-phase solvent-gradient HPLC retention indexes of drugs. *J. Anal. Toxicol.* 18(5) (1994) 233–242.

45. E. Kováts. Characterization of organic compounds by gas chromatography. Part 1. Retention. Indices of aliphatic halides, alcohols, aldehydes and ketones. *Helv. Chim. Acta.* 41(7) (1958) 1915–1932.

46. E. L. Schymanski, M. Meringer, W. Brack. Automated strategies to identify compounds on the basis of GC/EI-MS and calculated properties. *Anal. Chem.* 83(3) (2011) 903–912.

47. S. Wolf, S. Schmidt, M. Müller-Hannemann, S. Neumann. In silico fragmentation for computer assisted identification of metabolite mass spectra. *BMC Bioinformatics.* 11 (2010) 148.

48. W. Guo, Y. Lu, X. M. Zheng. The predicting study for chromatographic retention index of saturated alcohols by MLR and ANN. *Talanta.* 51(3) (2000) 479–488.

49. J. Acevedo-Martínez, J. C. Escalona-Arranz, A. Villar-Rojas, F. Téllez-Palmero, R. Pérez-Rosés, L. González, R. Carrasco-Velar. Quantitative study of the structure-retention index relationship in the imine family. *J. Chromatogr. A.* 1102(1–2) (2006) 238–244.

50. E. Konoz, M. H. Fatemi, R. Faraji. Prediction of Kovats retention indices of some ali-phatic aldehydes and ketones on some stationary phases at different temperatures using artificial neural network. *J. Chromatogr. Sci.* 46 (2008) 406–412.

51. P. Ertl, B. Rohde, P. Selzer. Fast calculation of molecular polar surface area as a sum of fragment-based contributions and its application to the prediction of drug transport properties. *J. Med. Chem.* 43(20) (2000) 3714–3717.

52. winMolconn, Version 1.4; Hall Associates Consulting: Quincy, MA, 2011.

53. L. H. Hall. A structure-information approach to prediction of biological activities and properties. *Chem. Biodivers.* 1 (2004) 183–201.

54. D. R Albaugh, L. M. Hall, D. W. Hill, T. M. Kertesz, M. Parham, L. H. Hall, D. F. Grant. Prediction of HPLC retention index using artificial neural networks and IGroup E-State indices. *J. Chem. Inf. Model.* 49(4) (2009) 788–799.

55. MDL QSAR, Version 2.2 b365; SymyxTechnology MDL: San Ramon, CA, 2003.

56. Emergent, Version 5.2; University of Colorado: Boulder, CO.

57. B. Aisa, B. Mingus, R. O'Reilly. The emergent neural modeling system. *Neural Netw.* 21 (2008) 1045–1212.

58. D. Daniel, F. S. Wood. *Fitting Equations to Data.* Wiley-Interscience/John Wiley & Sons, New York, 1980.

59. M. Bogusz. Influence of elution conditions on HPLC retention index values of selected acid and basic drugs measured in the 1-nitroalkane scale. *J. Anal. Toxicol.* 15 (1991) 174–178.

60. D. C. Montgomery. Experiments with random factors. In *Design and Analysis of Experiments*, 5th Edition, 511–556. John Wiley & Sons, Inc., New York (2008).
61. winMolconn, Version 1.2.2.1; Hall Associates Consulting: Quincy, MA, 2011.
62. D. R. Albaugh, L. M. Hall, D. W. Hill, T. M. Kertesz, M. Parham, L. H. Hall, D. F. Grant. Prediction of HPLC retention index using artificial neural networks and IGroup E-State indices. *J. Chem. Inf. Model.* 49(4) (2009) 788–799.
63. L. Prechelt. Automatic early stopping using cross-validation: Quantifying the criteria. *Neural Netw.* 11(4) (1998) 761–767.
64. L. Eriksson, E. Johansson, N. Kettaneh-Wold, S. Wold. *Multi- and Megavariate Data Analysis, Principles and Applications*, 107. Umetrics AB, Umeå, Sweden (2001).

V.G. Berezkin, S.N. Shtykov, and E.G. Sumina

CONTENTS

7.1 INTRODUCTION

The role and meaning of analytical chemistry in science and industry has been growing permanently in the recent years. It is connected with the permanent growth of the role and meaning of production control in the modern economy, the role of analytical methods in diagnostics of human health and environmental protection being expanded. In correspondence with the most common definition, "Analytical chemistry is a science developing general methodology, methods and means of studying of chemical composition of a substance and working out ways of analysis of different objects" [1]. As it follows from this definition, the development of analytical methods and corresponding means for its implementation is one of the principal tasks in the development of analytical chemistry. That is why many scientists repeatedly underlined the meaning of a method in science development. Thus, for example, a number of Nobel Prize laureates highly appreciated a role of a method in research. I.P. Pavlov wrote "A method holds in hands a fate of a research"; L.D. Landau outlined, "A method is more significant than a discovery because a correct method will lead to new, more valuable discoveries" (http://www.libaforizm.ru/content/750).

Naturally, developing different methods exerts different influence on science developing; however, undoubtedly, it is that development of research methods should be permanently paid much attention.

Programming of mobile-phase properties in thin layer chromatography (TLC) naturally, in many cases, influences substantially the separation of the analyzed mixtures. For example, if the programming of the mobile-phase composition (and of, consequently, its properties) is implemented before a chromatographic plate, there are methods and devices developed for their implementation, which are described in detail in the literature (see, e.g., Refs [2–11]). In the given review, the new variant of TLC with the programming of composition and properties of a mobile phase (MP) is described. The new method of programming is the result of dissolving the active volatile reagent being in the chamber atmosphere and dissolving in the MP migrating on the plate adsorption layer. The active volatile compounds (the modifiers) as a result of dissolving in the MP change the properties of the latter in a direction necessary for an experimenter. It is interesting to note that according to the generally accepted opinion, presence of a gas phase in contact with a TLC plate is considered as a negative factor of TLC by most of the researches [2–4]. However, a gas phase has been shown by us to play a positive role as well, improving separation of analyzed mixtures.

The given review is devoted to the description of the bases of the new TLC method with a controlled gas phase (TLC-CGP) [12–14]. The authors hope that the method described will arouse the reader's interest and wish to use this method in one's work and to propose new variants of its further development and practical application.

7.2 BASES OF THE NEW VARIANT OF TLC WITH A CONTROLLED GAS PHASE (TLC-CGP) INFLUENCING THE SEPARATION

TLC is the simplest version of modern liquid chromatography [2–5]. It can be applied for solving many analytical tasks not only in industry but also at home, at school, in village first-aid stations, and so on. Planar chromatography is currently widely used in pharmacy, scientific researches, medicine, environmental protection, industry, and so on. In Russia, TLC is used in more than 10,000 laboratories. The further development of TLC as a pilot method for modern liquid chromatography is rather prospective [4]. The wide practical usage of TLC in modern industry, science, and other spheres is explained by a number of its benefits in comparison with column liquid chromatography (CLC). One of the main drawbacks of traditional TLC is, in many famous researches' opinion, a free gas phase surrounding a plate. Presence of the gas phase in the experiment zone leads to poor repeatability of data obtained by the TLC method [15].

The role of gas phase in various variants of TLC is reflected in Figure 7.1. In the first work on TLC by N.A. Izmailov and M.S. Shraiber [16], the chromatographic process was conducted on a completely open sorbent layer.

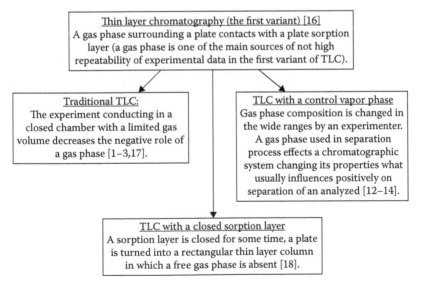

FIGURE 7.1   Schematic of TLC development (the main TLC variants reflecting the role of a gas phase contacting a plate sorption layer in a chromatographic experiment).

However, in the process of TLC development, it was proposed to conduct the experiment in a closed chamber (i.e., in closed space) for repeatability improvement, given that it was recommended previously to create the gas atmosphere in the chamber that contained saturated vapors of the used MP. For implementation of this main "chamber" variant of TLC, various chamber types have been proposed (see, e.g., Refs [2–4,17]).

The new "chamberless" TLC variant with a closed sorption layer in which a gas phase was completely absent, and a chamber was transformed into a distinctive flat column was proposed in the work [18]. A sorption layer is closed by tightly covering with another plate (for example polymer) and a TLC plate is turned into a rectangular thin layer column in which a free gas phase is absent.

However, a gas phase in contact with a plate adsorption layer can be given a new function; that is, it can also be used in a gas phase with the insertion of particular active components (e.g., acid or basic character) to influence expediently (positively in an analytical aspect) on selectivity of a used chromatographic system (i.e., on a sorbent and a MP of a plate on which separation takes place). As a result of such an influence, separation properties of a used chromatographic system may be purposefully changed as a result of what separation can be improved substantially.

The development of analytical chemistry is, as a rule, the development (like in other spheres of chemical science) of principally new methods of research. Thus, in the opinion of Yu. A. Zolotov, "a method of analysis is a sufficiently universal and, as a rule, theoretically based way of determining of chemical composition, usually irrelatively to a determined component and an analyzed object" [19].

The main task of our research was working out and researching the new variant of TLC with a gas phase controlled by an experimenter and changed in time and space (relatively to an TLC plate), which can actively interact and change properties of a MP and a sorbent directly on a plate [14].

The new variant is called "thin layer chromatography with a controlled gas phase" (TLC-CGP). TLC-CGP is the variant of TLC in which, with the purpose of separation, improving properties of a chromatographic system (a sorbent + a MP) are purposefully changed by insertion of active components (e.g., acid or basic) surrounding a plate in a gas medium, which are partially dissolved in a liquid MP migrating on a plate with respect to the change in MP properties (as well as adsorbent properties). This permits improving separation properties of an initial chromatographic system and correspondingly of chromatographed compounds.

When working out a new method, the following question arises among many analysts and chromatographists: What new positive features does a new analytical method possess? The answer to this significant question for the new variant of TLC is given in Table 7.1. As follows from the analysis of the proposed and traditional methods (see Table 7.1), TLC-CGP, in our opinion, is undoubtedly a useful and new analytical method [20].

Two chromatograms, as an example, that directly testify to the principal possibilities of the proposed method are provided in Figure 7.2 [13], which were obtained when separating the mixture of benzoic acid derivatives on the plates in the usual air atmosphere and in the carbon dioxide atmosphere.

While carrying out the comparison of the chromatograms A and B (Figure 7.2), it should be noted that, first, carrying out the separation in the atmosphere of carbon

TABLE 7.1
Comparison of the New Method of TLC-CGP with the Most Known Method of Programming of Mobile Phase Composition

No	Characteristic of Comparison	The Known Method of Programming of Mobile Phase Composition	The Proposed Method of TLC with a Control Vapor Phase (TLC-CVP) Contacting a Mobile Phase on a Plate
1	The main idea of the used method of mobile phase programming	The direct programming of composition of a liquid phase coming in contact with a plate during the process of separation (programming of composition of a liquid phase is implemented before its coming in contact with a plate on which an analyzed mixture is separated)	Programming of mobile phase composition is implemented during the process of separation of an analyzed mixture as a result of interaction (absorption or adsorption) of a gas phase containing active components with a mobile phase and an adsorption layer being on a plate
2	The principle of the method implementing	Purposeful changing of mobile phase composition before it coming in contact with a plate	Purposeful changing of composition of a gas phase placed or migrating above a plate (i.e., above a sorption layer partially wetted by a mobile phase with already partially separated compounds) as a result of a mobile phase contact with a liquid phase (a sorbent), given active components of a gas phase (or the part of a gas phase proper) are dissolved in a sorptionally active plate layer and change its properties (e.g., when dissolving CO_2 a mobile phase becomes more acidic and when dissolving NH_3 it becomes more basic); as a result separating properties of a used chromatographic system can be purposefully changed and separation can be improved
3	The main differences of compared methods	The principle of method implementation does not permit to change mobile phase properties directly on a plate (i.e., practically instantly) since substitution of a mobile phase takes place only on the one side of a plate (on the side of mobile phase coming), that is, for the known method appearance of a factor of delay is typical	The principle of the proposed method implementation permits to change mobile phase properties practically instantly since a contact of a gas phase containing active compounds with a mobile phase takes place on all the plate length simultaneously and, consequently, its properties change simultaneously. This distinctive feature of the proposed method of programming was not known earlier either in TLC or in column liquid chromatography, it is a principally new factor in liquid chromatography and it is used in TLC (TLC-CVP)

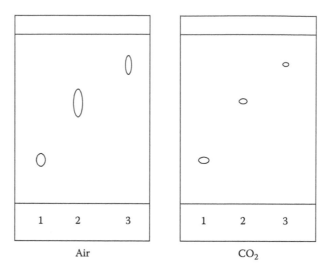

FIGURE 7.2 Chromatograms reflecting the influence of the gas atmosphere of carbon dioxide on separation. The conditions of separation include the mobile phase: isopropropanol–water (60:40, vol.); the sorbent: polyamide-II; and the separated compounds: (1) 2,4-Dihydroxybenzoic acid, (2) para-aminobenzoic acid; and (3) ortho-aminobenzoic acid [13].

dioxide leads to the sharp improvement of separation of the analyzed isomers (para-aminobenzoic acid and ortho-aminobenzoic acid) and, second, substantially improves the efficiency of separation. That is why the usage of the proposed method widens analytical possibilities of TLC, and it is expedient to be used when solving different practical tasks.

7.3 MAIN TYPES OF THE SPECIAL CHAMBERS USED FOR IMPLEMENTING OF THE NEW VARIANT (TLC-CGP)

The description of an analytical method is usually the integrated whole of two components. First, it includes the description of the proper idea of a new method (it is usually uncovered as the consequence [the totality] of particular analytical operations), and second the description of the idea of design of a new device that permits to implement a new analytical method sufficiently, simply, and reliably. Naturally, for the implementation of the main idea of a method, several designs (several methods of its implementation) can be proposed. Let us note that the device (the arrangement) permitting to implement the idea of a new method in practice plays a very significant role in implementing a new method. The main arrangements that were worked out and used when implementing the new method of TLC-CGP are described in Sections 7.3.1 and 7.3.2.

7.3.1 Gas-Flowing Chamber for TLC

The gas-flowing chamber that was used in our first work on the new method implementation is schematically shown in Figure 7.3 [12–14]. The flows of acidic (e.g., carbon dioxide) or basic (e.g., ammonia) gases was used for changing the

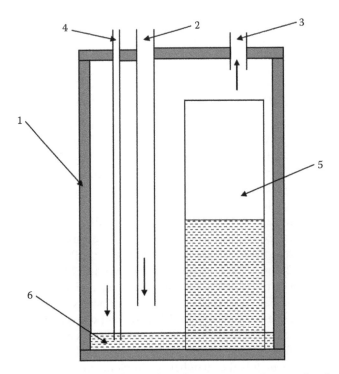

FIGURE 7.3 Schematic of the TLC chamber whose gas atmosphere can be changed during the process of separation. The designations: (1) The chamber; (2) The tube for supplying the required gas into the chamber; (3) The tube for the gas extraction; (4) The tube for supplying the mobile phase; (5) The plate for TLC; and (6) The liquid eluent.

atmosphere in the chromatographic chamber. During the process of chromatographic separation of the analyzed mixture, the above-mentioned gases with the characteristics different from neutral were dissolved in the film of the liquid MP migrating upward on the plate in the conditions of ascending chromatography. As a result of the processes described, the change in a pH of the film of the MP on the plate took place, which changed the form of chromatographed compounds; for example, changing from a molecular state to a ionic state or, on the contrary, it, naturally, manifested in the changing of adsorptive capacity of the chromatographed molecules and their chromatographic mobility on the plate.

Chromatograms of the derivatives of benzoic acid are provided, as examples, in the work by Berezkin et al.: (1) in the conditions of neutral air atmosphere and (2) in the conditions of the "acid" gas atmosphere of carbon dioxide. As it follows from these chromatograms, carbon dioxide when dissolved in the MP decreases the pH of the used MP and increases sorptive capacity of the chromatographed acids that are in the chromatographic system in the molecular form. Simultaneously, broadening of the zones of the analyzed compounds decreases.

As follows from the preceding examples [13], use of gas modifiers substantially improves separation and increases its efficiency.

Thereby, the application of the proposed method and the device (the chamber) broadens the possibilities of an analyst in control of chromatographic process in TLC and permits the implementation of more selective separation.

7.3.2 DIFFUSION CHAMBER FOR IMPLEMENTING THE TLC-CGP VARIANT

Our first chamber and some of the following ones proposed and used for implementing the new TLC variant were based on the supplying of an active agent (e.g., ammonia and carbon dioxide) to the plate working surface by the flow of an inert gas. Using this principle, different examples of the proposed TLC variant were implemented. Unfortunately, this TLC-CGP variant has one significant drawback: the gas flow leaving the chamber uncontrolledly takes away with it vapors of a volatile MP that can negatively and uncontrolledly influence the characteristics of the chromatographic experiment as the whole. To rule this out, it is expedient to use (1) a slow speed of the active gas flow affecting the MP to the working surface of a TLC plate or (2) low-volatile compounds as a MP in TLC. However, the drawback discussed can be overcome if the principle of the active compound supplying (e.g., of an acid or a base) to the plate working surface is changed; for example, it is possible to use the phenomenon of diffusion and convection for this purpose (since in the latter case the losses of a MP used in TLC can be sharply reduced) instead of the gas flow as a means of supplying an actively acting substance. The diffusion chamber, which is the chamber for implementing of one of the variants of TLC-CGP in which the diffusion-convection method of the modifier supplying to the plate working surface is used, is described in the following discussion [21].

The design of the diffusion-convection chamber is illustrated in Figure 7.4.

The variant of the diffusion-convection chamber used is schematically provided in Figure 7.4. The chemical agent modifying the working plate is positioned on the donor plate wetted by the donor liquid. The modifying agent is transferred from the donor plate to the working plate with respect to the changes of the MP properties.

The donor plate with the solution of the volatile agent is placed in the subchamber C-2, the sorption layer being positioned at the partition 5. The chamber is closed with the cover 6. The whole chamber is made of glass and glued with silicone glue.

Chromatographic research was carried out on the plates PTSKh-P-A (Sorbfil®,* Krasnodar, Russia) of size 10 × 10 cm. The following individual solvents and their solutions were used as the MPs: toluene, toluene + ethanol (in the ratio 8:2), and toluene + ethanol (in the ratio 1:1). The mixture of the dyes used as the separated mixture in toluene was Siba F II, Indophenol, Ariabel Red, Sudan Blue, Sudan II, and dimethylaminoazobenzene in the ratio 1:1:1:1:1:1.

For forming of the active gas phase in the special chamber, aqueous solution of either ammonia (2.5%) or acetic acid (5%) was used.

---

* Registered trademark of Close Corporation "Sorbpolimer," Krasnodar, Russia.

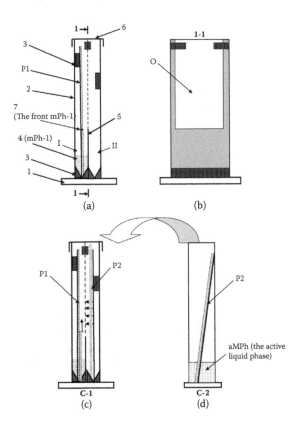

FIGURE 7.4 Schematic of conducting the experiment using the diffusion-convection chamber. (a) The working chamber (in the beginning of the experiment, it is used for traditional separation using the mobile phase MPh-1): (1) the basement of the chamber, (2) the chamber, (3) the clamps of the plate position in the chamber, (4) the mobile phase, (5) the partition with the "window," (6) the chamber cover, and (7) the front (upper boundary) of the developing mobile phase 1 (Mph-1). (b) The central partition of the chamber with the "window" O through which the diffusion-convection transfer of vapors of the active liquid phase (aMPh) from the surface of the donor plate 2 (after its transfer from the chamber C-2 into the working chamber C-1) takes place. (c) The working chamber after the installation of the donor plate (P-D): (P1) the TLC plate with the separated components and (P2) the donorplate, when implementing the TLC-CGP method, the vapors of the aMPh with which the plate P-D is impregnated as a result of diffusion and convection transfer through the "window" O to the plate P1. (d) The additional chamber used for and impregnation and saturation of the working surface of the donor plate with the phase aMPh.

Two to three milliliters of MP was poured into the sub-chamber C-4 by a syringe. After that when on the TLC plate (the size of 4.5–4.8 × 10 cm) at a distance of 15 mm from the plate edge, the analyzed samples (in our case, they were dye mixtures) were applied, the working plate was sunk in the sub-chamber C-4 by the tweezers carefully not touching the inner walls, in the MP, the adsorption layer being placed to the partition, watching attentively for the plate being in the position illustrated in Figure 7.4a. Then, the chamber was closed with cover. The MP front started ascending on the TLC plate separating the analyzed mixture. The MP front having traversed a distance of 3 cm, the donor plate (wetted, e.g., by the solution of ammonia or the solution of acetic acid) was installed in the second part of the chamber and the adsorption layer being positioned to the partition as well (see Figure 7.4c), while the distance between the working TLC plate and the donor plate did not exceed 10 mm.

Note that for increasing the capacity of the donor plate fasten an additional strip of filter paper (of the same size) to it. Once the experiment is finished, the working plate with the separated samples was taken away from the chamber by the tweezers, and then it was done with the donor plate.

It should be noted that it is extremely dangerous to leave the MP (especially toluene and other solvents) in the chamber tank for a long time (e.g., for a night or for more than 4 hours). This may lead to partial damage of a layer of used glue or to damage of the chamber integrity.

7.4 INFLUENCE OF THE ACTIVE GAS PHASE ON THE CHROMATOGRAPHIC CHARACTERISTICS OF AROMATIC AMINES AND HALOGEN DERIVATIVES OF BENZOIC ACIDS IN THE NEW VARIANT OF TLC-CGP

To study the features of the new variant of TLC-CGP, it seemed to be expedient to study the influence of the gas phase on the changing chromatographic characteristics of organic amines and acids using plates both with silica gel and with nonpolar phases (silica gel RP-18, polyamide).

Taking into account, first, that the "age" of the TLC-CGP method does not exceed several years, second, that the given review is primarily addressed [22] to qualified TLC specialists, and the most interesting factor for everybody in the given case is, acquaintance with the initial experimental material we used, which is, in this paper—the initial TLC-chromatograms as they present in original papers. That is why the given experimental material in the review is represented in such a form.

Chromatographic process of separation of organic acids and amines in the TLC-CGP conditions was implemented on the following plates:

1. "Sorbfil" plates [23] ("Sorbpolimer," Krasnodar, Russia) with polymeric or aluminum substrate; the sorption layer was silica gel containing luminiferous indicator.
2. "Silufol"®* plates [24] produced in Votitsa at the plant "Kavalier"; the substrate was aluminum foil; the sorption layer was silica gel containing admixtures of luminophore.

---

* Registered trademark of Sklarni Kavalier, Inc., Sazava, Czech Republic.

3. "Silica gel RP-18" plates [8], Merck®,*.
4. "Polyamide-11" plates [8], Merck, nonpolar organic polymeric sorbent with hydrophobic properties.

7.4.1 Thin Layer Chromatography of Aromatic Amines

7.4.1.1 Influence of Carbon Dioxide (Sorbent on TLC Plate Is Silica Gel)

In Figure 7.5, the chromatograms of the individual isomers of nitroanilines on the domestic plates "Sorbfil" with the polymeric substrate and on the plates "Silufol" with the aluminum substrate are given. The chromatograms were obtained by using binary solution, hexane–isopropanol (9.5:0.5), as the MP; air and carbon dioxide were used as the chamber gas atmosphere.

As follows from the given chromatograms, when using carbon dioxide, difference in the retention values for the studied compounds increases, the value of efficiency increases simultaneously, but the width of chromatographic zones, correspondingly, decreases.

7.4.1.2 Influence of Acetic Acid Vapors (Sorbent on TLC Plate Is Silica Gel)

Acetic acid vapors negatively influence the efficiency of the separation of individual nitroanilines. Chromatographic zones of nitroanilines broaden more substantially

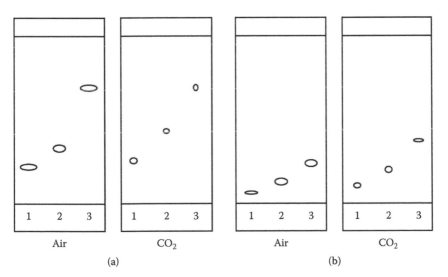

FIGURE 7.5 Chromatograms of some nitroanilines on the plates (a) "Sorbfil" and (b) "Silufol." The experiment conditions: the duration of the contact with the plates with CO_2 is 15 minutes. (1) Para-nitroaniline; (2) Meta-nitroaniline; and (3) Ortho-nitroaniline.

* Registered trademark of "Merck KGaA," Darmstadt, Germany.

(in comparison with zones of the same compounds when carrying out separation in air atmosphere); however, the order of elution does not change.

7.4.1.3 Influence of Ammonia Vapors (Sorbent on TLC Plate Is Silica Gel)

With regard to the influence of ammonia vapors on the TLC separation of nitroanilines, it should have been expected that the pH of the mobile phase under the considered conditions is to be increased. As a result, adsorption capacity of amines in these conditions should have increased, but the value of retardation factor (R_f) should have decreased. The given forecast of the influence of ammonia on chromatographic characteristics of nitroanilines completely agrees with the chromatogram (see Figure 7.6).

7.4.1.4 Influence of Ethanol Vapors (Sorbent on TLC Plate Is Silica Gel)

Even small changes in MP composition are known to influence substantially on selectivity of chromatographic separation [4,5]. That is why it was of interest to widen the range of conditioning solvents, the vapors of which can also noticeably influence the separation. In Figure 7.7, the chromatograms of separation of individual nitroanilines are obtained under the traditional conditions and in the conditions of ethanol vapors' influence (neutral nonionogenic compound) on the chromatographic process of separation on "Sorbfil" and "Silufol" plates.

As follows from the chromatograms given on both plates, ethanol vapors influenced positively on the selectivity (separation) as well as on the efficiency of zones of the separated compounds.

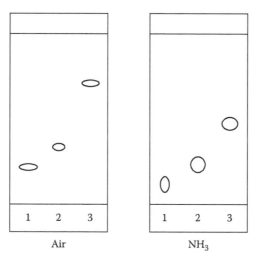

FIGURE 7.6 Chromatograms of some nitroanilines on the plate "Sorbfil" with polymeric substrate. The experiment conditions: the mobile phase: hexane–isopropanol (9.5:0.5); the duration of the contact with the plates "Sorbfil" with NH_3 is 15 minutes. (1) p-nitroaniline (2) m-nitroaniline (3) o-nitroaniline.

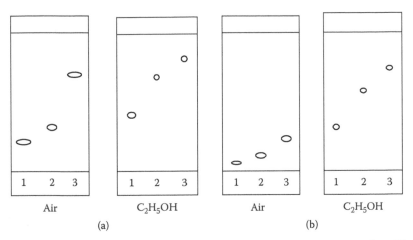

FIGURE 7.7 Chromatograms of some nitroanilines on the plates (a) "Sorbfil" and (b) "Silufol." The experiment conditions: the mobile phase: hexane–isopropanol (9.5:0.5); the duration of the contact with the plates "Sorbfil" with ethanol vapors is 15 minutes. (1) p-nitroaniline (2) m-nitroaniline (3) o-nitroaniline.

7.4.1.5   Influence of Carbon Dioxide and Ammonia (Sorbents Are Modified Silica Gel RP-18 and Polyamide)

Studying the separation of ortho-, meta-, and para-isomers of individual nitroanilines on the TLC plates with silica gel RP-18 and polyamide in the conditions of air atmosphere, carbon dioxide, and ammonia vapors showed that for the studied mixtures, the use of new gas atmospheres (carbon dioxide and ammonia) instead of traditional air in the separation conditions does not always lead to improved results.

However, when separating another mixture (diphenylamine, 1-naphthylamine, and aniline) in the gas atmosphere of carbon dioxide, separation with better chromatographic characteristics was obtained than when using traditional air atmosphere (see Figure 7.8).

As it follows from the data provided in Figure 7.8, when separating the mixture stated in the preceding discussion, an improvement in the chromatographic characteristics, especially the characteristics of efficiency, was seen. The positive results of the experimental estimate of the TLC-CGP variant obtained by using separation of aromatic amines testify to the expediency of further development of the given method and its usage for solving of various practical tasks.

In Section 7.4.2, usage of TLC-CGP method in separation of halogen derivatives of benzoic acid is considered.

7.4.2   THIN LAYER CHROMATOGRAPHY OF HALOGEN DERIVATIVES
OF BENZOIC ACID

In this section, results of separation by TLC method are given for the following benzoic acids:

1. o-chlorobenzoic (o-ClBA) acid
2. p-chlorobenzoic (p-ClBA) acid

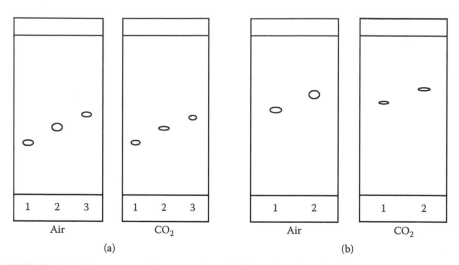

FIGURE 7.8 Chromatographic separation of diphenylamine (1), 1-naphthylamine (2), and aniline sulfate (3) on the plates with (a) silica gel RP-18 and with (b) polyamide. The experiment conditions: the mobile phase when using the plates RP-18: isopropanol–water (6:4); and for the plates with polyamide phase: acetonitrile–water (6:4).

3. o-bromobenzoic (o-BrBA) acid
4. p-bromobenzoic (p-BrBA) acid

7.4.2.1   Influence of Carbon Dioxide (Sorbent Is Silica Gel)

Typical TLC chromatograms of the individual halogen derivatives of benzoic acids when using the MP hexane–isopropanol (9.5:0.5) on the plates "Sorbfil" and "Silufol" under the usual conditions and with carbon dioxide effect (20 minutes) are provided in Figure 7.9.

The comparison of the obtained chromatograms shows that effect of carbon dioxide on the plate in the chromatographic process improves separation (see Figure 7.9). For example, if in the traditional air atmosphere, the zones of halogen derivatives of benzoic acid do not practically move on the plate, in the atmosphere of carbon dioxide all the compounds separate well (see Figure 7.9a). Selectivity of separation improves substantially also on the "Silufol" plates. Thereby, using carbon dioxide instead of the traditional (air) atmosphere improves the separation of the researched mixture when using both the plates "Sorbfil" "Silufol."

7.4.2.2   Influence of Acetic Acid Vapors (Sorbent Is Silica Gel)

The chromatograms of the individual halogen benzoic acids when using the MP: hexane–isopropanol (9.5:0.5) on the plates "Sorbfil" both under the usual conditions (the gas phase is air) and acetic acid vapor effect are provided in Figure 7.10.

Chromatograms obtained under traditional conditions (i.e., air atmosphere) and after 20 minutes of separation in the carbon dioxide atmosphere are provided in the figure. The results obtained also testify to the expediency of using the new TLC-CGP variant in the analytical practice.

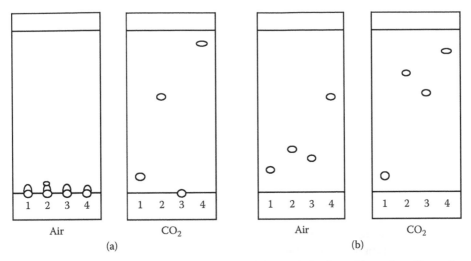

FIGURE 7.9 Chromatograms of the individual halogen derivatives of benzoic acids on the plates (a) "Sorbfil" and (b) "Silufol." The experiment conditions: The duration of the plates in contact with CO_2 is 20 minutes; the mobile phase: hexane–isopropanol (9.5:0.5). (1) o-ClBA (2) p-ClBA (3) o-BrBA (4) p-BrBA.

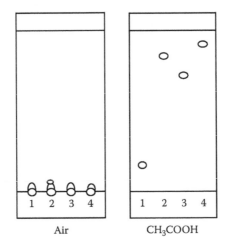

FIGURE 7.10 Chromatograms of the individual halogen derivatives of benzoic acids on the plates "Sorbfil." The experiment conditions: the mobile phase: hexane–isopropanol (9.5:0.5); the duration of the plates contact with acetic acid vapors is 20 minutes. (1) o-ClBA (2) p-ClBA (3) o-BrBA (4) p-BrBA.

7.4.2.3 Influence of Ammonia (Sorbent Is Silica Gel)

The influence of ammonia when using "Sorbfil" plate manifested only in the mobility of p-bromobenzoic acid, the value of which increased sharply. However, the zones of rest of the benzoic acids stayed on the start line. It is interesting to note that the unusual changing of chromatographic characteristics was observed, as well for p-BrBA for other cases.

7.4.2.4 Influence of Ethanol Vapors (Sorbent Is Silica Gel)

Studying the influence of gas atmosphere on the chromatographic characteristics of the separated compounds, it was natural to estimate the influence of not only compounds changing the pH of a MP in TLC but also of a neutral compound. For this purpose, the influence of ethanol vapors on the separation of halogen derivatives of benzoic acids was studied.

In Figure 7.11, the chromatograms of halogen derivatives of benzoic acid are provided in the conditions of the traditional (air) atmosphere and in the conditions when the gas atmosphere contacts the plate containing ethanol vapors.

As it follows from the chromatograms, using of vapors of nonionogenic modifier (ethanol) in the gas phase also permitted to substantially improve the separation on nonpolar sorbents (silica gel RP-18, polyamide).

7.4.2.5 Influence of Carbon Dioxide (Sorbents Are Modified Silica Gel RP-18 and Polyamide)

The chromatograms of the benzoic acid derivatives on the plates with the sorption polyamide and silica gel RP-18 layers in the conditions of the traditional separation and under the atmosphere of carbon dioxide are provided in Figure 7.12.

As it follows from the chromatograms in Figure 7.12, using the atmosphere of carbon dioxide in the conditions of the chromatographic separation of the benzoic acid derivatives permitted to sharply improve the efficiency and the selectivity of separation on the plate with polyamide and to increase the efficiency when separating on the plate with silica gel RP-18. The obtained results testify to the expediency of using nonstandard gas atmosphere in TLC as well as when using nonpolar sorbents.

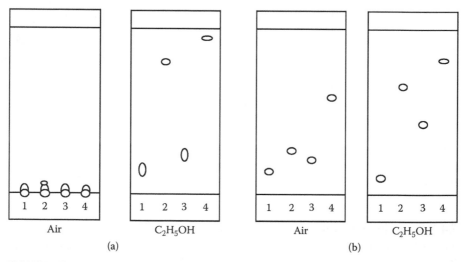

FIGURE 7.11 Chromatograms of individual derivatives of benzoic acids on the plates (a) "Sorbfil" and (b) "Silufol." The experiment conditions: the mobile phase: hexane–iso-propanol (9.5:0.5); the duration of the plates contact with ethanol is 20 minutes. (1) o-ClBA (2) p-ClBA (3) o-BrBA (4) p-BrBA.

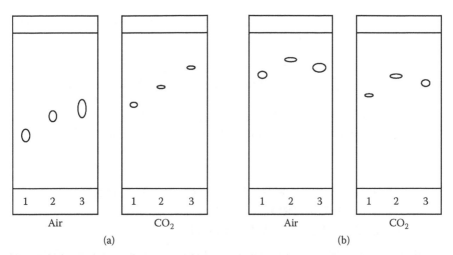

FIGURE 7.12 Chromatograms of the individual derivatives of benzoic acid on (a) polyamide and on (b) silica gel RP-18. The experiment conditions: the mobile phase: isopropanol–water (6:4); the duration of TLC plates in contact with carbon dioxide is 60 minutes. (1) m-nitrobenzoic acid (2 m-aminobenzoic acid (3) benzoic acid.

7.4.2.6   Influence of Ammonia (Nonpolar Sorbents Are Polyamide and Silica Gel RP-18)

When using the plates with nonpolar phases (polyamide and silica gel RP-18), the separation of m-nitrobenzoic, m-aminobenzoic, and benzoic acids in the conditions of ammonia atmosphere permits to substantially improve the efficiency of zones of the separated compounds (to compare the separation in the air atmosphere); however, the selectivity of separation in these conditions did not practically change.

Thereby, the obtained results testify to the expediency of using the semipolar polyamide-11 and nonpolar RP-18 plates, as well, when practical usage of the TLC-CGP mode is proposed.

7.4.2.7   Conclusion

The theoretical considerations and the obtained experimental results of the conducted research permit the recommendation of using the new TLC variant with a CGP for solving many practical tasks, since in many cases the new method substantially improves the main chromatographic characteristics: selectivity and efficiency. The quantitative interpretation of the influence of carbon dioxide and ammonia on the changing of retention of such significant classes of organic compounds as amines and acids is a complicated task. However, when researching the given problem, it seems expedient, beforehand, to pay attention to acid–base equilibriums in the MP and to their connection with adsorption of chromatographed compounds. Let us note that when it comes in contact with carbon dioxide, the pH of the MP (isopropanol–water 6:4) decreases from 6.6 to 4.0, and when in contact with ammonia, it increases from 6.6 to 10.2. Using the approach considered in the preceding discussion when interpreting the phenomena revealed in the work taking into account acid–base equilibriums seemed to be one of the most prospective ones [10,25].

The new proposed variant of TLC is undoubtedly expedient to be used in analytical practice. By using the proposed method of TLC-CGP, an experimenter obtains the new tool permitting to improve separation of an analyzed mixture.

It should be underlined that the proposed method is distinguished by simplicity and the implementation of the new method does not require the use of complex and expensive equipment.

7.5 ELECTROOSMOTIC PLANAR CHROMATOGRAPHY BY USING A CONTROLLED (CHANGED) GAS PHASE CONTACTING A TLC PLATE

Traditional TLC widely used in science, medicine, environmental monitoring, and so on is one of the simplest and the most effective methods in modern analytical chemistry. However, this method has particular drawbacks, one of which is velocity of mobile liquid phase on a sorbent layer steadily decreasing in time. Electroosmotic TLC, which has been actively developing in the recent years, does not have this drawback (see, e.g., [26, 27]).

Let us consider the new variant of electroosmotic TLC in which MP properties (e.g., pH) change in the process of electroosmotic separation as a result of the substitution of the neutral gas medium above the TLC plate adsorption layer to the new gas medium that contains the active (e.g., acid or base) components. These active components of the gas medium are dissolved in the MP and change its pH and, consequently, its chromatographic characteristics. It is also possible that in some cases, the active components of the gas phase can chemically react with separated compounds. The new variant of electroosmotic TLC [26] described in the preceding discussion is expedient to be considered and designated as ETLC-CGP.

The arrangement for the implementing of the new variant of ETLC-CGP used in our work is given in Figure 7.13.

A voltage of 3 kV was applied to the electrodes, the electrical current making up 250 μA. When adding gas ammonia, the voltage decreased to 2 kV and the electrical current increased to 350 μA.

When conducting the experiment, the samples of sorbates were applied on the TLC plate in 0.2 μL at a distance of 2 cm from the edges and at the center for controlling uniformity of migration of the MP on the whole working surface of the plate. Then, the plate prewetted by the MP was placed in the arrangement, the sorption layer being down (see Figure 7.13), so that two of its edges touched the strips of the filter paper and the central (working) area was above the central tank that contained the ammonia solution (i.e., it was the source of gas ammonia, which was dissolved in the sorption system of the TLC plate).

In Figure 7.14, the chromatograms obtained by the method of electroosmotic chromatography by using different gas phases are provided.

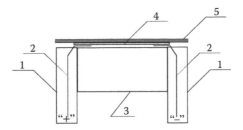

FIGURE 7.13   Schematic of the arrangement for electroosmotic TLC with the changed and controlled composition of the gas phase that contacts with the plate sorption layer. (1) The tank for supplying the plate with the mobile phase; (2) the filter paper for supplying (and removal) the mobile phase to the plate; (3) the rectangular central chamber for creating and changing the gas phase contacting with the plate sorption layer; (4) the TLC plate in the chamber that usually contained the volatile reagent; (5) the glass for closing the chromatographic plate from above (the cover). (From Berezkin, V.G., Nekhoroshev, G.A., *Sorption and Chromatographic Processes*, 7, 2, 2007. With permission.)

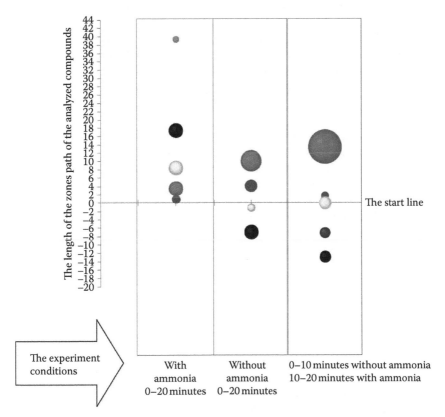

FIGURE 7.14   The changing of the position of the chromatographic zones of the analyzed compounds in different experiment conditions. (From Berezkin, V.G., Nekhoroshev, G.A., Sorption and Chromatographic Processes [*Sorbtsionnye i khromatograficheskiye processy*], 7, 2, 2007. With permission.)

As follows from the provided chromatograms, the gas phases contacting the TLC plate substantially influence the separation of the chromatographed dyes.

In conclusion, it should be noted that when implementing the new method, changes of the selectivity of the separation of the artificial dyes mixture and increase in the separation coefficient up to 50% were obtained. Thereby, the experimental research of the new variant of electroosmotic TLC-CGP showed the prospects of its practical using for separation for the mixtures of charged substances that change the polarity when changing a pH of medium.

7.6 VARIANT OF THIN LAYER CHROMATOGRAPHY BY USING AN INERT GAS FLOW MOVING ABOVE A PLATE SORPTION LAYER

With the purpose of the further development of the effective methods based on the use of the special properties of a gas phase and on its using in TLC, beforehand, as a factor improving separation, we proposed to use during the process of sample separation the gas flow moving above the plate sorption layer under special conditions when the distance between the TLC plate and the upper chamber wall is not large (~0.1–0.8 mm) [28].

In the examples given, the use of chemically active gas in contact with liquid MP is not described, although the implementation of such a method of controlling the separation is undoubtedly possible. That is why the authors believed that the description of the given method can be undoubtedly considered as one of the variants of TLC-CGP.

In Figure 7.15, the general schematic of the arrangement and the principle of its operation including the direction of the gas flow movement above the plate sorption layer and the migration of the volatile MP are shown.

The use of two independent flows, of the gas (above the plate) and of the liquid MP, migrating on the plate adsorption layer substantially increases the efficiency of separation and its velocity. The experiments were conducted on 100×50 mm plates.

In Figure 7.16, the chromatograms of the separation of the dye mixture in toluene obtained using the arrangement shown in Figure 7.15 are provided.

Analyzing the obtained chromatograms, it is necessary to note the following: Better separation was obtained when using the unsaturated chamber (Figure 7.16b) to compare the separation of the same mixture in the chamber previously saturated with the MP vapors (Figure 7.16a).

Improvement of chromatographic characteristics of the analyzed compounds is observed when separating by using the air flow moving above the plate in the direction reversed to the direction of the liquid MP migration (see Figure 7.15). Under the conditions of the counterflow of air and the liquid MP, (1) the MP front as a result of evaporation starts migrating in the direction reversed to the initial one and (2) the zones of the separated compounds in these conditions are "compressed" in the longitudinal direction. As a result of the zones "compression," the increase of the coefficient of the peaks resolution (R_s) takes place. Thus, for example, the value R_s increases 1.8 times for indophenol and 1.3 times increases for Sudan blue. On average, for all the zones on the chromatogram, the values of R_s and N increase by factors 1.5 and 3.6, respectively, when the reverse migration of the MP front takes place.

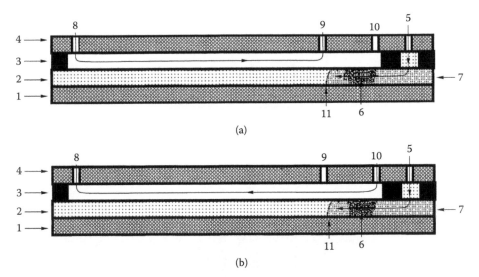

FIGURE 7.15 General schematic of the arrangement for implementing of the new variant of TLC. (1) The plate substrate; (2) the layer of the adsorbent on the plate; (3) the Teflon®* gasket; (4) the covering glass plate (the cover); (5) the hole for the supplying the mobile phase (the eluent) in the narrow rectangular canal in the gasket; (6) the place of the analyzed sample application on the plate; (7) the part of the plate sorption layer impregnated with the mobile phase 8, 9, and (10); the longitudinal narrow cuts in the upper chamber cover for supplying the gas flow to the chamber and its removal from the chamber; (11) the position of the migrating front of the liquid mobile phase on the plate adsorption layer. (a) The counterflow direction of the flows: of the gas (8→9) and of the liquid phase (5→11). (b) The forward-flow direction of the flows: of the gas (10→8) and of the liquid mobile phase (5→11). (From Berezkin, V.G., Sednev, K.V., *Doklady Chem.*, 419, 1, 2008. With permission.)

The observed positive effects can be explained by compression of zones because of reversed moving of MP front that occurs as a result of evaporation of the volatile MP from the plate surface under the effect of counter-current moving of the air flow. Thus, for example, when the MP front touches the upper (front) border of the substance zone, the lower border of its zone under the action of the eluate flow continues to move in the direction to the front border of the zone that has already stopped moving on the adsorption layer. In the conditions of the described process, the longitudinal zone sizes begin to decrease automatically (the zone is compressed) until the "reversed" MP front "meets" the lower borders of the chromatographic zone.

To improve "the effect of the zones compression," the procedure of the effect described above (Figure 7.16) may be repeated many times to obtain more degree of bands (zones) compression. In Figure 7.17, the chromatograms obtained at the multiple zones compression during the process of the chromatographic separation of the mixture on the TLC plate are provided.

In Figure 7.17, the chromatograms of the separation of the same mixture in the conditions permitting to estimate the efficiency of the method and the arrangement

---

* Registered trademark of E.I. duPont de Nemours & Company, Inc., Wilmington, Delaware.

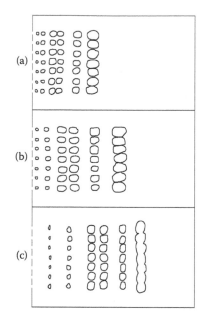

FIGURE 7.16 Chromatograms of the separation of the dyes in the flow of toluene on the plates for TLC of the company "Sorbfil" (the adsorbent is silica gel, CC "Sorbpolimer," Krasnodar, Russia). The chromatographic zones of the used dyes when separating were positioned in the following order (beginning with the most retained): Siba FII, Indophenol, Ariabel Red, Sudan blue, Sudan II, and dimethylaminoazobenzene. The given chromatograms were obtained in the following experiment conditions: (a) The chamber saturated with the vapors of the mobile phase (the previous saturation of the chamber with the vapors of the mobile phase was carried out by passing of the air flow through the external bubbler filled with toluene), the air flow was not used in the process of separation. (b) The saturated chamber (the duration of the mobile phase front migration is 22 minutes), the air flow was not used in the process of separation. (c) The chamber not saturated with the toluene vapors (the first stage of the separation), then the air flow whose velocity was exponentially increased for 30 minutes (the second stage of the separation) was passed through the chamber for 30 minutes in the counterflow (regarding to the direction of the liquid mobile phase flow). (From Berezkin, V.G., Sednev, K.V., *Doklady Chem.*, 419, 1, 2008. With permission.)

proposed are provided. In Figure 7.17a, the final chromatogram obtained after the fourfold separation (the positions of the MP front after 1, 2, 3, 4 separation are indicated by a dashed line) is given. As a result, the elution duration increased: 4, 9, 16, and 25 minutes.

When using the new method proposed in this work, the first development took 4 minutes, after that using the gas counterflow for 9 minutes, the MP front was returned to the start line. At the second development, the MP front migrated to the end of the plate (23 minutes) with the slow movement of the gas flow in the same direction, then the plate was dried using the reversed increasing counterflow for 16 minutes.

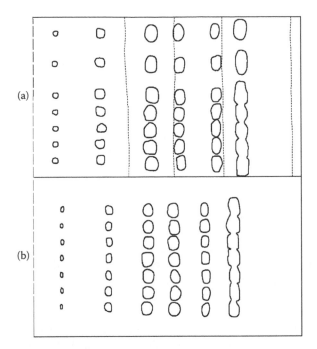

FIGURE 7.17 Chromatograms of TLC obtained in the saturated chamber at the fourfold (traditional) separation with the intermediate drying of the plate (a) and at the twofold separation by drying of the plate at the end of every separation in the conditions of using the reversed gas flow (b). (From Berezkin, V.G., and Sednev, K.V. *Doklady Chem.*, 419, 1, 2008. With permission.)

The chromatograms given in Figure 7.17 permit to draw the conclusion that the twofold use of the described method with the reversed gas flow permits to obtain better separation than in the classical method with fourfold (i.e., multiple) separation.

Taking into account the expenditures of time when using the classical multiple separation method also including the expenditures of time for the additional operations (drying, etc.), it should be believed that the described method with the reversed gas flow takes substantially less expenditures of time. Let us note that for the method with the reversed gas flow (on the whole chromatogram), the coefficient of the zone resolution increases 1.25 times and the efficiency N_1 increases 1.46 times (in comparison with the traditional method).

It should be noted that the results obtained by us permit to draw the conclusion that the method worked out by us substantially (in efficiency and rapidness) exceeds the classical method of multiple elution [2,29,30]. The essence of the traditional method "is in the multiple elution of the chromatogram by the same solvent." In this method, a separated sample is eluted by a selected solvent, a plate is taken away from a chamber, a plate is dried, then a plate is again put in a chamber, and a sample is eluted for the second time. Process of elution can be repeated a number of times until the required separation is achieved [28]. However, the method proposed by us permits to obtain better separation, to increase the rapidness, and to simplify the implementation of the method.

Using the method described in the preceding discussion can be widened if it is used as the method of TLC with CGP influencing the separation [31].

In the work by Sednev and Berezkin [31], separation of polar and ionic substances in carbon dioxide and ammonia atmospheres on unmodified silica gel TLC plates with reduced moisture content has been studied by single and multiple development chromatography with counterflow drying.

The use of active gas flows over the adsorbent layer in TLC has significant advantages in the separation of polar ionic substances. Activation of the adsorbent, stabilization of MP pH and ionic strength, the saturation of the adsorbent with the MP, evaporation of the MP to perform multiple development, and creating gradients and preventing their undesired creation can all be done by simple passage of gases in one or other direction over the adsorbent layer. All these features can also be programmed to change continuously during a separation. Programmed methods are well known and successfully used in chromatography and the method reported in this paper can be regarded as a new variant of programming.

The apparatus is very economical in relation to solvent use compared to traditional chambers.

Potentially, many other operations in addition to those being reported in this chapter can be carried out by use of gas flows over the TLC plate, so this field of research will expand.

7.7 CONCLUSION

The new TLC-CGP method is based on the purposeful changing of physicochemical properties of a used chromatographic system as a result of a contact of an active gas (vapor) phase with a plate during process of chromatographic separation. It was experimentally shown that usage of the given method permits, as a rule, to substantially improve separation. The new method is sufficiently general; it was successfully applied in electroosmotic TLC as well.

It should be underlined that for the time being, practically only the bases of the new TLC-CGP method have been worked out. The following development of the method is of common interest and this method will hopefully arouse interest among chromatographists, and some of them might take part both in the practical usage of the method described in the given review and in its further development.

REFERENCES

1. Zolotov, Yu. A. and Vershinin, V.I. *Istoriya i Metodologiya Analiticheskoy Khimii* (*The History and Methodology of Analytical Chemistry*). Akademiya, Moscow, 2007, 378.
2. Sherma, J. and Fried, B. (Eds.). *Handbook of Thin-Layer Chromatography*. Marcel Dekker, New York, 2003, 1016.
3. Nyiredy, Sz. (Ed.). *Planar Chromatography*. Springer Scientific Publisher, Budapest, 2001, 614.
4. Geiss, F. *Fundamentals of Thin Layer Chromatography* (*Planar Chromatography*). Huethig Verlag, Heidelberg, 1987, 482.

5. Zlatkis, A. and Kaiser, R.E. (Eds.). *HPLC High Performance Thin-Layer Chromatography, Bad Duerkheim*. Institute of Chromatography, Amsterdam, 1977, 240.
6. Berezkin, V.G. and Bochkov, A.S. *Kolichestvennaya tonkosloynaya khromatografiya. Instrumental'nyye metody. (Quantitative Thin Layer Chromatography. Tool Methods)*. Nauka, Moscow, 1980, 183.
7. Poole, C.F. *The Essence of Chromatography*. Elsevier, Amsterdam, 2003, 925.
8. Sumina, E.G., Shtykov, S.N., and Tyurina, N.V. *Tonkosloynaya Khromatografiya. Teoreticheskiye Osnovy i Praktichesloye Primeneniye (Thin Layer Chromatography. The Theoretical Bases and the Practical Application)*. Saratovsky University Publishing House, Saratov, 2006, 112.
9. Krasikov, V.D. *Osnovy Planarnoy Khromatografii (The Bases of Planar Chromatography)*. Khimizdat, St. Petersburg, 2005, 232.
10. Hahn-Deinstrop, E. *Applied Thin-Layer Chromatography*. Wiley-VCH Verlag, Weinheim, 2007, 314.
11. Kowalska, T. and Sherma, J. (Eds.). *Preparative Thin-Layer Chromatography*. CRC Press, Boca Raton, 2006, 424.
12. Berezkin, V.G., Sumina, E.G., Shtykov, S.N., Atayan, V.Z., Tyurina, N.V., Zagniboroda, D.A., and Gaidamakin, D.V. New Variant of Planar Chromatography with Time-Dependent Properties of Chromatographic System upon Contact with Acidic and Basic Gas Stream. *Proceedings of the International Symposium on Planar Separations: "Planar Chromatography" 2005*, 29–31 May. Hungary, 2005, 273.
13. Berezkin, V.G., Sumina, E.G., Shtykov, S.N., Atayan, V.Z., Zagniboroda, D.A., and Nekhoroshev, G.A. Effect of Chamber Gas Phase on Mobile Phase pH and on Separation Efficiency in TLC. A New Mode of Chromatography. *Chromatographia*, 64 (1/2), 105, 2006.
14. Berezkin, V.G., Sumina, E.G., Shtykov, S.N., Zagniboroda, D.A., and Atayan, V.Z. New thin layer chromatography method for ionizable compounds based on the change in the mobile phase acidity during elution. *Doklady Physical Chemistry*, 407 (1), 77, 2006.
15. Siouffi, A.M. and In Issaq, H.J. (Eds.). *A Century of Separation Science*. Marcel Dekker, New York, 2002, 65.
16. Berezkin, V.G. (Complier and scientific editor), N.A. Izmailov and M.S. Shraiber *Otkrytiye Tonkosloynoy Khromatografii (The Discovery of Thin Layer Chromatography)*. GEOS, Moscow, 2007, 128.
17. Gocan, S. TLC Sandwich Chamber. In: J. Cazes (Ed.), *Encyclopedia of Chromatography*. Marcel Dekker, New York, 2001, 251.
18. Berezkin, V.G. and Kormishkina E.V. Version of conventional thin-layer chromatography with a contact-covered sorption layer (Chamberless TLC). *Journal of Analytical Chemistry*, 61 (10), 991, 2006
19. Zolotov, Yu. A. *Analiticheskaya Khimiya: Fragmenty i Kartiny (Analytical Chemistry: Fragments and Pictures)*. Geokhi, Moscow, 1999, 41.
20. Berezkin, V.G., Sumina, E.G., Shtykov, S.N., Zagniboroda, D.A., Atayan, V.Z. and Nekhoroshev, G.A. Sposob razdeleniya soedineniy metodom tonkosloynoy khromatografii i ustroystvo dlya yego realizatcii (The way of separation of compounds by the method of thin layer chromatography and the arrangement for its implementing), The solution of Rospatent, Federal Institute of Industrial Property, department 98, November, 21, 2007, on issue of the patent for invention (the patent).
21. Berezkin, V.G., Chausov, A.V., Shtykov, S.N., and Sumina, Ye.G. Vliyanie izmeneniya sostava okrujayutchey plastinku gazovoy fazy na uderjivaniye razdelyamykh komponentov v TSKh (The influence of the changing of composition of a gas phase surrounding a plate on retention of separated components in TLC). *Sorbtsionnye i khromatograficheskiye processy* (Sorption and Chromatographic Processes), 7 (1), 106, 2007.

22. Berezkin, V.G., Sumina, E.G., Shtykov, S.N., Atayan, V.Z., Zagniboroda, D.A., Nekhoroshev, G.A., and Chausov, A.V. *Novy Variant Tonkosloynoy Khromatografii s Upravlyayemoy Gazovoy Fazoy, Vliyayutchey na Razdeleniye* (*The New Variant of Thin Layer Chromatography with a Controlled Gas Phase Influencing the Separation*). INKhS RAN, Moscow, 2008.

23. Kogan, Yu.D. Development of Modern Quatitative Thin Layer Chromatography. In: *Khromatografiya na blago Rossiyi* (*Chromatography for the Benefit of Russia*). Granitsa, Moscow, 2007, 480.

24. Sharshunova, M., Shvarts, V., and Mikhalets, Ch. *Chromatografia na Tenkych Vrstvachvo Farmacil a v Klinickej Biochemiiю* (*Thin Layer Chromatography in Pharmacy and Clinical Biochemistry*). Translated from Slovak, Mir, Moscow, 1980, 52.

25. Kraus, Lj., Koch, A., and Hoffstetter-Kuhn, S. *Duhnschichtchromatographie* (Thin Layer Chromatography). Springer-Verlag, Berlin, 1995.

26. Berezkin, V.G. and Nekhoroshev, G.A. Novy metod elektroosmoticheskoy TSKh, osnovanny na izmenenii kislotnosti podvijnoy fazy v rezul'tate yeyo vzaimodeistviya s aktivnoy gazovoy fazoy v processe eluirovaniya (The new method of electroosmotic TLC based on the changing of acidity of the mobile phase as the result of its interaction with the active gas phase during the process of elution). *Sorbtsionnye i khromatograficheskiye processy* (Sorption and Chromatographic Processes), 7 (2), 224, 2007.

27. Berezkin, V.G., Litvin, Ye. F., Balushkin, A.O., Rozhilo, Ya. K., and Malinovska, I., Elektroosmoticheskaya tonkosloynaya khromatografiya (obzor) (Electroosmotic thin layer chromatography (survey)). *Zavodskaya Laboratoriya. Diagnostika Materialov* (*Industrial Laboratory Diagnostics of Materials*), 12 (70), 3, 2004.

28. Berezkin, V.G. and Sednev, K.V. New thin layer chromatography technique using a gas stream above the plate sorption layer. *Doklady Chemistry*, 419 (1), 65, 2008.

29. Kirchner, J.G. *Thin-Layer Chromatography*. Interscience, New York, 1967.

30. Cazes, J. (Ed.). *Encyclopedia of Chromatography*. Marcel Dekker, New York, 2001, 927.

31. Sednev, K.V. and Berezkin V.G. Thin-layer chromatography of polar and ionic compounds using active gas flow over the silica gel adsorbent layer. *Journal of Planar Chromatography*, 23 (3),181, 2011.

Index